RIEMANNIAN GEOMETRY

PRINCETON LANDMARKS IN MATHEMATICS AND PHYSICS

Non-standard Analysis,
by Abraham Robinson

General Theory of Relativity,
by P.A.M Dirac

Angular Momentum in Quantum Mechanics,
by A. R. Edmonds

Mathematical Foundations of Quantum Mechanics,
by John von Neumann

Introduction to Mathematical Logic,
by Alonzo Church

Convex Analysis,
by R. Tyrrell Rockafellar

Riemannian Geometry,
by Luther Pfahler Eisenhart

RIEMANNIAN GEOMETRY

BY

LUTHER PFAHLER EISENHART

PRINCETON UNIVERSITY PRESS
PRINCETON, NEW JERSEY

Published by Princeton University Press, 41 William Street,
Princeton, New Jersey 08540
In the United Kingdom: Princeton University Press, Chichester,
West Sussex

ISBN 0-691-08026-7

Princeton University Press books are printed on acid-free paper
and meet the guidelines for permanence and durability of
the Committee on Production Guidelines for Book
Longevity of the Council on Library Resources

Second Printing, 1950
Third Printing, 1952
Fourth Printing, 1960
Fifth Printing, 1964
Seventh Printing, 1993
Eighth Printing, for the Princeton Landmarks in
Mathematics and Physics series, 1997

Printed in the United States of America

9 11 13 15 14 12 10 8

Preface

The recent physical interpretation of intrinsic differential geometry of spaces has stimulated the study of this subject. Riemann proposed the generalization, to spaces of any order, of the theory of surfaces, as developed by Gauss, and introduced certain fundamental ideas in this general theory. From time to time important contributions to the theory were made by Bianchi, Beltrami, Christoffel, Schur, Voss and others, and Ricci coördinated and extended the theory with the use of tensor analysis and his Absolute Calculus. Recently there has been an extensive study and development of Riemannian Geometry, and this book aims to present the existing theory.

Throughout the book constant use is made of the methods of tensor analysis and the Absolute Calculus of Ricci and Levi-Civita. The first chapter contains an exposition of tensor analysis in form and extent sufficient for the reader of the book who has not previously studied this subject. However, it is not intended that the exposition shall give an exhaustive foundational treatment of the subject.

Most, if not all, of the contributors to the theory of Riemannian Geometry have limited their investigations to spaces with a metric defined by a positive definite quadratic differential form. However, the theory of relativity deals with spaces with an indefinite fundamental form. Consequently the former restriction is not made in this book. Although many results of the older theory have been modified accordingly, much remains to be done in this field.

The theory of parallelism of vectors in a general Riemannian manifold, as introduced by Levi-Civita and developed by others, is set forth in the second chapter and is applied in other parts of the book. The extensions of this theory to non-Riemannian geometries are not developed in this book, since it is my intention to present some of them in a later book.

Of the many exercises in the book some involve merely direct applications of the formulas of the text, but most of them constitute extensions of the theory which might properly be included as portions of a more extensive treatise. References to the sources of these exercises are given for the benefit of the reader. All references in the book are to the papers listed in the Bibliography.

In the writing of this book I have had invaluable assistance and criticism by four of my students, Dr. Arthur Bramley, Dr. Harry Levy, Dr. J. H. Taylor and Dr. J. M. Thomas. I desire also to express my appreciation of the courtesies extended by the printers Lütcke & Wulff and by the Princeton University Press.

October, 1925.

Luther Pfahler Eisenhart.

Contents

CHAPTER I

Tensor analysis

CHAPTER II

Introduction of a metric

CHAPTER III

Orthogonal ennuples

Chapter IV

The geometry of sub-spaces

Chapter V

Sub-spaces of a flat space

CHAPTER VI

Groups of motions

CHAPTER I

Tensor analysis

1. Transformation of coördinates. The summation convention. Any n independent variables x^i, where i takes the values 1 to n, may be thought of as the coördinates of an n-dimensional space V_n in the sense that each set of values of the variables defines a point of V_n. Unless stated otherwise it is understood that the coördinates are real.

Suppose that we have n independent real functions φ^i of the variables $x^1, x^2, \cdots, x^n$.* A necessary and sufficient condition that the functions be independent is that the Jacobian does not vanish identically;† that is,

$$(1.1) \qquad \left|\frac{\partial \varphi^i}{\partial x^j}\right| = \begin{vmatrix} \frac{\partial \varphi^1}{\partial x^1} & \cdots & \frac{\partial \varphi^n}{\partial x^1} \\ \cdot & \cdot & \cdot \\ \cdot & \cdot & \cdot \\ \frac{\partial \varphi^1}{\partial x^n} & \cdots & \frac{\partial \varphi^n}{\partial x^n} \end{vmatrix} \neq 0.$$

If we put

$$(1.2) \qquad x'^i = \varphi^i(x^1, \cdots, x^n) \qquad (i = 1, \cdots, n),$$

the quantities x'^i are another set of coördinates of the space; when in the right-hand members of (1.2) we substitute the coördinates x^i of any point P, these equations give the coördinates x'^i of P. Thus equations (1.2) define a *transformation of coördinates* of the space V_n. In consequence of the assumption (1.1) the x's are expressible in terms of the x''s, say

$$(1.3) \qquad x^i = \psi^i(x'^1, \cdots, x'^n) \qquad (i = 1, \cdots, n).$$

---

* When we consider any function, it is understood that it is real and continuous, as well as its derivatives of such order as appear in the discussion, in the domain of the variables considered, unless stated otherwise.

† *Goursat*, 1904, 1, p. 57; *Wilson*, 1911, 1, p. 133.

If we think of the x's as functions of the x''s, then by the rules for differentiation

$$\frac{\partial x^k}{\partial x^j} = \sum_i^{1,\cdots,n} \frac{\partial x^k}{\partial x'^i} \frac{\partial x'^i}{\partial x^j}.$$

However, since the x's are independent, the left-hand member of the above equation is zero unless $k = j$, in which case it is unity. Accordingly we can write

$$\sum_i^{1,\cdots,n} \frac{\partial x^k}{\partial x'^i} \frac{\partial x'^i}{\partial x^j} = \delta_j^k, \tag{1.4}$$

where by definition

$$\delta_j^k = 1 \text{ or } 0, \quad \text{as } k = j \text{ or } k \neq j. \tag{1.5}$$

These are called the *Kronecker deltas* and are used frequently throughout this work. In like manner we have

$$\sum_i^{1,\cdots,n} \frac{\partial x'^k}{\partial x^i} \frac{\partial x^i}{\partial x'^j} = \delta_j^k. \tag{1.6}$$

If in (1.4) we hold k fixed and let j take the values 1 to n, we have n equations linear in $\frac{\partial x^k}{\partial x'^i}$ for $i = 1, \cdots, n$. Solving for these quantities, we obtain

$$\frac{\partial x^k}{\partial x'^i} = \frac{\textit{cofactor of } \frac{\partial x'^i}{\partial x^k} \textit{ in } \left| \frac{\partial x'^i}{\partial x^j} \right|}{\left| \frac{\partial x'^i}{\partial x^j} \right|}. \tag{1.7}$$

Any direction at a point P of the space is determined by the differentials dx^i and the same direction is determined in another set of coördinates x'^i by the differentials dx'^i, where from (1.2)

$$dx'^i = \sum_j^{1,\cdots,n} \frac{\partial \varphi^i}{\partial x^j} dx^j = \sum_j^{1,\cdots,n} \frac{\partial x'^i}{\partial x^j} dx^j. \tag{1.8}$$

It is desirable now to introduce a convention which will be used throughout this book, namely that when the same letter appears in any term as a subscript and superscript, it is understood that this letter is summed for all the values, say n, which this letter takes and consequently the one term stands for the sum of n terms. Thus we write (1.8) in the form

$$dx'^i = \frac{\partial x'^i}{\partial x^j} dx^j \qquad (i, j = 1, \cdots, n). \tag{1.9}$$

Since j appears twice in the right-hand member in the manner indicated and i appears only once, the right-hand member stands for the sum

$$\frac{\partial x'^i}{\partial x^1} dx^1 + \frac{\partial x'^i}{\partial x^2} dx^2 + \cdots + \frac{\partial x'^i}{\partial x^n} dx^n.$$

When the same index appears twice and has the significance just defined, we call it a *dummy index*, since the letter used for such an index is immaterial. However, a letter appearing as another index must not also be used for a dummy index, otherwise an ambiguity would be introduced. Thus i in (1.9) could not be used also in place of j, but the right-hand member of (1.9) could be written in such forms as

$$\frac{\partial x'^i}{\partial x^k} dx^k, \qquad \frac{\partial x'^i}{\partial x^l} dx^l \qquad (k, l = 1, \cdots, n).$$

It should be remarked that (1.9) represents n equations obtained by giving i the values from 1 to n.

Using the summation convention, we write (1.4) and (1.6) in the forms

$$\frac{\partial x^k}{\partial x'^i} \frac{\partial x'^i}{\partial x^j} = \delta^k_j; \qquad \frac{\partial x'^k}{\partial x^i} \frac{\partial x^i}{\partial x'^j} = \delta^k_j. \tag{1.10}$$

2. Contravariant vectors. Congruences of curves. Let λ^j be any n functions of the x's and let n functions λ'^i be defined by

$$\lambda'^i = \lambda^j \frac{\partial x'^i}{\partial x^j} \qquad (i, j, = 1, \cdots, n). \tag{2.1}$$

We observe that equations (1.9) are of this form. If equations (2.1) be multiplied by $\frac{\partial x^k}{\partial x'^i}$ and i be summed from 1 to n, we have in consequence of (1.10)

$$\frac{\partial x^k}{\partial x'^i}\lambda'^i = \lambda^j \frac{\partial x'^i}{\partial x^j}\frac{\partial x^k}{\partial x'^i} = \lambda^j \delta^k_j .$$

The right-hand member is the sum of n terms each of which is zero by (1.5) unless $j = k$, and consequently the right-hand member reduces to the single term λ^k. Accordingly we have

$$\lambda^k = \lambda'^i \frac{\partial x^k}{\partial x'^i}. \tag{2.2}$$

The same result is obtained if we solve (2.1) for λ^j by algebraic processes and make use of (1.7). However, the process used above is very simple and will be used frequently. From (2.1) and (2.2) it is seen that the relation between the λ's and λ''s is entirely reciprocal.

Suppose now that we have a set of functions λ''^i in another coördinate system x''^i defined by equations of the form (2.1), thus

$$\lambda''^i = \lambda^k \frac{\partial x''^i}{\partial x^k}.$$

Then by means of (2.2) we have

$$\lambda''^i = \lambda'^l \frac{\partial x^k}{\partial x'^l}\frac{\partial x''^i}{\partial x^k} = \lambda'^l \frac{\partial x''^i}{\partial x'^l}.$$

Observe that we have changed the dummy index i in (2.2) to l, since i appears already. The above equations and (2.2) being similar to (2.1), we see that the relations (2.1) possess what may be called the *group property*.

When two sets of functions λ^i and λ'^i are related as in (2.1), we say that λ^i are the *components* of a *contravariant vector* in the system x^i and λ'^i the components of the same vector in the system x'^i. From this definition it follows that any n functions of the x's in one coördinate system may be taken as the components of a contravariant vector, whose components in any other system are defined by (2.1). From (1.9) we see that the first differentials of the coördinates in any system are the components of a contravariant vector whose components in any other system are the first differentials of the coördinates of that system.

A contravariant vector as defined determines a direction at each point of the space, that is, a field of vectors in the ordinary sense that a vector is a direction at a point. However, we will use interchangeably the terms vector and *vector-field.**

If λ^i are the components of any contravariant vector, a displacement in the direction of the vector at a point satisfies the equations

$$\frac{dx^1}{\lambda^1} = \frac{dx^2}{\lambda^2} = \cdots = \frac{dx^n}{\lambda^n}. \tag{2.3}$$

From the theory of differential equations of this form we have that these equations admit $n-1$ independent solutions

$$\varphi^j(x^1, x^2, \ldots, x^n) = c^j \qquad (j = 1, \cdots, n-1), \tag{2.4}$$

where the c's are arbitrary constants and the matrix $\left\| \dfrac{\partial \varphi^j}{\partial x^i} \right\|$ is of rank $n-1$. The functions φ^j are solutions of the partial differential equation†

$$\lambda^i \frac{\partial \varphi}{\partial x^i} = 0. \tag{2.5}$$

If now we effect the transformation of coördinates (1.2) in which for φ^j, where $j = 1, \cdots, n-1$, we take the above solutions and for φ^n any function such that (1.1) is satisfied, we have from (2.1)

$$\lambda'^j = 0 \quad (j = 1, \cdots, n-1), \qquad \lambda'^n \neq 0. \tag{2.6}$$

Hence:

When a contravariant vector is given, a system of coördinates can be chosen in terms of which all the components but one of the vector are equal to zero.

* Many of the ideas developed in this chapter were studied first by *Christoffel*, 1869, 1, and by *Ricci*, whose development was presented by him and *Levi-Civita* in their paper, 1901, 1.

† *Goursat*, 1891, 1, p. 29.

If the coördinates of any point P are substituted in (2.4), the values of c^j are determined and the $n-1$ equations (2.4) for these values of c^j define a *curve* through P, that is, the locus of points whose coordinates satisfy $n-1$ equations, or, what is equivalent, whose coordinates are expressible as functions of a single parameter. Thus equations (2.4) define a *congruence* of curves, one of which passes through each point of the space V_n. We say that the congruence is determined by the vector-field λ^i and that the vector λ^i at a point is *tangent* to the curve of the congruence through the point. Thus we identify the differentials for a curve with components of the tangent vector.

3. Invariants. Covariant vectors. If a function f of the x's and a function f' of the x''s are such that they are reducible to one another by the equations of the transformations of the variables, they are said to define an *invariant*. In this sense an invariant is a *scalar* as defined in vector analysis, and is so called by some writers on tensor analysis. It should be remarked that the term invariant as thus used has a different connotation from its definition in the field of algebraic invariants. In fact, any function of the x's can be taken as an invariant and then its definition in any other coordinate system is determined by the transformation of coördinates.

If f be any function, we have

$$\frac{\partial f}{\partial x'^i} = \frac{\partial f}{\partial x^j}\frac{\partial x^j}{\partial x'^i} \qquad (i, j = 1, \cdots, n). \tag{3.1}$$

These equations are a special case of the equations

$$\lambda'_i = \lambda_j \frac{\partial x^j}{\partial x'^i}, \tag{3.2}$$

where λ_j are any functions of the x's and the λ''s are functions of the x''s defined by (3.2). As in § 2 it can be shown that (3.2) are equivalent to

$$\lambda_j = \lambda'_i \frac{\partial x'^i}{\partial x^j}; \tag{3.3}$$

also that the relation (3.2) possesses the group property (§ 2). When two sets of functions λ_i and λ'_i are in the relation (3.2), we say that the λ's are the components of a *covariant vector* in the x's and the λ''s are the components of the same vector in the x''s. Evidently a covariant vector is defined uniquely by choosing any set of n functions in one coördinate system. In particular, it follows from (3.1) that if the first derivatives of a function f are taken as the components of a covariant vector, the components of the same vector in any other system are the first derivatives of the function with respect to the new coördinates. Such a covariant vector is called the *gradient* of f.

It should be observed that the index of a contravariant vector is written as a superscript and of a covariant vector as a subscript; this is done so that the summation convention can be used in (2.1), (2.2), (3.2) and (3.3).

If λ^i and μ_i are the components of any contravariant and covariant vectors respectively, from equations of the forms (2.1) and (3.2) and from (1.10) we have

$$\lambda'^i \mu'_i = \lambda^j \frac{\partial x'^i}{\partial x^j} \mu_k \frac{\partial x^k}{\partial x'^i} = \lambda^j \mu_k \delta^k_j.$$

If in the right-hand member we sum first for k, all the terms for any j vanish except when $k = j$, and consequently

$$\lambda'^i \mu'_i = \lambda^j \mu_j = \lambda^i \mu_i. \tag{3.4}$$

Each member of this equation consists of the sum of n terms, and the members being equal because of (1.2), it follows that $\lambda^i \mu_i$ is an invariant.

Suppose conversely that we have an equation such as (3.4) in which it is assumed that λ^i are the components of a contravariant vector. In consequence of (2.1) we have

$$\lambda^j \left(\frac{\partial x'^i}{\partial x^j} \mu'_i - \mu_j \right) = 0.$$

From this equation it can be concluded that μ_i is a covariant vector, if λ^j is an arbitrary vector and only in this case. Hence:

If the quantity $\lambda^i \mu_i$ is an invariant and either λ^i or μ_i are the components of an arbitrary vector, the other set are components of a vector.

Let λ_i be the components of a covariant vector and consider the equation

$$\lambda_i \, dx^i = 0. \tag{3.5}$$

This equation admits $n-1$ linearly independent sets of values of the differentials dx^i in terms of which any other set is linearly expressible. The totality of directions at a point satisfying (3.5) constitute what may be called an *elemental* V_{n-1} at the point. Hence a covariant vector field may be considered geometrically as defining an elemental V_{n-1} at each point. In general, equation (3.5) does not admit a family of solutions of the form $f(x^1, \cdots, x^n) = c$, where c is a constant; when it does, that is, when (3.5) is completely integrable, the elemental V_{n-1}'s at all points of such a *hypersurface* $f = c$ coincide with the hypersurface.

Exercises

1. If $\lambda^1 = \varphi$, $\lambda^j = 0\ (j \neq 1)$, where φ is an arbitrary function of the x's, are taken as the components of a contravariant vector in the x's, the components λ'^i in any other coordinate system x'^i are given by

$$\lambda'^i = \varphi \frac{\partial x'^i}{\partial x^1}.$$

2. If $\lambda_1 = \varphi$, $\lambda_j = 0\ (j \neq 1)$, where φ is an arbitrary function of the x's, are taken as the components of a covariant vector in the x's, the components in any other coordinate system x'^i are given by

$$\lambda'_i = \varphi \frac{\partial x^1}{\partial x'^i}.$$

3. If $\lambda_{\alpha|}{}^i$ are the components of n vector-fields in a V_n, where i for $i = 1, \cdots, n$ indicates the component and α for $\alpha = 1, \cdots, n$ the vector, and these vectors are independent, that is, the determinant $|\lambda_{\alpha|}{}^i| \neq 0$, then any vector-field λ^i is expressible in the form

$$\lambda^i = a^\alpha \lambda_{\alpha|}{}^i,$$

where the a's are invariants.

4. If μ_i are the components of a given vector-field, any vector-field λ^i satisfying $\lambda^i \mu_i = 0$ is expressible linearly in terms of $n-1$ independent vector-fields $\lambda_{\alpha|}{}^i$ for $\alpha = 1, \cdots, n-1$ which satisfy the equation. (The vectors $\lambda_{\alpha|}{}^i$ are independent, if the rank of the matrix $\|\lambda_{\alpha|}{}^i\|$ is $n-1$).

5. For a linear transformation of the form $x'^i = a^i_j x^j$, where the a's are constants and the determinant $a = |a^i_j| \neq 0$, the coördinates are components of a contravariant vector-field in both coördinate systems. If we put

$$u'_i x'^i = u_j x^j,$$

we have an induced transformation on the u's given by $u'_i = A^j_i u_j$, where A^j_i is the cofactor of a^i_j in the determinant a divided by a. Show that u'_i and u_i are components of a covariant vector in the x's and x''s for the given transformation.

4. Tensors. Symmetric and skew-symmetric tensors.

Let λ^i, μ^i be the components of two contravariant vectors and ξ_i, η_i the components of two covariant vectors. If we put

$$(4.1) \qquad a^{ij} = \lambda^i \mu^j, \qquad a_{ij} = \xi_i \eta_j, \qquad a^i_j = \lambda^i \xi_j,$$

and denote by a'^{ij}, a'_{ij} and a'^i_j the same functions in the components λ'^i, μ'^i, ξ'_i, η'_i for a coördinate system x'^i, it follows from equations of the form (2.1) and (3.2) that

$$(4.2) \qquad a'^{ij} = a^{kl} \frac{\partial x'^i}{\partial x^k} \frac{\partial x'^j}{\partial x^l},$$

$$(4.3) \qquad a'_{ij} = a_{kl} \frac{\partial x^k}{\partial x'^i} \frac{\partial x^l}{\partial x'^j},$$

$$(4.4) \qquad a'^i_j = a^k_l \frac{\partial x'^i}{\partial x^k} \frac{\partial x^l}{\partial x'^j}.$$

If we have any two sets of functions in two coördinate systems satisfying equations of one of these forms, we say that

a^{kl} are the components of a *contravariant tensor of the second order*,

a_{kl} are the components of a *covariant tensor of the second order*,

a^k_l are the components of a *mixed tensor of the second order*.

It should be observed that as thus defined any tensor of one of these types is not necessarily obtainable from vectors as in (4.1).

From this definition it follows that any set of n^2 quantities can be taken as the components of a tensor of the second order of any type and the components of the tensor in any other coördinate system are defined by (4.2), (4.3) or (4.4), according as the tensor is to be contravariant, covariant or mixed.

As an example we consider the case $a^k_l = \delta^k_l$, where δ^k_l are the Kronecker deltas defined by (1.5). From (4.4) we have

$$a'^i_j = \delta^k_l \frac{\partial x'^i}{\partial x^k} \frac{\partial x^l}{\partial x'^j} = \frac{\partial x'^i}{\partial x^k} \frac{\partial x^k}{\partial x'^j} = \delta^i_j.$$

Hence:

If the Kronecker deltas are taken as the components of a mixed tensor of the second order in one set of coördinates, they are the components of the tensor in any set of coördinates.

Tensors of any order are defined by generalizing (4.2), (4.3), (4.4). Thus the equations

$$a'^{r_1 \cdots r_m} = a^{s_1 \cdots s_m} \frac{\partial x'^{r_1}}{\partial x^{s_1}} \cdots \frac{\partial x'^{r_m}}{\partial x^{s_m}} \tag{4.5}$$

define a *contravariant tensor of the mth order*;

$$a'_{r_1 \cdots r_m} = a_{s_1 \cdots s_m} \frac{\partial x^{s_1}}{\partial x'^{r_1}} \cdots \frac{\partial x^{s_m}}{\partial x'^{r_m}} \tag{4.6}$$

a *covariant tensor of the mth order*;

$$a'^{r_1 \cdots r_m}_{p_1 \cdots p_q} = a^{s_1 \cdots s_m}_{t_1 \cdots t_q} \frac{\partial x'^{r_1}}{\partial x^{s_1}} \cdots \frac{\partial x'^{r_m}}{\partial x^{s_m}} \frac{\partial x^{t_1}}{\partial x'^{p_1}} \cdots \frac{\partial x^{t_q}}{\partial x'^{p_q}} \tag{4.7}$$

a *mixed tensor of the $m+q$ order* which is *contravariant of the mth order* and *covariant of the qth order*.*

Concerning these definitions we make the following observations and deductions:

(1) A superscript indicates contravariant character, a subscript covariant;

(2) Any set of functions in sufficient number can be taken as the components of a tensor of any type and order in one coördinate system and the components in any other system are defined by equations (4.5), (4.6) or (4.7) as the case may be;

(3) A contravariant vector is a contravariant tensor of the first order; a covariant vector is a covariant tensor of the first order;

(4) An invariant is a tensor of zero order. The latter designation is a more appropriate term than invariant because of the possible ambiguity of the term invariant;

(5) From (4.5), (4.6) and (4.7) it follows that if the components of a tensor in one coördinate system are zero at a point, they are

* It can be shown as in § 2 that these definitions possess the group property.

zero at this point in every coördinate system; in particular, if the components are identically zero in one coördinate system, they are identically zero in every coördinate system.

From the form of equations (4.5), (4.6) and (4.7) it is clear that the order of the indices plays a rôle in these equations. Suppose, however, that the relative position in the a's of two or more indices, either contravariant or covariant, is immaterial, which means that the a's with these indices interchanged are equal. Then from the form of these equations it follows that the order of the corresponding indices in the a''s is immaterial. For example, suppose that in (4.5) $a^{s_1 s_2 \cdots s_m} = a^{s_2 s_1 \cdots s_m}$, then we have

$$\begin{aligned} a'^{r_1 \cdots r_m} &= a^{s_1 s_2 \cdots s_m} \frac{\partial x'^{r_1}}{\partial x^{s_1}} \frac{\partial x'^{r_2}}{\partial x^{s_2}} \cdots \frac{\partial x'^{r_m}}{\partial x^{s_m}} \\ &= a^{s_2 s_1 \cdots s_m} \frac{\partial x'^{r_2}}{\partial x^{s_2}} \frac{\partial x'^{r_1}}{\partial x^{s_1}} \cdots \frac{\partial x'^{r_m}}{\partial x^{s_m}} = a'^{r_2 r_1 \cdots r_m}. \end{aligned}$$

When the relative position of two or more indices, either contravariant or covariant, in the components of a tensor is immaterial, the tensor is said to be *symmetric with respect to these indices.* If the order of all the indices is immaterial, the tensor is said to be *symmetric.*

A general tensor of the second order has n^2 components, whereas, if the tensor is symmetric, there are only $n(n+1)/2$ different components. Similar formulas for the number of components can be obtained for symmetric tensors of higher order or tensors symmetric with respect to certain indices.

When for a tensor two components obtained from one another by the interchange of two particular indices, either contravariant or covariant, differ only in sign, the tensor is said to be *skew-symmetric with respect to these indices.* When the interchange of any two indices, either contravariant or covariant, produces only a change in sign in the components, the tensor is said to be *skew-symmetric.* It can shown as above, that if a tensor has the property of skew-symmetry in one system of coördinates, it has it in every system.

If a_{ij} is skew-symmetric, then $a_{ii} = 0$ and there are only $n(n-1)/2$ different components. Also, if $a_{r_1 \cdots r_n}$ is skew-symmetric in an n-dimensional space, all the components are zero or equal to within sign. For a four-dimensional space there are 6 different components of a skew-symmetric tensor a_{ij} (it is sometimes called a *six-vector*).

5. Addition, subtraction and multiplication of tensors. Contraction. From the form of equations (4.5), (4.6) and (4.7) it follows that the sum or difference of two tensors of the same type and order is a tensor of the same type and order. The same is true of any linear combination of tensors of the same type and order whose coefficients are constants or invariants. As an example, we consider any tensor a_{ij}. If we write

$$a_{ij} = \frac{1}{2}(a_{ij} + a_{ji}) + \frac{1}{2}(a_{ij} - a_{ji}), \tag{5.1}$$

the first term on the right is a symmetric tensor and the second is skew-symmetric. Hence any covariant (or contravariant) tensor of the second order can be written as the sum of a symmetric tensor and a skew-symmetric tensor.

The process which was used in (4.1) to obtain tensors from vectors is not limited to the case of combining vectors. Thus if a_{ij} and b^{rst} are the components of two tensors in coördinates x^i, we have

$$a'_{\alpha\beta}\, b'^{\mu\nu\sigma} = a_{ij}\, b^{rst} \frac{\partial x^i}{\partial x'^\alpha} \frac{\partial x^j}{\partial x'^\beta} \frac{\partial x'^\mu}{\partial x^r} \frac{\partial x'^\nu}{\partial x^s} \frac{\partial x'^\sigma}{\partial x^t}, \tag{5.2}$$

and consequently $a_{ij} b^{rst}$ are the components of a tensor of the fifth order, covariant of order 2 and contravariant of order 3. This process is general, so that by multiplying the components of any number of tensors, we obtain a tensor, called the *product* of the given tensors, which is covariant and contravariant of the orders obtained by adding the covariant orders and contravariant orders respectively. This is sometimes called the *outer* product.

For any mixed tensor a^{ij}_{rst} the expression a^{ij}_{rsj} is the sum of n components of this tensor. We shall show that it is a tensor of the third order. For we have

$$\begin{aligned} a'^{\alpha\beta}_{\mu\nu\beta} &= a^{ij}_{rst} \frac{\partial x'^\alpha}{\partial x^i} \frac{\partial x'^\beta}{\partial x^j} \frac{\partial x^r}{\partial x'^\mu} \frac{\partial x^s}{\partial x'^\nu} \frac{\partial x^t}{\partial x'^\beta} \\ &= a^{ij}_{rst} \frac{\partial x'^\alpha}{\partial x^i} \frac{\partial x^r}{\partial x'^\mu} \frac{\partial x^s}{\partial x'^\nu} \delta^t_j = a^{ij}_{rsj} \frac{\partial x'^\alpha}{\partial x^i} \frac{\partial x^r}{\partial x'^\mu} \frac{\partial x^s}{\partial x'^\nu}. \end{aligned}$$

Hence from (4.7) it follows that a^{ij}_{rsj} is a tensor, covariant of the second order and contravariant of the first order. This process by means of which from a mixed tensor of order r we obtain a tensor of order $r-2$ is called *contraction*. Observe that in applying contraction any superscript may be used with any subscript.*

In particular, from the tensor a^i_j we obtain an invariant a^i_i by contraction. In § 4 we saw that the Kronecker deltas δ^i_j are the components of a mixed tensor; by contraction we get the sum of n terms each of which is 1, and thus the invariant δ^i_i is n.

This process may be repeated, thus from the above tensor we have by two contractions a vector, such as any of the following: a^{ij}_{isj}, a^{ij}_{rij}, a^{ij}_{rji}.

Multiplication and contraction may be combined to give tensors. Thus from the tensors a_{ij} and b^{rst} we may obtain a tensor of the third order, such as $a_{ij}\,b^{jst}$, or $a_{ij}\,b^{rit}$, or a vector as $a_{ij}\,b^{ijt}$. This combined process is referred to by some writers as *inner multiplication*. We remark that this process was used in (3.4).

Let a^{ij}_{klm} be a set of functions of x^i and $a'^{\alpha\beta}_{\mu\nu\sigma}$ be a set of functions of x'^i such that $a^{ij}_{klm}\,\lambda^l$ and $a'^{\alpha\beta}_{\mu\nu\sigma}\,\lambda'^\nu$ are the components of a tensor, when λ^l is an arbitrary vector. From this hypothesis and in consequence of (4.7) and (2.2) we have

$$\begin{aligned} a'^{\alpha\beta}_{\mu\nu\sigma}\,\lambda'^\nu &= a^{ij}_{klm}\,\lambda^l\,\frac{\partial x'^\alpha}{\partial x^i}\,\frac{\partial x'^\beta}{\partial x^j}\,\frac{\partial x^k}{\partial x'^\mu}\,\frac{\partial x^m}{\partial x'^\sigma} \\ &= a^{ij}_{klm}\,\lambda'^\nu\,\frac{\partial x^l}{\partial x'^\nu}\,\frac{\partial x'^\alpha}{\partial x^i}\cdots\frac{\partial x^m}{\partial x'^\sigma}. \end{aligned}$$

Since λ'^ν is arbitrary, we have

$$a'^{\alpha\beta}_{\mu\nu\sigma} = a^{ij}_{klm}\,\frac{\partial x'^\alpha}{\partial x^i}\,\frac{\partial x'^\beta}{\partial x^j}\,\frac{\partial x^k}{\partial x'^\mu}\,\frac{\partial x^l}{\partial x'^\nu}\,\frac{\partial x^m}{\partial x'^\sigma},$$

and consequently a^{ij}_{klm} and $a'^{\alpha\beta}_{\mu\nu\sigma}$ are the components of a mixed tensor of the fifth order. This proof applies equally well when any of the subscripts is used for contraction with λ^l; also a similar result can be established if the arbitrary vector is covariant. Since

---

* *Ricci* and *Levi-Civita*, 1901, 1, p. 133 call the process *composition*, and German writers, *Verjüngung*.

the proof is not conditioned by the number of indices of the functions a, we have the following theorem of which the theorem of § 3 is a particular case:

Given a set of functions $a^{r_1\cdots r_m}_{p_1\cdots p_q}$ *of* x^i *and a set* $a'^{s_1\cdots s_m}_{t_1\cdots t_q}$ *of* x'^i, *if* $a^{r_1\cdots r_m}_{p_1\cdots p_l\cdots p_q}\lambda^{p_l}$ *and* $a'^{s_1\cdots s_m}_{t_1\cdots t_l\cdots t_q}\lambda'^{t_l}$ *are components of a tensor in the coördinates* x^i *and* x'^i *respectively, when* λ^i *and* λ'^i *are components of an arbitrary vector in these respective coördinates, the given functions are components of a tensor of one higher order.*

A similar theorem holds if λ^i is replaced by a tensor of any type and one of the indices is contracted. This is sometimes called the *quotient law* of tensors.

6. Conjugate symmetric tensors of the second order. Associate tensors. Let g_{ij} be the components of a symmetric covariant tensor of the second order, that is, $g_{ij} = g_{ji}$. We denote by g the determinant of the g_{ij}'s, that is,

$$g = \begin{vmatrix} g_{11} & \cdots & g_{1n} \\ \cdot & \cdot & \cdot \\ g_{n1} & \cdots & g_{nn} \end{vmatrix}. \tag{6.1}$$

If g^{ij} denotes the cofactor of g_{ij} divided by g, we have

$$g^{ij}g_{kj} = \delta^i_k, \tag{6.2}$$

where δ^i_k have the values (1.5). For it follows from the definition of g^{ij} that when $i \neq k$ the left-hand member of (6.2) is the sum of the product of the terms of one row (or column) of (6.1) by the cofactors of another row (or column) divided by g; and when $i = k$, this sum is equal to g/g.

Let λ^i be the components of an arbitrary vector, then $g_{ij}\lambda^i$ is an arbitrary vector, say μ_j. Now by (6.2)

$$g^{kj}\mu_j = g^{kj}g_{ij}\lambda^i = \delta^k_i\lambda^i = \lambda^k.$$

Since μ_j is an arbitrary vector, we have as a consequence of the last theorem of § 5:

If g is the determinant of a symmetric covariant tensor g_{ij}, the cofactors of g_{ij} divided by g and denoted by g^{ij} are the components of a symmetric contravariant tensor.

It is clear that in like manner if g^{ij} are the components of a symmetric contravariant tensor, the cofactors of g^{ij} in the determinant of the g^{ij}'s divided by the determinant are the components of a symmetric covariant tensor of the second order. In either case we say that the tensor obtained by this process is the *conjugate* of the given one.

As a consequence of the above result and (6.2) we have that δ^i_k are the components of a mixed tensor, which was proved directly in § 4.

If in (6.2) we replace k by i and sum for i, we get n terms each of which is unity. Hence for the invariant obtained from a symmetric tensor of the second order and its conjugate we have

$$g^{ij}\, g_{ij} = n. \tag{6.3}$$

If we denote by $\overline{g}$ the determinant of g^{ij}, we have by the rule for multiplying determinants and (6.2)

$$g\overline{g} = \begin{vmatrix} g_{11} & \cdots & g_{1n} \\ \cdot & \cdot & \cdot \\ \cdot & \cdot & \cdot \\ g_{n1} & \cdots & g_{nn} \end{vmatrix} \cdot \begin{vmatrix} g^{11} & \cdots & g^{1n} \\ \cdot & \cdot & \cdot \\ \cdot & \cdot & \cdot \\ g^{n1} & \cdots & g^{nn} \end{vmatrix} = \begin{vmatrix} 1 & 0 & 0 \cdots 0 \\ 0 & 1 & 0 \cdots 0 \\ \cdot & \cdot & \cdot \\ 0 & 0 & \cdots 1 \end{vmatrix} = 1, \tag{6.4}$$

and from (6.2) it follows that g_{ij} is the cofactor of g^{ij} in $\overline{g}$ divided by $\overline{g}$.

By means of a symmetric tensor g_{ij} and its conjugate g^{ij} we can obtain from a given tensor, by means of the methods of § 5, tensors of the same order but different character. Thus, if a_{ijk} are the components of a tensor the following expressions are components of tensors of the character indicated by their indices:

$$\begin{aligned} &a^l{}_{jk} = g^{li}\, a_{ijk}; \qquad a_i{}^l{}_k = g^{lj}\, a_{ijk}; \qquad a_{ij}{}^l = g^{lk}\, a_{ijk}; \\ &a^{lm}{}_k = g^{li}\, g^{mj}\, a_{ijk}; \quad a^l{}_j{}^m = g^{li}\, g^{mk}\, a_{ijk}; \quad a^{lmp} = g^{li}\, g^{mj}\, g^{pk}\, a_{ijk}. \end{aligned} \tag{6.5}$$

In similar manner from the tensor of components b^{ijk} we obtain tensors of the following types of components:

$$b_l{}^{jk} = g_{li}\, b^{ijk}; \qquad b_{lm}{}^k = g_{li}\, g_{mj}\, b^{ijk}; \qquad b_{lmp} = g_{li}\, g_{mj}\, g_{pk}\, b^{ijk}. \tag{6.6}$$

We say that these tensors are *associate* to the given tensor by means of g_{ij}. Similarly we find tensors associate to any mixed tensor. We speak of this process as *raising* the subscripts by means of g^{ij} and *lowering* superscripts by means of g_{ij}. We might write the first of (6.5) thus a^l_{jk}, but we use the notation in (6.5) to indicate which index has been raised or lowered.

We remark that this process is reversible. Thus multiplying the first of (6.5) by g_{lm} and summing for l, we have

$$g_{lm}\, a^l{}_{jk} = g_{lm}\, g^{li}\, a_{ijk} = \delta^i_m\, a_{ijk} = a_{mjk},$$

which is the tensor from which $a^l{}_{jk}$ was obtained.

Exercises

1. If a $a^{ij}_k \lambda_i \mu_j \nu^k$ is an invariant for λ_i, μ_i and ν^i arbitrary vectors, then a^{ij}_k are the components of a tensor.

2. If $a_{ij}\lambda^i\lambda^j$ is an invariant for λ^i an arbitrary vector, then $a_{ij}+a_{ji}$ are the components of a tensor; in particular, if $a_{ij}\lambda^i\lambda^j = 0$, then $a_{ij}+a_{ji} = 0$.

3. If $a_{ijk}\,dx^i dx^j dx^k = 0$ for arbitrary values of the differentials, then

$$a_{ijk} + a_{jki} + a_{kij} + a_{ikj} + a_{kji} + a_{jik} = 0.$$

See App. 1 **4.** If $a_{ij}\lambda^i\lambda^j = 0$ for all vectors λ^i such that $\lambda^i\mu_i = 0$, where μ_i is a given covariant vector, if ν^i is a vector not satisfying this condition, and by definition

$$a_{ij}\nu^i = \sigma_j, \qquad \nu^i\mu_i = \tau,$$

then $\left(a_{ij} - \frac{1}{\tau}\mu_i\sigma_j\right)\xi^i\xi^j = 0$ is satisfied by every vector-field ξ^i (cf. Ex. 4, p. 8), and consequently

$$a_{ij} + a_{ji} = \frac{1}{\tau}(\mu_i\sigma_j + \mu_j\sigma_i).$$

Schouten, 1924, 1, p. 59.

5. If a_{rs} are the components of a tensor and b and c are invariants, show that if $b\,a_{rs} + c\,a_{sr} = 0$, then either $b = -c$ and a_{rs} is symmetric, or $b = c$ and a_{rs} is skew-symmetric.

6. Let b_{ij} be a set of functions of x^i $(i = 1, \cdots, n)$ such that the determinant $|b_{ij}| = 0$, and λ^i the set of functions defined by the equations $b_{ij}\lambda^j = 0$; if b_{ij} and λ^j are taken as the components of a tensor and vector in the x's, in accordance with the theorem of § 2 a coordinate system x'^i can be chosen for which $b'_{1j} = 0\ (j = 1, \cdots, n)$.

7. By definition the *rank* of a tensor of the second order a_{ij} is the rank of the determinant $|a_{ij}|$. Show that the rank is invariant under all transformations of coördinates.

8. Show that the rank of the tensor of components $a_i b_j$, where a_i and b_j are the components of two vectors, is one; show that for the symmetric tensor $a_i b_j + a_j b_i$ the rank is two.

7. The Christoffel 3-index symbols and their relations. We consider any symmetric covariant tensor of the second order g_{ij} and the conjugate tensor g^{ij} and define two expressions, due to Christoffel, which will be of frequent use. They are

$$[ij, k] = \frac{1}{2}\left(\frac{\partial g_{ik}}{\partial x^j} + \frac{\partial g_{jk}}{\partial x^i} - \frac{\partial g_{ij}}{\partial x^k}\right), \tag{7.1}$$

$$\left\{ {l \atop ij} \right\} = g^{lk}[ij, k].* \tag{7.2}$$

Observe that from their definition $[ij, k]$ and $\left\{ {l \atop ij} \right\}$ are symmetric in i and j. The symbols defined by (7.1) and (7.2) are called the *Christoffel symbols* of the *first* and *second kinds* respectively. From (7.2) and (6.2) we have

$$g_{lh}\left\{ {l \atop ij} \right\} = g_{lh}\, g^{lk}[ij, k] = \delta^k_h [ij, k] = [ij, h]. \tag{7.3}$$

Again from (7.1) we have

$$\frac{\partial g_{ik}}{\partial x^j} = [ij, k] + [kj, i]. \tag{7.4}$$

Differentiating (6.2) with respect to x^l, we have

$$g^{ij}\frac{\partial g_{kj}}{\partial x^l} + g_{kj}\frac{\partial g^{ij}}{\partial x^l} = 0.$$

Multiplying by g^{km} and summing for k, we obtain

$$\frac{\partial g^{im}}{\partial x^l} = -g^{ij} g^{km} \frac{\partial g_{kj}}{\partial x^l}. \tag{7.5}$$

Substituting in the right-hand member from (7.4), we find in consequence of (7.2)

$$\frac{\partial g^{im}}{\partial x^l} = -\left(g^{ij}\left\{ {m \atop lj} \right\} + g^{jm}\left\{ {i \atop lj} \right\}\right). \tag{7.6}$$

* The historical forms of these respective symbols are $\left[{ij \atop k} \right]$ and $\left\{ {ij \atop l} \right\}$ but we have adopted the above forms because they are in keeping with the summation convention. Cf. *Christoffel*, 1869, 1, p. 49.

From (6.3) we have by differentiation

$$g^{ij}\frac{\partial g_{ij}}{\partial x^l} + g_{ij}\frac{\partial g^{ij}}{\partial x^l} = 0. \tag{7.7}$$

Applying the rule for the differentiation of a determinant and the definition of g^{ij}, we have

$$\frac{\partial g}{\partial x^l} = g g^{ij}\frac{\partial g_{ij}}{\partial x^l} = -g g_{ij}\frac{\partial g^{ij}}{\partial x^l}, \tag{7.8}$$

the last expression being a consequence of (7.7). Substituting from (7.4) or (7.6) in (7.8), we have

$$\frac{\partial \log \sqrt{g}}{\partial x^l} = \left\{ {i \atop i\,l} \right\}, \tag{7.9}$$

the right-hand member being summed for i.

The Christoffel symbols of either kind are not components of a tensor as will be seen from the following results. If g_{ij} and g'_{ij} are components of the given tensor in coördinate systems x^i and x'^i, it follows from (4.3) that

$$g'_{\mu\nu} = g_{ij}\frac{\partial x^i}{\partial x'^\mu}\frac{\partial x^j}{\partial x'^\nu}. \tag{7.10}$$

Differentiating with respect to x'^σ, we have

$$\frac{\partial g'_{\mu\nu}}{\partial x'^\sigma} = \frac{\partial g_{ij}}{\partial x^k}\frac{\partial x^k}{\partial x'^\sigma}\frac{\partial x^i}{\partial x'^\mu}\frac{\partial x^j}{\partial x'^\nu} + g_{ij}\left(\frac{\partial x^i}{\partial x'^\mu}\frac{\partial^2 x^j}{\partial x'^\nu\,\partial x'^\sigma} + \frac{\partial x^j}{\partial x'^\nu}\frac{\partial^2 x^i}{\partial x'^\mu\,\partial x'^\sigma}\right). \tag{7.11}$$

The first of the following equations is obtained from (7.11) by interchanging μ and σ throughout and the dummy indices i and k in the first term of the right-hand member, the second by interchanging ν and σ throughout and the dummy indices j and k in the first term of the right-hand member:

$$\frac{\partial g'_{\sigma\nu}}{\partial x'^\mu} = \frac{\partial g_{kj}}{\partial x^i}\frac{\partial x^k}{\partial x'^\sigma}\frac{\partial x^i}{\partial x'^\mu}\frac{\partial x^j}{\partial x'^\nu} + g_{ij}\left(\frac{\partial x^i}{\partial x'^\sigma}\frac{\partial^2 x^j}{\partial x'^\mu\partial x'^\nu} + \frac{\partial x^j}{\partial x'^\nu}\frac{\partial^2 x^i}{\partial x'^\mu\partial x'^\sigma}\right),$$

$$\frac{\partial g'_{\mu\sigma}}{\partial x'^\nu} = \frac{\partial g_{ik}}{\partial x^j}\frac{\partial x^k}{\partial x'^\sigma}\frac{\partial x^i}{\partial x'^\mu}\frac{\partial x^j}{\partial x'^\nu} + g_{ij}\left(\frac{\partial x^i}{\partial x'^\mu}\frac{\partial^2 x^j}{\partial x'^\nu\,\partial x'^\sigma} + \frac{\partial x^j}{\partial x'^\sigma}\frac{\partial^2 x^i}{\partial x'^\mu\,\partial x'^\nu}\right).$$

If from the sum of these two equations we subtract (7.11) and divide by 2, we have in consequence of (7.1)

$$(7.12)\qquad [\mu\nu,\sigma]' = [ij,k]\frac{\partial x^i}{\partial x'^\mu}\frac{\partial x^j}{\partial x'^\nu}\frac{\partial x^k}{\partial x'^\sigma} + g_{ij}\frac{\partial x^i}{\partial x'^\sigma}\frac{\partial^2 x^j}{\partial x'^\mu \partial x'^\nu},$$

where $[\mu\nu,\sigma]'$ is formed with respect to the tensor $g'_{\mu\nu}$. Since these equations are not of the form (4.5), (4.6) or (4.7), it follows that the functions $[ij, k]$ are not components of a tensor. The same is true of $\left\{ {k \atop ij} \right\}$ as follows from (7.2), or from the following equation obtained by multiplying (7.12) by $g'^{\sigma\lambda}\frac{\partial x^l}{\partial x'^\lambda}$, summing for σ and making use of

$$(7.13)\qquad g'^{\sigma\lambda}\frac{\partial x^k}{\partial x'^\sigma}\frac{\partial x^l}{\partial x'^\lambda} = g^{kl}$$

and (6.2):

$$(7.14)\qquad \left\{ {\lambda \atop \mu\nu} \right\}' \frac{\partial x^l}{\partial x'^\lambda} = \left\{ {l \atop ij} \right\}\frac{\partial x^i}{\partial x'^\mu}\frac{\partial x^j}{\partial x'^\nu} + \frac{\partial^2 x^l}{\partial x'^\mu \partial x'^\nu}.$$

8. Riemann symbols and the Riemann tensor. The Ricci tensor. We consider now equation (7.14) and the similar equation

$$(8.1)\qquad \frac{\partial^2 x^l}{\partial x'^\mu \partial x'^\sigma} = \left\{ {\lambda \atop \mu\sigma} \right\}' \frac{\partial x^l}{\partial x'^\lambda} - \left\{ {l \atop ij} \right\}\frac{\partial x^i}{\partial x'^\mu}\frac{\partial x^j}{\partial x'^\sigma}.$$

If we differentiate this equation with respect to x'^ν and (7.14) with respect to x'^σ and eliminate $\frac{\partial^3 x^l}{\partial x'^\mu \partial x'^\nu \partial x'^\sigma}$, the resulting equation is reducible by means of equations of the form (8.1) to

$$(8.2)\qquad R'^\lambda{}_{\mu\sigma\nu}\frac{\partial x^l}{\partial x'^\lambda} = R^l{}_{ijk}\frac{\partial x^i}{\partial x'^\mu}\frac{\partial x^j}{\partial x'^\sigma}\frac{\partial x^k}{\partial x'^\nu},$$

where

$$(8.3)\qquad R^l{}_{ijk} = \frac{\partial}{\partial x^j}\left\{ {l \atop ik} \right\} - \frac{\partial}{\partial x^k}\left\{ {l \atop ij} \right\} + \left\{ {m \atop ik} \right\}\left\{ {l \atop mj} \right\} - \left\{ {m \atop ij} \right\}\left\{ {l \atop mk} \right\},$$

and $R'^{\lambda}{}_{\mu\sigma\nu}$ is the similar expression in the symbols for $g'_{\mu\nu}$. If (8.2) be multiplied by $\dfrac{\partial x'^{\alpha}}{\partial x^l}$ and summed for l, we have

$$R'^{\alpha}{}_{\mu\sigma\nu} = R^l{}_{ihj} \frac{\partial x'^{\alpha}}{\partial x^l} \frac{\partial x^i}{\partial x'^{\mu}} \frac{\partial x^h}{\partial x'^{\sigma}} \frac{\partial x^j}{\partial x'^{\nu}}. \tag{8.4}$$

Hence $R^l{}_{ihj}$, which are called *Riemann symbols of the second kind*, are the components of a tensor contravariant of the first order and covariant of the third order. It is called the *mixed Riemann tensor of the fourth order*. From (8.3) it follows that the tensor is skew-symmetric in j and k. The components R_{hijk} of the associate covariant tensor of the fourth order, defined by

$$R_{hijk} = g_{lh} R^l{}_{ijk}, \qquad R^l{}_{ijk} = g^{lh} R_{hijk}, \tag{8.5}$$

are called the *Riemann symbols of the first kind*.

If (8.2) be multiplied by $g_{lh} \dfrac{\partial x^h}{\partial x'^{\tau}}$ and summed for l, we have, in consequence of (7.10) and (8.5),

$$R'_{\tau\mu\sigma\nu} = R_{hijk} \frac{\partial x^h}{\partial x'^{\tau}} \frac{\partial x^i}{\partial x'^{\mu}} \frac{\partial x^j}{\partial x'^{\sigma}} \frac{\partial x^k}{\partial x'^{\nu}}. \tag{8.6}$$

From (7.3) and (7.4) we have

$$\begin{aligned} g_{lh} \frac{\partial}{\partial x^j} \begin{Bmatrix} l \\ ik \end{Bmatrix} &= \frac{\partial}{\partial x^j} \left(g_{lh} \begin{Bmatrix} l \\ ik \end{Bmatrix} \right) - \begin{Bmatrix} l \\ ik \end{Bmatrix} \frac{\partial g_{lh}}{\partial x^j} \\ &= \frac{\partial}{\partial x^j} [ik, h] - \begin{Bmatrix} l \\ ik \end{Bmatrix} ([lj, h] + [hj, l]). \end{aligned} \tag{8.7}$$

Hence from (8.3), (8.5) and (8.7) we obtain

$$R_{hijk} = \frac{\partial}{\partial x^j} [ik, h] - \frac{\partial}{\partial x^k} [ij, h] + \begin{Bmatrix} l \\ ij \end{Bmatrix} [hk, l] - \begin{Bmatrix} l \\ ik \end{Bmatrix} [hj, l]. \tag{8.8}$$

In consequence of (7.1) and (7.2) this is reducible to

$$\begin{aligned} R_{hijk} = \frac{1}{2} &\left(\frac{\partial^2 g_{hk}}{\partial x^i \partial x^j} + \frac{\partial^2 g_{ij}}{\partial x^h \partial x^k} - \frac{\partial^2 g_{hj}}{\partial x^i \partial x^k} - \frac{\partial^2 g_{ik}}{\partial x^h \partial x^j} \right) \\ &+ g^{lm} ([ij, m] [hk, l] - [ik, m] [hj, l]). \end{aligned} \tag{8.9}$$

From (8.9) we find that the symbols of the first kind satisfy the following identities:

$$
\begin{aligned}
R_{hijk} &= -R_{ihjk},\\
R_{hijk} &= -R_{hikj}, \qquad (8.10)\\
R_{hijk} &= R_{jkhi},
\end{aligned}
$$

and

$$R_{hijk} + R_{hjki} + R_{hkij} = 0. \tag{8.11}$$

From (8.10) it follows that not more than two of the indices can be alike without the components vanishing; the same is true if the first two or second two indices are alike. Because of (8.10) there are $n(n-1)/2$ ($\equiv n_2$) ways in which the first pair of indices are like the second pair, and $n_2(n_2-1)/2$ ways in which the first pair and second pair are unlike; hence there is a total of $n_2(n_2+1)/2$ distinct symbols as regards (8.10). However, there are $n(n-1)(n-2)(n-3)/4!$ ($\equiv n_4$) equations of the form (8.11). Consequently there are $n_2(n_2+1)/2 - n_4 = n^2(n^2-1)/12$ distinct symbols of the first kind.*

In consequence of (8.10) we have from (8.8)

$$R_{ihkj} = \frac{\partial}{\partial x^j}[ik, h] - \frac{\partial}{\partial x^k}[ij, h] + \left\{ {l \atop ij} \right\}[hk, l] - \left\{ {l \atop ik} \right\}[hj, l]. \tag{8.12}$$

Also from (8.10) and (8.5) we have

$$R^l_{\ ijk} = -R_i{}^l{}_{jk}.\dagger \tag{8.13}$$

If $R^l_{\ ijk}$ be contracted for l and k, we have, in consequence of (7.9), the tensor R_{ij} whose components are given by

$$
\begin{aligned}
R_{ij} = R^k_{\ ijk} = \frac{\partial^2 \log \sqrt{g}}{\partial x^i \partial x^j} - \frac{\partial}{\partial x^k}\left\{ {k \atop ij} \right\} + \left\{ {m \atop ik} \right\}\left\{ {k \atop mj} \right\}& \\
- \left\{ {m \atop ij} \right\} \frac{\partial \log \sqrt{g}}{\partial x^m}&,
\end{aligned} \tag{8.14}
$$

* Cf., *Christoffel*, 1869, 1, p. 55.

† *Ricci* and *Levi Civita*, 1901, 1, p. 142 denote R_{ihkj} as defined by (8.12) by $a_{ih,kj}$, and *Bianchi*, 1902, 1, p. 73 denotes it by (ih, kj). Also the latter puts $\{il, kj\} = g^{lh}(ih, kj)$; hence $\{il, kj\}$ is equal to $-R^l_{\ ijk}$ by (8.13).

which evidently is symmetric. We call the tensor R_{ij} the *Ricci tensor*, as it was first considered by Ricci who gave it a geometrical interpretation in case g_{ij} is the fundamental tensor of a Riemann space (cf. § 34).*

Exercises

1. If R^l_{ijk} in (8.3) is contracted for l and i, the resulting tensor is a zero tensor.

2. If $R_{ij} = \varrho g_{ij}$, then $\varrho = \frac{1}{n} R$, where $R = g^{ij} R_{ij}$.

3. Show from (7.14) that for transformations $x'^i = \varphi^i(x^1, \cdots, x^{n-1})$, $x'^n = x^n$ the Christoffel symbols $\left\{ {n \atop ij} \right\}$, where $i, j = 1, \cdots, n-1$, are the components of a symmetric covariant tensor in a variety $x^n = \text{const.}$; likewise $\left\{ {i \atop nj} \right\}$ and $\left\{ {n \atop ni} \right\}$ are the components of a mixed tensor and a covariant vector respectively.

4. Show that the tensor equation $a^i_j \lambda_i = \alpha \lambda_j$, where α is an invariant, can be written in the form $(a^i_j - \alpha \delta^i_j) \lambda_i = 0$. Show also that $a^i_j = \delta^i_j \alpha$, if the equation is to hold for an arbitrary vector λ_i.

5. If $a^i_j \lambda_i = \alpha \lambda_j$ holds for all vectors λ_i such that $\mu^i \lambda_i = 0$, where μ^i is a given vector, then

$$a^i_j = \alpha \delta^i_j + \sigma_j \mu^i.$$

Schouten, 1924, 1, p. 59.

9. Quadratic differential forms.

If g_{ij} are the components of a tensor, the *quadratic differential form* $g_{ij}\, dx^i\, dx^j$ is an invariant, that is (§§ 2, 3),

$$g'_{\mu\nu}\, dx'^\mu\, dx'^\nu = g_{ij}\, dx^i\, dx^j. \tag{9.1}$$

Conversely if this condition is satisfied for arbitrary values of the differentials, it follows from equations similar to (1.9) that

$$\left(g'_{\mu\nu} - g_{ij} \frac{\partial x^i}{\partial x'^\mu} \frac{\partial x^j}{\partial x'^\nu} \right) dx'^\mu\, dx'^\nu = 0,$$

and consequently

$$g'_{\mu\nu} + g'_{\nu\mu} = (g_{ij} + g_{ji}) \frac{\partial x^i}{\partial x'^\mu} \frac{\partial x^j}{\partial x'^\nu}.$$

If we assume that g_{ij} is symmetric this reduces to (7.10). However, if in (9.1) we put $\bar{g}_{ij} = \frac{1}{2}(g_{ij} + g_{ji})$, we have a quadratic form whose coefficients are symmetric. Hereafter we assume that we deal with symmetric forms.

---

* *Ricci*, 1904, 2, p. 1234.

At any point of space $g_{ij}\, dx^i\, dx^j$ is an algebraic quadratic form in the differentials, and the transformation (1.9) is a linear transformation with constant coefficients. Hence we can apply the algebraic theory of transformations at a point. In particular, we know that the values of $\frac{\partial x'^i}{\partial x^j}$ can be chosen at a point so that $g'_{\mu\nu} = 0$ for $\mu \neq \nu$. If the transformation is to be real, it is not always possible to choose the transformation so that all of the quantities $g'_{\mu\mu}$ are positive. But according to Sylvester's law of inertia the difference between the number of positive coefficients and the number of negative coefficients is invariant for real transformations; this difference is called the *signature* of the form. Thus by a real transformation a quadratic form at a point is reducible to

$$(9.2)\qquad (dx^1)^2 + \cdots + (dx^p)^2 - (dx^{p+1})^2 - \cdots - (dx^n)^2,$$

where the integer $2p - n$ is the signature of the form.* In particular, if the signature is n for each point of space, the quadratic form is said to be *positive definite*.

If g' denotes the determinant $|g'_{\mu\nu}|$, from the rule for multiplication of determinants and (7.10) it follows that

$$(9.3)\qquad g' = gJ^2,$$

where J is the Jacobian $\left|\frac{\partial x^i}{\partial x'^\mu}\right|$. Thus if g and g' differ in sign at a point, the transformation is imaginary.

10. The equivalence of symmetric quadratic differential forms. We have seen that equations (7.10) are a necessary consequence of the equivalence of two symmetric quadratic forms (9.1). We seek further conditions upon the g's and the g''s in order that (7.10) may admit a set of n independent solutions $x^i = \psi^i(x'^1, \cdots, x'^n)$ for $i = 1, \cdots, n$, by means of which the forms (9.1) are transformable into one another.

If we put

$$(10.1)\qquad \frac{\partial x^i}{\partial x'^\mu} = p^i_\mu,$$

* Cf. *Bôcher*, 1907, 1, p. 146.

equations (8.1) become

$$(10.2)\qquad \frac{\partial p^l_\mu}{\partial x'^\sigma} = \left\{ {\lambda \atop \mu\,\sigma} \right\}' p^l_\lambda - \left\{ {l \atop i\,j} \right\} p^i_\mu p^j_\sigma .$$

Hence the problem reduces to the determination of $n(n+1)$ functions x^i, p^i_μ satisfying these differential equations and also the $n(n+1)/2$ finite equations

$$(10.3)\qquad g'_{\mu\nu} - g_{ij} p^i_\mu p^j_\nu = 0,$$

which follow from (7.10).

The conditions of integrability of (10.1) are satisfied identically in consequence of (10.2), and the conditions of integrability of (10.2) are

$$(10.4)\qquad R'_{\tau\mu\sigma\nu} = R_{hijk}\, p^h_\tau\, p^i_\mu\, p^j_\sigma\, p^k_\nu ,$$

as follows from (8.6) which is equivalent to (8.2).

From the manner in which equations (7.14) were obtained from (7.10) it follows that for any set of solutions of (10.1) and (10.2) the left-hand member of (10.3) is constant, and consequently, if the initial values are chosen to satisfy (10.3), the solutions will satisfy (10.3). This imposes $n(n+1)/2$ conditions on the constants of integration of (10.1) and (10.2). Hence the solution, if it exists, admits at most $n(n+1)/2$ arbitrary constants, and then only, if (10.4) is satisfied identically or as a consequence of (10.3). For otherwise equations (10.4) impose further conditions, as may also the equations obtained by differentiating them and substituting the expressions for the first derivatives from (10.2). This result may be stated as follows:

The general transformation of a quadratic differential form in n variables into another form contains at most $n(n+1)/2$ arbitrary constants.

From the results of § 9 it follows that for the transformations to be real at a point the signature of the two forms must be equal at the point.

Consider in particular the case of two sets of functions g_{ij} and $g'_{\mu\nu}$ for which the Riemann symbols of the first kind for both sets vanish. Then (10.4) is satisfied identically and consequently the differential forms $g_{ij}\,dx^i\,dx^j$ and $g'_{\mu\nu}\,dx'^\mu\,dx'^\nu$ are transformable into one another by a transformation involving $n(n+1)/2$ constants. The Riemann symbols of the first kind for the g''s are zero, if the quantities $g'_{\mu\nu}$ are constants, as follows from (7.1) and (8.8), and these symbols for the g's must be zero, if the two forms are equivalent. Hence:

A necessary and sufficient condition that a quadratic differential form $g_{ij}\,dx^i\,dx^j$ be reducible to a form with constant coefficients is that the components of the Riemann tensor vanish; the transformation involves $n(n+1)/2$ arbitrary constants.

From the results of § 9 it follows that any quadratic form satisfying the conditions of the theorem is reducible by real transformations to the form (9.2), where p is determined by the signature of the given form.

Returning to the consideration of (10.4), we remark that, if (10.4) is to be a consequence of (10.3), the tensor R_{hijk} must be the sum of tensors of the fourth order whose terms are products of two g's. Since g_{ij} is symmetric, the most general form is

$$R_{hijk} = a\,g_{hi}\,g_{jk} + b\,g_{hj}\,g_{ik} + c\,g_{hk}\,g_{ij},$$

where a, b, c are invariants. Interchanging j and k and subtracting the resulting equation from the above, we have, in consequence of (8.10) and on replacing $\frac{1}{2}(b-c)$ by b,

$$R_{hijk} = b(g_{hj}\,g_{ik} - g_{hk}\,g_{ij}). \tag{10.5}$$

It is readily shown that (8.10) and (8.11) are satisfied, whatever be b. However, it will be shown in § 26 that b must be a constant. A quadratic differential form possessing the property (10.5) is said to have *constant curvature* b; the significance of this term will appear in § 26.

When two given quadratic forms satisfy (10.5) for the same constant b, the equations (10.4) are satisfied identically. Hence:

Two irreducible quadratic differential forms which have the same constant curvature admit a transformation into one another involving $n(n+1)/2$ arbitrary constants; conversely, unless this condition is satisfied by two irreducible forms the number of parameters is less than $n(n+1)/2$.

It is beyond the scope of this work to consider further the equivalence of two quadratic differential forms. Christoffel* has given the solution of the general problem.

11. Covariant differentiation with respect to a tensor g_{ij}. In § 3 it was seen that the derivatives of an invariant are the components of a covariant vector. It will be shown that this is the only case for a general system of coordinates in which the derivatives of the components of a tensor are the components of a tensor, but at the same time we shall find expressions involving the first derivatives which are components of a tensor.

Let λ^i and λ'^μ be the components in two coordinate systems of a contravariant vector, and differentiate with respect to x^j the equation

$$\lambda^i = \lambda'^\mu \frac{\partial x^i}{\partial x'^\mu} = \lambda'^\sigma \frac{\partial x^i}{\partial x'^\sigma}; \tag{11.1}$$

with the aid of (8.1), (2.1) and (2.2), we obtain

$$\begin{aligned}
\frac{\partial \lambda^i}{\partial x^j} &= \frac{\partial \lambda'^\mu}{\partial x'^\nu}\frac{\partial x'^\nu}{\partial x^j}\frac{\partial x^i}{\partial x'^\mu} + \lambda'^\sigma \frac{\partial^2 x^i}{\partial x'^\sigma \partial x'^\nu}\frac{\partial x'^\nu}{\partial x^j} \\
&= \frac{\partial \lambda'^\mu}{\partial x'^\nu}\frac{\partial x'^\nu}{\partial x^j}\frac{\partial x^i}{\partial x'^\mu} + \lambda'^\sigma \frac{\partial x'^\nu}{\partial x^j}\left(\begin{Bmatrix}\mu\\ \sigma\nu\end{Bmatrix}'\frac{\partial x^i}{\partial x'^\mu} - \begin{Bmatrix}i\\ hk\end{Bmatrix}\frac{\partial x^h}{\partial x'^\sigma}\frac{\partial x^k}{\partial x'^\nu}\right) \\
&= \left(\frac{\partial \lambda'^\mu}{\partial x'^\nu} + \lambda'^\sigma \begin{Bmatrix}\mu\\ \sigma\nu\end{Bmatrix}'\right)\frac{\partial x'^\nu}{\partial x^j}\frac{\partial x^i}{\partial x'^\mu} - \lambda^h \begin{Bmatrix}i\\ hj\end{Bmatrix}.
\end{aligned}$$

If we put

$$\lambda^i{}_{,j} = \frac{\partial \lambda^i}{\partial x^j} + \lambda^h \begin{Bmatrix}i\\ hj\end{Bmatrix}, \tag{11.2}$$

the above equation becomes

$$\lambda^i{}_{,j} = \lambda'^\mu{}_{,\nu}\frac{\partial x'^\nu}{\partial x^j}\frac{\partial x^i}{\partial x'^\mu}.$$

Hence $\lambda^i{}_{,j}$ are the components of a mixed tensor of the second order. The components $\lambda^i{}_{,j}$ as defined by (11.2) are said to be

* 1869, 1, p. 60.

obtained from the vector λ^i by *covariant differentiation with respect to the tensor* g_{ij}. We speak also of the tensor as the covariant derivative of the vector with respect to g_{ij}. Throughout the remainder of this chapter it is understood that covariant differentiation is with respect to g_{ij}.

If we proceed in similar manner with equations (3.3), we find that $\lambda_{i,j}$, defined by

$$\lambda_{i,j} = \frac{\partial \lambda_i}{\partial x^j} - \lambda_h \begin{Bmatrix} h \\ ij \end{Bmatrix}, \tag{11.3}$$

are the components of a covariant tensor of the second order. The components $\lambda_{i,j}$ are said to be obtained from the vector λ_i *by covariant differentiation with respect to the tensor* g_{ij}.

From (11.3) we have

$$\lambda_{i,j} - \lambda_{j,i} = \frac{\partial \lambda_i}{\partial x^j} - \frac{\partial \lambda_j}{\partial x^i},$$

which is the *curl* of the vector λ_i. For $\lambda_{i,j}$ to be symmetric, λ_i must be a gradient (§ 3). Hence:

A necessary and sufficient condition that the first covariant derivative of a covariant vector be symmetric is that the vector be a gradient.

If we differentiate with respect to x'^σ the equation

$$a'_{\mu\nu} = a_{ij} \frac{\partial x^i}{\partial x'^\mu} \frac{\partial x^j}{\partial x'^\nu}$$

and substitute for the second derivatives of x^i and x^j expressions of the form (8.1), the resulting equation is reducible to

$$\frac{\partial a'_{\mu\nu}}{\partial x'^\sigma} - a'_{\mu\lambda} \begin{Bmatrix} \lambda \\ \nu\sigma \end{Bmatrix}' - a'_{\lambda\nu} \begin{Bmatrix} \lambda \\ \mu\sigma \end{Bmatrix}'$$
$$= \left(\frac{\partial a_{ij}}{\partial x^k} - a_{ih} \begin{Bmatrix} h \\ jk \end{Bmatrix} - a_{hj} \begin{Bmatrix} h \\ ik \end{Bmatrix} \right) \frac{\partial x^i}{\partial x'^\mu} \frac{\partial x^j}{\partial x'^\nu} \frac{\partial x^k}{\partial x'^\sigma}.$$

Hence $a_{ij,k}$, defined by

$$a_{ij,k} = \frac{\partial a_{ij}}{\partial x^k} - a_{ih} \begin{Bmatrix} h \\ jk \end{Bmatrix} - a_{hj} \begin{Bmatrix} h \\ ik \end{Bmatrix}, \tag{11.4}$$

are the components of a covariant tensor of the third order. The components $a_{ij,k}$ are called the first covariant derivatives of a_{ij} with respect to g_{ij}. In like manner it can be shown that the covariant derivatives of a^{ij} and a^i_j, defined by

$$a^{ij}{}_{,k} = \frac{\partial a^{ij}}{\partial x^k} + a^{ih}\left\{ {j \atop hk} \right\} + a^{hj}\left\{ {i \atop hk} \right\}, \tag{11.5}$$

and

$$a^i_{j,k} = \frac{\partial a^i_j}{\partial x^k} + a^h_j \left\{ {i \atop hk} \right\} - a^i_h \left\{ {h \atop jk} \right\}, \tag{11.6}$$

are mixed tensors of the second order. Observe that covariant differentiation is indicated by a subscript preceded by a comma. In particular, the covariant derivative of an invariant f is the ordinary derivative of the function, and is indicated by $f_{,i}$.

The general rule for covariant differentiation is

$$\begin{aligned} a^{r_1 \cdots r_m}_{s_1 \cdots s_p, i} = \frac{\partial a^{r_1 \cdots r_m}_{s_1 \cdots s_p}}{\partial x^i} &+ \sum_{\alpha}^{1 \cdots m} a^{r_1 \cdots r_{\alpha-1} j r_{\alpha+1} \cdots r_m}_{s_1 \cdots s_p} \left\{ {r_\alpha \atop j\,i} \right\} \\ &- \sum_{\beta}^{1 \cdots p} a^{r_1 \cdots r_m}_{s_1 \cdots s_{\beta-1} l s_{\beta+1} \cdots s_p} \left\{ {l \atop s_\beta\, i} \right\}.* \end{aligned} \tag{11.7}$$

From (11.4), (7.4) and (11.5), (7.6) we have

$$g_{ij,k} = 0, \qquad g^{ij}{}_{,k} = 0. \tag{11.8}$$

Also from (1.5) and (11.6)

$$\delta^i_{j,k} = 0. \tag{11.9}$$

In consequence of the form of (11.7) it follows that the covariant derivative of the sum (or difference) of two tensors of the same order and kind is the sum (or difference) of their covariant derivatives.

If we effect the covariant derivative of the tensor $a_{ij}\, b^{kl}$, we have

$$\begin{aligned} (a_{ij}\, b^{kl})_{,m} &= \frac{\partial}{\partial x^m}(a_{ij}\, b^{kl}) - b^{kl}\left(a_{hj}\left\{ {h \atop im} \right\} + a_{ih}\left\{ {h \atop jm} \right\}\right) \\ &\qquad + a_{ij}\left(b^{kh}\left\{ {l \atop hm} \right\} + b^{hl}\left\{ {k \atop hm} \right\}\right) \\ &= b^{kl}\, a_{ij,m} + a_{ij}\, b^{kl}{}_{,m}, \end{aligned}$$

* The tensor character of covariant derivatives was first established by *Christoffel*, 1869, 1, p. 56.

which is the same as the rule of the differential calculus. Since a tensor formed by multiplication and contraction is a sum of products, we have also

$$(a_{ij}\, b^{jl})_{,k} = a_{ij,k}\, b^{jl} + a_{ij}\, b^{jl}{}_{,k}.$$

Hence we have the general rule:

Covariant differentiation of the sum, difference, outer and inner multiplication of tensors obeys the same rules as in ordinary differentiation.

From (11.8) and (11.9) follows also the rule:

The tensors g_{ij}, g^{ij} and δ^i_j behave as though they were constants in covariant differentiation with respect to g_{ij}.

Thus if λ^i and μ_i are any vectors and λ_i and μ^i are their respective associates by means of g_{ij} (§ 6), the derivatives of the invariant

$$I = \lambda^i \mu_i = g^{il} \lambda_l \mu_i \tag{11.10}$$

are given by

$$I_{,k} = g^{il}(\lambda_{l,k}\, \mu_i + \lambda_l\, \mu_{i,k}) = \mu^l \lambda_{l,k} + \lambda^i \mu_{i,k}. \tag{11.11}$$

If λ_i in (11.3) is the gradient $f_{,i}$ of an invariant f, we have

$$f_{,ij} - f_{,ji} = \frac{\partial}{\partial x^j}\left(\frac{\partial f}{\partial x^i}\right) - \frac{\partial}{\partial x^i}\left(\frac{\partial f}{\partial x^j}\right) = 0, \tag{11.12}$$

$f_{,ij}$ denoting the first covariant derivative of $f_{,i}$ and the *second* of f.

It will be found that this is the only case in which the order of covariant differentiation is immaterial.

If we differentiate covariantly the tensor $\lambda_{i,j}$ defined by (11.3), we have

$$\begin{aligned}
\lambda_{i,jk} &= \frac{\partial}{\partial x^k}\left(\frac{\partial \lambda_i}{\partial x^j} - \lambda_l \begin{Bmatrix} l \\ ij \end{Bmatrix}\right) - \left(\frac{\partial \lambda_h}{\partial x^j} - \lambda_l \begin{Bmatrix} l \\ hj \end{Bmatrix}\right)\begin{Bmatrix} h \\ ik \end{Bmatrix} \\
&\qquad - \left(\frac{\partial \lambda_i}{\partial x^h} - \lambda_l \begin{Bmatrix} l \\ ih \end{Bmatrix}\right)\begin{Bmatrix} h \\ jk \end{Bmatrix} \\
&= \frac{\partial^2 \lambda_i}{\partial x^j\, \partial x^k} - \frac{\partial \lambda_l}{\partial x^k}\begin{Bmatrix} l \\ ij \end{Bmatrix} - \frac{\partial \lambda_h}{\partial x^j}\begin{Bmatrix} h \\ ik \end{Bmatrix} - \frac{\partial \lambda_i}{\partial x^h}\begin{Bmatrix} h \\ jk \end{Bmatrix} \\
&\qquad - \lambda_l\left(\frac{\partial}{\partial x^k}\begin{Bmatrix} l \\ ij \end{Bmatrix} - \begin{Bmatrix} l \\ ih \end{Bmatrix}\begin{Bmatrix} h \\ jk \end{Bmatrix} - \begin{Bmatrix} h \\ ik \end{Bmatrix}\begin{Bmatrix} l \\ hj \end{Bmatrix}\right).
\end{aligned} \tag{11.13}$$

Consequently we have

(11.14) $$\lambda_{i,jk} - \lambda_{i,kj} = \lambda_l R^l_{ijk},$$

where R^l_{ijk} is given by (8.3).

In like manner for a tensor a_{ij} we find

(11.15) $$a_{ij,kl} - a_{ij,lk} = a_{ih} R^h_{jkl} + a_{hj} R^h_{ikl},$$

and in general

(11.16) $$a_{r_1 \cdots r_m, kl} - a_{r_1 \cdots r_m, lk} = \sum_{\alpha}^{1 \cdots m} a_{r_1 \cdots r_{\alpha-1} h r_{\alpha+1} \cdots r_m} R^h_{r_\alpha kl}.$$

This result is due to Ricci and is called the *Ricci identity*.* When covariant differentiation is used in place of ordinary differentiation, this identity must be used in place of the ordinary condition of integrability. Thus (11.14) follows from (11.13) as a consequence of

$$\frac{\partial}{\partial x^k}\left(\frac{\partial \lambda_i}{\partial x^j}\right) = \frac{\partial}{\partial x^j}\left(\frac{\partial \lambda_i}{\partial x^k}\right).$$

The corresponding formulas for contravariant tensors follow on raising indices by means of g^{ij} and noting that the latter behave like constants in covariant differentiation. Thus, if (11.14) be multiplied by g^{ih} and summed for i, we have

$$(g^{ih}\lambda_i)_{,jk} - (g^{ih}\lambda_i)_{,kj} = g^{ih}\lambda^l R_{lijk} = -g^{ih}\lambda^l R_{iljk},$$

and consequently

(11.17) $$\lambda^h_{,jk} - \lambda^h_{,kj} = -\lambda^l R^h_{ljk}.$$

In general

(11.18) $$a^{r_1 \cdots r_m}_{s_1 \cdots s_p, jk} - a^{r_1 \cdots r_m}_{s_1 \cdots s_p, kj} = \sum_{\alpha}^{1 \cdots p} a^{r_1 \cdots r_m}_{s_1 \cdots s_{\alpha-1} l s_{\alpha+1} \cdots s_p} R^l_{s_\alpha jk} - \sum_{\beta}^{1 \cdots m} a^{r_1 \cdots r_{\beta-1} l r_{\beta+1} \cdots r_m}_{s_1 \cdots s_p} R^{r_\beta}_{ljk}.$$

A necessary and sufficient condition that the Christoffel symbols be zero is that all of the g_{ij}'s be constant, as follows from (7.1) and (7.4). Combining this result with the second theorem of § 10, we have the theorem:

* *Ricci* and *Levi-Civita*, 1901, 1, p. 143.

In order that there exist a coördinate system in which the first covariant derivatives with respect to a tensor g_{ij} reduce to ordinary derivatives at every point in space, it is necessary and sufficient that the Riemann symbols formed with respect to g_{ij} be zero and that the x's be those for which g_{ij} are constants. (Cf. § 18.)

Exercises

1. The second theorem of § 11, and the identities (11.16) and (11.18) are consequences of the definitions of covariant differentiation and do not involve an assumption that the quantities differentiated are components of tensors.

2. By applying the general rule of covariant differentiation of § 11 to the invariant $\lambda^i \mu_i$ show that this rule implies that the covariant derivative of an invariant is the ordinary derivative.

3. The tensor defined by

$$a^{\alpha_1 \cdots \alpha_r, i}_{\beta_1 \cdots \beta_s} = g^{il} a^{\alpha_1 \cdots \alpha_r}_{\beta_1 \cdots \beta_s, l}$$

is called the *contravariant* derivative of $a^{\alpha_1 \cdots \alpha_r}_{\beta_1 \cdots \beta_s}$ with respect to g_{ij}. Show that $g^{ij,k} = 0$. *Ricci* and *Levi-Civita*, 1901, 1, p. 140.

4. If a_{ij} is the curl of a covariant vector, show that

$$a_{ij,k} + a_{jk,i} + a_{ki,j} = 0,$$

and that this is equivalent to

$$\frac{\partial a_{ij}}{\partial x^k} + \frac{\partial a_{jk}}{\partial x^i} + \frac{\partial a_{ki}}{\partial x^j} = 0.$$

Is this condition sufficient as well as necessary that a skew-symmetric tensor a_{ij} be the curl of a vector? *Eisenhart*, 1922, 1.

5. By definition a^{ijk}_{lm} are the components of a *relative tensor* of *weight p*, if the equations connecting the components in two coördinate systems are of the form

$$a'^{\alpha\beta\gamma}_{\delta\varepsilon} = J^p a^{ijk}_{lm} \frac{\partial x'^\alpha}{\partial x^i} \frac{\partial x'^\beta}{\partial x^j} \frac{\partial x'^\gamma}{\partial x^k} \frac{\partial x^l}{\partial x'^\delta} \frac{\partial x^m}{\partial x'^\varepsilon},$$

where J is the Jacobian $\left| \frac{\partial x^i}{\partial x'^\alpha} \right|$. Show that if a_{ij} is a covariant tensor, then the cofactor of a_{ij} in the determinant $|a_{ij}|$ is a relative contravariant tensor of weight two.

6. If $a_{\alpha\beta}$ is a covariant tensor of rank $n-1$ (cf. Ex. 7, p. 16), there exist two relative vectors λ^α and μ^α, both of weight one, such that the cofactor $A^{\alpha\beta}$ of $a_{\alpha\beta}$ is of the form $A^{\alpha\beta} = \lambda^\alpha \mu^\beta$. When $a_{\alpha\beta}$ is symmetric, λ^α and μ^α are the same relative vectors.

7. When a relative tensor is of weight one it is called a *tensor density*. Show that if the components of any tensor are multiplied by the square root of the non-vanishing determinant of a covariant tensor, they are the components of a tensor density.

8. The invariant $\lambda^i_{,i}$ is called the *divergence* of the vector λ^i with respect to the symmetric tensor g_{ij}. Show that

$$\lambda^i_{,i} = \frac{1}{\sqrt{g}} \frac{\partial}{\partial x^i}(\lambda^i \sqrt{g}).$$

9. Show that the divergence of the tensor a^{ij} with respect to the symmetric tensor g_{ij}, that is, $a^{ij}_{,j}$, has the expression

$$a^{ij}_{,j} = \frac{1}{\sqrt{g}} \frac{\partial}{\partial x^j}(a^{ij}\sqrt{g}) + a^{jk}\left\{ {i \atop jk} \right\},$$

and that the last term vanishes, if a^{ij} is skew-symmetric.

10. The divergence of a mixed tensor a_i^j is reducible to

$$a_{i,j}^j = \frac{1}{\sqrt{g}} \frac{\partial}{\partial x^j}(a_i^j \sqrt{g}) - a_l^j \left\{ {l \atop ij} \right\}.$$

Show that if the associate tensor a^{ij} is symmetric,

$$a_{i,j}^j = \frac{1}{\sqrt{g}} \frac{\partial}{\partial x^j}(a_i^j\sqrt{g}) - \frac{1}{2} a^{jk} \frac{\partial g_{jk}}{\partial x^i} = \frac{1}{\sqrt{g}} \frac{\partial}{\partial x^j}(a_i^j \sqrt{g}) + \frac{1}{2} a_{jk} \frac{\partial g^{jk}}{\partial x^i}.$$

Einstein, 1916, 1, p. 799.

11. When g_{ij} and a_{ij} are the components of two symmetric tensors, if

$$g_{ij}\, a_{kl} - g_{il}\, a_{jk} + g_{jk}\, a_{il} - g_{kl}\, a_{ij} = 0 \qquad (i, j, k, l = 1, \cdots, n),$$

then $a_{ij} = \varrho\, g_{ij}$.

12. If a_{ijkl} is a tensor satisfying the conditions (8.10) and for a vector λ^i we have $\lambda^i a_{ijkl} = 0$, a coördinate system x'^i can be chosen for which a'_{ijkl} are zero, when one or more of the indices is n.

13. Let $\lambda_{\alpha|}{}^i$ for $i = 1, \cdots, n$ denote the components of n independent contravariant vectors, where the value of α for $\alpha = 1, \cdots, n$ indicates the vector (cf. Ex. 3, p. 8), and let Λ_i^α denote the cofactor of $\lambda_{\alpha|}{}^i$ in the determinant $\Lambda = |\lambda_{\alpha|}{}^i|$ divided by Λ. Show that the quantities Λ_i^α for each coördinate system are the components of a covariant vector, α indicating the vector and i the component.

14. Show that if $a_{hijk}\, \lambda_{1|}{}^h \lambda_{2|}{}^i \lambda_{1|}{}^j \lambda_{2|}{}^k = 0$ for any two arbitrary vectors $\lambda_{1|}{}^i$ and $\lambda_{2|}{}^i$, then

$$a_{hijk} + a_{hkji} + a_{jihk} + a_{jkhi} = 0;$$

also when a_{hijk} possesses the properties (8.10) and (8.11), then $a_{hijk} = 0$.

15. Show that when in a V_3 the coördinates can be chosen (Cf. § 15) so that the components of a tensor g_{ij} are zero when $i \neq j$, then

$$R_{hj} = \frac{1}{g_{ii}} R_{hiij},$$

$$R_{hh} = \frac{1}{g_{ii}} R_{hiih} + \frac{1}{g_{jj}} R_{hjjh},$$

$$R_{hiih} - g_{hh}\, R_{ii} - g_{ii}\, R_{hh} + \frac{1}{2} R\, g_{hh}\, g_{ii} = 0 \qquad (h, i, j \neq),$$

where $R = g^{ij} R_{ij}$. Hence the tensor C_{hijk}, defined by

$$C_{hijk} = R_{hijk} + g_{jh} R_{ik} - g_{hk} R_{ij} + g_{ik} R_{hj} - g_{ij} R_{hk} + \frac{R}{2} (g_{hk} g_{ij} - g_{hj} g_{ik}),$$

is a zero tensor (Cf. § 28).

16. If $a_{r_1 \cdots r_m}$ and $\bar{a}_{r_1 \cdots r_m}$ are the components of a tensor in V_n for coördinate systems in the relation

$$\bar{x}^1 = x^1, \quad \bar{x}^j = \varphi^j(x^2, \cdots, x^n) \qquad (j = 2, \cdots, n)$$

and $a_{r_1 \cdots r_m}$ and $\bar{a}_{r_1 \cdots r_m}$, where $r_1, \cdots, r_m = 2, \cdots, n$, are developed in power series in x^1, the coefficients of any power of x^1 in these developments are components of the same tensor in any hypersurface $x^1 =$ constant. *Levy*, 1925, 1.

17. If $a_{r_1 \cdots r_m}$ and $\bar{a}_{r_1 \cdots r_m}$ are the components of a tensor in V_n for coördinate systems x^i and $\bar{x}^i$ in the relation

$$\bar{x}^j = x^j \; (j = 1, \cdots, p), \quad \bar{x}^k = \varphi^k(x^{p+1}, \cdots, x^n) \qquad (k = p + 1, \cdots, n),$$

the functions $a_{r_1 \cdots r_m}$ and $\bar{a}_{r_1 \cdots r_m}$ for which $r_1, \cdots, r_m$ take the values $p + 1, \cdots, n$ and in which we put

(1) $$\bar{x}^j = x^j = a^j,$$

where the a's are constants, are components of the same tensor in the V_{n-p} defined by (1). *Levy*, 1925, 1.

18. If g_{ij} and $\bar{g}_{ij}$ are the components of two symmetric tensors, and $\left\{ {l \atop ij} \right\}$ and $\overline{\left\{ {l \atop ij} \right\}}$ are the corresponding Christoffel symbols, then b^l_{ij} defined by

$$\overline{\left\{ {l \atop ij} \right\}} = \left\{ {l \atop ij} \right\} + b^l_{ij}$$

are the components of a tensor. If $a^{\alpha_1 \cdots \alpha_r}_{\beta_1 \cdots \beta_s, i}$ and $a^{\alpha_1 \cdots \alpha_r}_{\beta_1 \cdots \beta_s, \bar{i}}$ denote the covariant derivatives of $a^{\alpha_1 \cdots \alpha_r}_{\beta_1 \cdots \beta_s}$ with respect to g_{ij} and $\bar{g}_{ij}$, then

$$a^{\alpha_1 \cdots \alpha_r}_{\beta_1 \cdots \beta_s, \bar{i}} - a^{\alpha_1 \cdots \alpha_r}_{\beta_1 \cdots \beta_s, i} = \sum_k^{1, \cdots, r} a^{\alpha_1 \cdots \alpha_{k-1} \sigma \alpha_{k+1} \cdots \alpha_r}_{\beta_1 \cdots \beta_s} b^{\alpha_k}_{\sigma i} - \sum_l^{1, \cdots, s} a^{\alpha_1 \cdots \alpha_r}_{\beta_1 \cdots \beta_{l-1} \sigma \beta_{l+1} \cdots \beta_s} b^{\sigma}_{\beta_l i}.$$

Also if R^i_{jkl} and $\bar{R}^i_{jkl}$ denote the corresponding Riemann symbols of the second kind, we have

$$\bar{R}^i_{jkl} - R^i_{jkl} = b^i_{jl,k} - b^i_{jk,l} + b^h_{jl} b^i_{hk} - b^h_{jk} b^i_{hl},$$

where the covariant derivatives are with respect to the tensor g_{ij}.

CHAPTER II

Introduction of a metric

12. Definition of a metric. The fundamental tensor. The geometry which has been considered thus far in the development of the ideas and processes of tensor analysis is geometry of position. In this geometry there is no basis for the determination of magnitude nor for a comparison of directions at two different points. In this chapter we define magnitude and parallelism, and develop consequences of these definitions.

We recall that the element of length of euclidean space of three dimensions, referred to cartesian coördinates, is given by

$$ds^2 = (dx^1)^2 + (dx^2)^2 + (dx^3)^2, \tag{12.1}$$

and for polar coördinates by

$$ds^2 = dr^2 + r^2(d\theta^2 + \sin^2\theta \, d\varphi^2). \tag{12.2}$$

This idea was generalized and applied to n-dimensions by *Riemann*,* who defined element of length by means of a quadratic differential form, thus $ds^2 = g_{ij}\, dx^i\, dx^j$, where the g's are functions of the x's. As thus defined ds is real for arbitrary values of the differentials only in case the quadratic form is assumed to be positive definite (§ 9). Much of the subsequent geometric development of this idea has been based on this assumption. However, the general theory of relativity has introduced a quadratic form which is not definite, and consequently it is advisable not to make the above assumption in the development of geometric ideas which are based on a quadratic differential form.

We take as the basis of the metric of space a real *fundamental quadratic* form

$$\varphi = g_{ij}\, dx^i\, dx^j, \tag{12.3}$$

---

* *Riemann*, 1854, 1.

where the g's are functions of the x's subject only to the restriction

$$(12.4) \qquad g = |g_{ij}| \neq 0.*$$

Element of length ds is defined by

$$(12.5) \qquad ds^2 = e g_{ij} \, dx^i \, dx^j,$$

where e is plus or minus one so that the right-hand member shall be positive, unless it is zero. The letter e will be used frequently and will always have this significance.

Since ds must be an invariant, it follows from § 9 that g_{ij} are the components of a covariant tensor of the second order which without loss of generality is assumed to be symmetric. It is called the *fundamental tensor* of the metric, and also is referred to as the fundamental tensor of the space. The metric defined by (12.5) is called the *Riemannian metric* and a geometry based upon such a metric is called a *Riemannian geometry*. Also we say that the space whose geometry is based upon such a metric is called a *Riemannian space*, just as a space with the metric (12.1) is called euclidean.

The significance of equation (12.5), as defining the element of length, is that ds is the magnitude of the contravariant vector of components dx^i. If λ^i are the components of any contravariant vector-field, then λ given by

$$(12.6) \qquad \lambda^2 = e g_{ij} \lambda^i \lambda^j$$

is an invariant, which is defined to be the *magnitude* of the vector (at each point of space). If λ_i are the components of any covariant vector and λ^i are the components of the associate vector (§ 6) by means of g^{ij}, the conjugate of g_{ij}, that is,

$$(12.7) \qquad \lambda^i = g^{ij} \lambda_j, \qquad \lambda_i = g_{ij} \lambda^j,$$

then

$$(12.8) \qquad g^{ij} \lambda_i \lambda_j = g^{ij} g_{ik} \lambda^k g_{jl} \lambda^l = g_{kl} \lambda^k \lambda^l = e\lambda^2.$$

Hence the invariant $g^{ij} \lambda_i \lambda_j$ is the square of the magnitude of the associate vector.

---

* Unless stated otherwise it is assumed that the coördinates are real.

If $\lambda = 0$ in (12.6) or (12.8), that is,

$$g_{ij}\lambda^i\lambda^j = 0 \quad \text{or} \quad g^{ij}\lambda_i\lambda_j = 0 \quad \text{or} \quad \lambda_i\lambda^i = 0, \tag{12.9}$$

at a point, we say that the vector is *null* at the point, and if (12.9) holds everywhere we have a *null* vector-field. If the fundamental form is definite at a point, at least one of the components of a null vector is imaginary at the point, in consequence of § 9.

If (12.9) is not satisfied, it follows from (12.6) and (12.8) that the components can be chosen so that respectively

$$g_{ij}\lambda^i\lambda^j = e, \qquad g^{ij}\lambda_i\lambda_j = e, \tag{12.10}$$

where, to use the above mentioned notation, e is plus or minus one according as the left-hand members are positive or negative. When the first of (12.10) is satisfied, we say that λ^i are the components of a *unit* contravariant vector; similarly the second of (12.10) is the condition for a *unit* covariant vector.

Any real curve C is defined by the x's as functions of a real parameter t (§ 2). Unless (12.3) is definite there may be portions of C for which, when dx^i in the right-hand member is replaced by $\frac{dx^i}{dt}dt$, this quantity is positive, negative, or zero. Let t_1 and t_2 be values of t at ends, or at interior points, of a portion for which this quantity is not zero. The *length* of the curve between these points is by definition

$$s = \int_{t_1}^{t_2} \sqrt{e g_{ij} \frac{dx^i}{dt} \frac{dx^j}{dt}}\, dt. \tag{12.11}$$

If we replace t_2 by t, equation (12.11) defines s as a function of t, and consequently the curve may be defined by the x's as functions of the fundamental parameter s, in which case we have

$$g_{ij}\frac{dx^i}{ds}\frac{dx^j}{ds} = e. \tag{12.12}$$

If for a portion of a curve, or for a whole curve,

$$g_{ij}\frac{dx^i}{dt}\frac{dx^j}{dt} = 0, \tag{12.13}$$

we say that it is of *length zero*, or *minimal*. We recall that in the space-time continuum of relativity certain lines of length zero are identified as the world-lines of light.

From continuity considerations it follows that a general curve consists of portions the length of which is thus defined, and hence we can speak of the length of a curve between any two of its points.

13. Angle of two vectors. Orthogonality. Let $\lambda_{1|}^{i}$ and $\lambda_{2|}^{i}$ be the components of two unit vectors, that is,

$$g_{ij}\,\lambda_{\alpha|}^{i}\,\lambda_{\alpha|}^{j} = e_{\alpha}, \qquad \alpha = (1,\,2).* \tag{13.1}$$

If we put

$$\cos\theta = g_{ij}\,\lambda_{1|}^{i}\,\lambda_{2|}^{j}, \tag{13.2}$$

it is clear that the right-hand member is an invariant determined by the two vectors. For euclidean space with the fundamental form (12.1) this is the cosine of the angle between the lines, and since it is an invariant it has the same meaning when polar coördinates, or any other, are used.

In the general case we define the *measure of the angle* by (13.2). Evidently $\cos\theta$ as thus defined is merely a symbol, unless the right-hand member is not greater than one in absolute value. In the latter case we give it the usual interpretation and thus the angle can be found. We shall show that this is always possible, if (12.3) is definite. In fact, $r\lambda_{1|}^{i} + t\lambda_{2|}^{i}$ are the components of a vector in the pencil determined by $\lambda_{1|}^{i}$ and $\lambda_{2|}^{i}$. The null vectors of this pencil, determined by the values of r/t for which

$$g_{ij}(r\,\lambda_{1|}^{i} + t\,\lambda_{2|}^{i})\,(r\,\lambda_{1|}^{j} + t\,\lambda_{2|}^{j}) = 0,$$

must be imaginary for this case. Hence we must have

$$(g_{ij}\,\lambda_{1|}^{i}\,\lambda_{2|}^{j})^{2} < 1,$$

and consequently $|\cos\theta|$ as defined by (13.2) is not greater than one.

---

* When dealing with more than one vector, we usually make use of the notation $\lambda_{\alpha|}^{i}$ and $\lambda_{\alpha|i}$ to denote the contravariant and covariant components of one of several vectors, where the value of α indicates the vector and i the component. In the present case α takes the values 1 and 2.

When the components are not chosen so that the vectors be unit vectors, we have

$$(13.3)\qquad \cos\theta = \frac{g_{ij}\,\lambda_{1|}^{i}\,\lambda_{2|}^{j}}{\sqrt{(e_1\,g_{ij}\,\lambda_{1|}^{i}\,\lambda_{1|}^{j})\,(e_2\,g_{kl}\,\lambda_{2|}^{k}\,\lambda_{2|}^{l})}},$$

as follows from (12.6). If dx^i and δx^i denote differentials for two curves through a point, neither of which is a curve of length zero, we have

$$(13.4)\qquad \cos\theta = \frac{g_{ij}\,dx^i\,\delta x^j}{\sqrt{(e_1\,g_{ij}\,dx^i\,dx^j)\,(e_2\,g_{kl}\,\delta x^k\,\delta x^l)}}.$$

When (12.3) is definite, a necessary and sufficient condition that two non-null vectors at a point be orthogonal is

$$(13.5)\qquad g_{ij}\,\lambda_{1|}^{i}\,\lambda_{2|}^{j} = 0,$$

and when the form is indefinite this is taken as the definition of *orthogonality*. The problem of determining vector-fields orthogonal to a given field will be treated later.

When one, or both, of the given vectors is a null vector, the right-hand member of (13.2) involves an indeterminate factor, since there is no analogue to unit vectors in this case. Accordingly in retaining (13.2) as the definition of angle, this indeterminateness is understood. Furthermore, we take (13.5) as the definition of orthogonality when one or both of the vectors is null. As a consequence, a null vector is self-orthogonal.

For the curves of parameter x^i of the space we have $dx^i \neq 0$, $dx^j = 0$, $(j \neq i)$. Hence, when they are not minimal, the components of the contravariant unit tangent vector are $\lambda^i = 1/\sqrt{e_i g_{ii}}$, $\lambda^j = 0\,(j \neq i)$. From this and (13.3) it follows that the angle ω_{ij} between the curves of parameters x^i and x^j at a point, when neither is a curve of length zero at the point, is given by

$$(13.6)\qquad \cos\omega_{ij} = \frac{g_{ij}}{\sqrt{e_i\,e_j\,g_{ii}\,g_{jj}}}.$$

In § 3 we saw that for a covariant vector-field λ_i the equation

$$(13.7)\qquad \lambda_i\,dx^i = 0$$

determines at each point an elemental V_{n-1}, which may be taken as the geometrical interpretation of the vector. In terms of the associate contravariant vector this becomes

$$g_{ij}\,\lambda^j\,dx^i = 0, \tag{13.8}$$

and consequently the vector λ^j at a point is orthogonal to any direction in the V_{n-1} at the point, and thus is normal to the V_{n-1}. Since either the normal or the V_{n-1} determines the other, we may look upon a vector of either type and its associate as defining the same geometrical configuration, and thus speak of λ^i and λ_i as the contravariant and covariant components of the same vector-field.

By means of (12.7) it is readily shown that from (13.2) we have

$$\cos\theta = g^{ij}\,\lambda_{1|i}\,\lambda_{2|j} \tag{13.9}$$

for the determination of the angle, when the covariant components of the vectors are given.* Likewise, the condition of orthogonality in this case is

$$g^{ij}\,\lambda_{1|i}\,\lambda_{2|j} = 0. \tag{13.10}$$

From (13.5) it is seen that at any point P the components of two orthogonal vectors may be interpreted as the homogeneous coördinates in a projective space of $n-1$ dimensions of two points harmonic with respect to the non-singular hyperquadric

$$g_{ij}\,y^i\,y^j = 0, \tag{13.11}$$

in which the g's are evaluated at the point. The problem of finding mutually orthogonal vectors at P is that of finding the vertices of polyhedra self-polar with respect to (13.11). Consider, for example, the case $n = 4$, that is, when (13.11) defines for P a non-singular quadric surface Q. One vertex, P_1, of such a tetrahedron can be chosen arbitrarily in the space but not on Q; a second vertex, P_2, arbitrarily in the polar plane of P_1, but not on Q; a third, P_3, arbitrarily on the intersection of the polar planes of P_1 and P_2, but not on Q. Then P_4 is determined as the intersection of the

* It is understood that the vectors are unit vectors, unless one or both are null vectors.

polar planes of P_1, P_2 and P_3. Since P_1, P_2 and P_3 can be chosen thus in ∞^3, ∞^2 and ∞^1 ways respectively, there are $\infty^6[=\infty^{n(n-1)/2}]$ sets of 4 mutually orthogonal non-null vectors at a point in a V_4.

We call n mutually orthogonal non-null vector-fields in a V_n an *orthogonal ennuple.* The analytical process of finding them is analogous to the above, the difference being that instead of choosing a point for P_1, we choose n arbitrary functions $\lambda_{1|}{}^i$ not satisfying (13.11) and so on.

Hence we have the theorem:

There exist $\infty^{n(n-1)/2}$ orthogonal ennuples in a Riemannian n-space.

Also we have:

A given non-null vector-field forms part of $\infty^{(n-1)(n-2)/2}$ orthogonal ennuples.

A null vector corresponds to a point P on the hyperquadric (13.11) and any non-null vector orthogonal to it to a point in the tangent hyperplane to (13.11) at P. Since this hyperplane is of $n-2$ dimensions, we have the theorem:

A null vector is orthogonal to $n-1$ linearly independent non-null vectors in terms of which it is linearly expressible.

From geometric considerations it is seen that these $n-1$ vectors cannot be chosen so as to be mutually orthogonal.

In like manner we have also:

Any vector orthogonal to a null vector is expressible linearly in terms of it and $n-2$ non-null vectors orthogonal to it.

If a null vector is orthogonal to $n-1$ linearly independent vectors, it is a linear function of them.

If $\lambda_{h|}{}^i$ are the components of the unit vectors of an orthogonal ennuple, where h for $h = 1, \cdots, n$ indicates the vector and i for $i = 1, \cdots, n$ the component, we have

$$(13.12) \qquad g_{ij}\,\lambda_{h|}{}^i\,\lambda_{h|}{}^j = e_h, \qquad g_{ij}\,\lambda_{h|}{}^i\,\lambda_{k|}{}^j = 0 \qquad (h \neq k).$$

Any other unit vector-field of components λ^i is defined by

$$(13.13) \quad \lambda^i = e_1 \cos\alpha_1\,\lambda_{1|}{}^i + e_2 \cos\alpha_2\,\lambda_{2|}{}^i + \cdots + e_n \cos\alpha_n\,\lambda_{n|}{}^i,$$

where in accordance with (13.2) $\cos\alpha_k = g_{ij}\,\lambda^i\,\lambda_{k|}{}^j$. If we put

$$(13.14) \qquad \xi_{h|}{}^i = t_h{}^l\,\lambda_{l|}{}^i \qquad (h, i, l = 1, \cdots, n),$$

where the t's are functions satisfying the conditions

$$\sum_l e_l (t_h^l)^2 \neq 0, \qquad \sum_l e_l t_h^l t_k^l = 0 \qquad (h \neq k), \tag{13.15}$$

the ξ's are components of an orthogonal ennuple. The determination of n^2 quantities t satisfying (13.15) is the problem of finding the self-polar polyhedra with respect to the hyperquadric $\sum_l e_l (y^l)^2 = 0$, and consequently there are $\infty^{n(n-1)/2}$ sets of solutions.

14. Differential parameters. The normals to a hypersurface. If f and φ are any functions of the x's, the functions defined by

$$\Delta_1 f = g^{ij} \frac{\partial f}{\partial x^i} \frac{\partial f}{\partial x^j} = g^{ij} f_{,i} f_{,j} \tag{14.1}$$

$$\Delta_1 (f, \varphi) = g^{ij} \frac{\partial f}{\partial x^i} \frac{\partial \varphi}{\partial x^j} = g^{ij} f_{,i} \varphi_{,j} \tag{14.2}$$

are invariants. They are called *differential parameters of the first order*. In like manner the invariant defined by

$$\Delta_2 f = g^{ij} f_{,ij} = g^{ij} \left(\frac{\partial^2 f}{\partial x^i \partial x^j} - \frac{\partial f}{\partial x^k} \begin{Bmatrix} k \\ ij \end{Bmatrix} \right) \tag{14.3}$$

is called a *differential parameter of the second order*.

An equation of the form $f(x^1, \cdots, x^n) = 0$ determines a V_{n-1} in V_n; we call it a *hypersurface*. For any displacement in this hypersurface we have

$$\frac{\partial f}{\partial x^i} dx^i = 0.$$

Consequently the quantities $\frac{\partial f}{\partial x^i}$ are the covariant components of the vector-field of *normals* to the V_{n-1}. From (14.1) and (12.9) it follows that

A necessary and sufficient condition that the normals to a hypersurface $f(x^1, \cdots, x^n) = 0$ form a null vector-field is that f be a solution of the differential equation

$$\Delta_1 f = 0. \tag{14.4}$$

If f_1 and f_2 are any functions not satisfying (14.4), the angle θ between the normals to two hypersurfaces $f_1 = 0$ and $f_2 = 0$ at a common point, the angle between the hypersurfaces, is given by

$$\cos\theta = \frac{\Delta_1(f_1, f_2)}{\sqrt{e_1 e_2 \Delta_1 f_1 \cdot \Delta_1 f_2}}, \tag{14.5}$$

as follows from (13.3), (13.9), (14.1) and (14.2). If either one or both of the functions f_1, f_2 is a solution of (14.4), we take

$$\cos\theta = \Delta_1(f_1, f_2) \tag{14.6}$$

as the measure of the angle between the hypersurfaces.

From the definitions of § 13 it follows that

$$\Delta_1(f_1, f_2) = 0 \tag{14.7}$$

is the condition that the hypersurfaces be orthogonal at each common point. Since

$$\Delta_1(x^i, x^j) = g^{ij}, \tag{14.8}$$

we have that a necessary and sufficient condition that the hypersurfaces $x^i =$ const., $x^j =$ const. at every point of space be orthogonal is that

$$g^{ij} = 0. \tag{14.9}$$

If $f^1(x^1, \cdots, x^n)$ is any real function, the differential equation

$$\Delta_1(f^1, f) = 0 \tag{14.10}$$

admits $n-1$ independent solutions.* If $f^2, \cdots, f^n$ denote such solutions, and if we introduce new coördinates defined by $x'^i = f^i$ for $i = 1, \cdots, n$, then from the equations $\Delta_1(x'^1, x'^j) = 0$ for $j = 2, \cdots, n$ expressed in terms of the fundamental form $g'_{ij}\, dx'^i\, dx'^j$ we have

$$g'^{1j} = 0 \qquad (j = 2, \cdots, n). \tag{14.11}$$

Since we have assumed that the determinant g' of the above form is not zero, it follows from (6.4) that $g'^{11} \neq 0$ and hence from the identity $g'^{1j} g'_{kj} = \delta^1_k$ we have

$$g'_{1j} = 0, \qquad g'_{11} \neq 0 \qquad (j = 2, \cdots, n). \tag{14.12}$$

* *Goursat*, 1891, 1, p. 29.

Hence the fundamental form is

$$\varphi = g'_{11}(dx^1)^2 + g'_{jk}\, dx'^j\, dx'^k \quad (j, k = 2, \cdots, n). \tag{14.13}$$

The geometrical interpretation of these results is that the hypersurfaces $f^j = \text{const.}$ for $j = 2, \cdots, n$ are orthogonal to the hypersurfaces $f^1 = \text{const.}$ and the former intersect in a congruence of curves orthogonal to the latter.

15. N-tuply orthogonal systems of hypersurfaces in a V_n. From (14.7) it follows that the condition that there exist in a V_n n families of hypersurfaces $f_i = \text{const.}$ $(i = 1, \cdots, n)$ such that every two hypersurfaces $f_i = \text{const.}$, $f_j = \text{const.}$ for $i, j = 1, \cdots, n (i \neq j)$ are orthogonal at every point is that the $n(n-1)/2$ simultaneous differential equations

$$\Delta_1(f_i, f_j) = 0 \tag{15.1}$$

admit n solutions. Evidently this is not possible for $n > 3$, when the fundamental form (12.3) is any whatever. When it is possible, we say that the Riemannian space admits an *n-tuply orthogonal system of hypersurfaces.*

If this condition is satisfied and these hypersurfaces are taken for the coördinate hypersurfaces $x^i = \text{const.}$, we have from (15.1)

$$g^{ij} = 0 \qquad (i, j = 1, \cdots, n;\ i \neq j). \tag{15.2}$$

Since we have assumed that the determinant g of the form (12.3) is not zero, it follows from (6.4) that none of the components g^{ii} is equal to zero.

Hence from the identities

$$g_{ij}\, g^{ik} = \delta^k_j$$

we have

$$g_{ij} = 0 \qquad (i, j = 1, \cdots, n;\ i \neq j). \tag{15.3}$$

Consequently the fundamental form is

$$\varphi = g_{11}(dx^1)^2 + g_{22}(dx^2)^2 + \cdots + g_{nn}(dx^n)^2. \tag{15.4}$$

Conversely when the fundamental form is reducible to (15.4), we have (15.2) and consequently the parametric hypersurfaces form an n-tuply orthogonal system.

Since in this case

$$g^{ii} = \frac{1}{g_{ii}}, \tag{15.5}$$

we have from (7.1), (15.2), (15.3) and (15.5), the following expressions for the Christoffel symbols formed with respect to (15.4):

$$[ij, k] = 0, \quad [ij, i] = -[ii, j] = \frac{1}{2}\frac{\partial g_{ii}}{\partial x^j}, \quad [ii, i] = \frac{1}{2}\frac{\partial g_{ii}}{\partial x^i} \qquad (i, j, k \neq), \tag{15.6}$$

$$\left\{ {k \atop ij} \right\} = 0, \quad \left\{ {j \atop ii} \right\} = -\frac{1}{2g_{jj}}\frac{\partial g_{ii}}{\partial x^j}, \quad \left\{ {i \atop ij} \right\} = \frac{1}{2}\frac{\partial \log g_{ii}}{\partial x^j}, \quad \left\{ {i \atop ii} \right\} = \frac{1}{2}\frac{\partial \log g_{ii}}{\partial x^i}. \tag{15.7}$$

From (8.9) we have in this case

$$
\begin{aligned}
R_{hijk} &= 0 \qquad (h, i, j, k \neq), \\
R_{hiik} &= \sqrt{g_{ii}}\left(\frac{\partial^2 \sqrt{g_{ii}}}{\partial x^h \partial x^k} - \frac{\partial \sqrt{g_{ii}}}{\partial x^h}\frac{\partial \log \sqrt{g_{hh}}}{\partial x^k} - \frac{\partial \sqrt{g_{ii}}}{\partial x^k}\frac{\partial \log \sqrt{g_{kk}}}{\partial x^h}\right) \qquad (h, i, k \neq), \\
R_{hiih} &= \sqrt{g_{ii}}\sqrt{g_{hh}}\left[\frac{\partial}{\partial x^h}\left(\frac{1}{\sqrt{g_{hh}}}\frac{\partial \sqrt{g_{ii}}}{\partial x^h}\right) + \frac{\partial}{\partial x^i}\left(\frac{1}{\sqrt{g_{ii}}}\frac{\partial \sqrt{g_{hh}}}{\partial x^i}\right) + \sum_m{}' \frac{1}{g_{mm}}\frac{\partial \sqrt{g_{ii}}}{\partial x^m}\frac{\partial \sqrt{g_{hh}}}{\partial x^m}\right] \qquad (h \neq i),
\end{aligned}
\tag{15.8}
$$

where $\sum_m{}'$ indicates the sum for $m = 1, \cdots, n$ excluding $m = h$ and $m = i$.

16. Metric properties of a space V_n immersed in a V_m. Consider a space V_m referred to coördinates y^α and with the fundamental form

$$\varphi = a_{\alpha\beta}\, dy^\alpha\, dy^\beta.* \tag{16.1}$$

If we put

$$y^\alpha = f^\alpha(x^1, \cdots, x^n), \tag{16.2}$$

* In this section Greek indices are supposed to take the values $1, \cdots, m$ and Latin indices $1, \cdots, n$.

where the f's are analytic functions of the x's such that the matrix $\left\| \frac{\partial f^\alpha}{\partial x^i} \right\|$ is of rank n, equations (16.2) define a space V_n immersed in V_m. If we write

$$(16.3) \qquad a_{\alpha\beta} \frac{\partial y^\alpha}{\partial x^i} \frac{\partial y^\beta}{\partial x^j} = g_{ij},$$

then from the definition of linear element for V_m, namely

$$(16.4) \qquad ds^2 = e\, a_{\alpha\beta}\, dy^\alpha\, dy^\beta,$$

we have for the linear element of V_n

$$(16.5) \qquad ds^2 = e g_{ij}\, dx^i\, dx^j.$$

Thus when a metric is defined for a space V_m, the metric of a subspace is in general determined (cf. Ex. 8, p. 48). This is an evident generalization of the case of a surface $x^i = f^i(u, v)$ (for $i = 1, 2, 3$) in a euclidean space with the linear element (12.1); in this case (16.5) assumes the well-known form $ds^2 = E\,du^2 + 2F\,du\,dv + G\,dv^2$ in the notation of Gauss.

The formula for V_m analogous to (13.4) is

$$(16.6) \qquad \cos\theta = \frac{a_{\alpha\beta}\, dy^\alpha\, \delta y^\beta}{\sqrt{(e_1\, a_{\alpha\beta}\, dy^\alpha\, dy^\beta)\,(e_2\, a_{\alpha\beta}\, \delta y^\alpha\, \delta y^\beta)}}.$$

From (16.2) we have

$$(16.7) \qquad dy^\alpha = \frac{\partial y^\alpha}{\partial x^i}\, dx^i.$$

Substituting in (16.6) and making use of (16.3), we obtain (13.4). Thus the invariant $\cos\theta$ of two directions at a point of V_n has the same value whether determined by the formula for V_n or for the enveloping space V_m. Later (§ 55) it will be shown that when the fundamental form of a space is positive definite there exists a euclidean space V_m, where $m \leq n(n+1)/2$ in which V_n can be considered as immersed. Consequently angle as defined by (13.4) for V_n is equal to the angle in the euclidean sense as determined in the enveloping V_m. In fact, in the differential geometry of a surface in euclidean 3-space, the angle between two directions on a surface is determined in the euclidean space and its expression in terms of the metric of the surface is derived therefrom; this gives a form of which (13.4) is an immediate generalization.*

If λ^i are the components of any contravariant vector-field in V_n, along any curve of the congruence of curves for which these are the tangent vectors we have $\frac{d x^i}{d t} = \lambda^i$. From (16.7) we have for this curve in V_m

$$\frac{d y^\alpha}{d t} = \frac{\partial y^\alpha}{\partial x^i} \frac{d x^i}{d t} = \frac{\partial y^\alpha}{\partial x^i} \lambda^i.$$

Hence the components in the y's of this vector-field are given by

$$\xi^\alpha = \frac{\partial y^\alpha}{\partial x^i} \lambda^i. \tag{16.8}$$

Conversely, if we have any vector-field ξ^α in V_m, for those vectors of the field in V_n, that is, tangential to V_n, the components λ^i in the x's are obtained by taking any† n of equations (16.8), replacing the y's by the expressions (16.2) and solving for the λ's.

From (16.8) and (16.3) we have

$$a_{\alpha\beta}\, \xi^\alpha \xi^\beta = g_{ij}\, \lambda^i \lambda^j, \tag{16.9}$$

and from (13.3) for two non-null vector-fields

$$\begin{aligned} \cos\theta &= \frac{a_{\alpha\beta}\, \xi_{1|}{}^\alpha\, \xi_{2|}{}^\beta}{\sqrt{(e_1\, a_{\alpha\beta}\, \xi_{1|}{}^\alpha\, \xi_{1|}{}^\beta)\,(e_2\, a_{\alpha\beta}\, \xi_{2|}{}^\alpha\, \xi_{2|}{}^\beta)}} \\ &= \frac{g_{ij}\, \lambda_{1|}{}^i\, \lambda_{2|}{}^j}{\sqrt{(e_1\, g_{ij}\, \lambda_{1|}{}^i\, \lambda_{1|}{}^j)\,(e_2\, g_{ij}\, \lambda_{2|}{}^i\, \lambda_{2|}{}^j)}}. \end{aligned} \tag{16.10}$$

From (16.7) it follows that $\frac{\partial y^\alpha}{\partial x^i}$ for $\alpha = 1, \cdots, m$ and a given i are the components in the y's of the tangents to the curves of

* Cf. *Eisenhart*, 1909, 1, p. 78.

† n suitable equations.

parameter x^i in V_n. Since the matrix $\left\| \frac{\partial y^\alpha}{\partial x^i} \right\|$ is of rank n by hypothesis, there are n such independent vector-fields in V_n in terms of whose components the components of any vector-field in V_n are linearly expressible. From this it follows that any m functions ξ^β satisfying the n equations

$$a_{\alpha\beta} \frac{\partial y^\alpha}{\partial x^i} \xi^\beta = 0 \tag{16.11}$$

are the components in the y's of a vector-field at points of V_n, such that the vector at a point of V_n is orthogonal to every vector in V_n at the point. Accordingly we say that a vector of components ξ^β satisfying (16.11) is *normal* to V_n. If (16.11) is written in the form

$$\frac{\partial y^\alpha}{\partial x^i} \xi_\alpha = 0, \tag{16.12}$$

we see that there are $m - n$ linearly independent vector-fields normal to V_n.

Exercises.

1. Show that a real coördinate system can be found for which $g = 1$ or -1. In this coördinate system the divergence of a vector λ^i (Ex. 8, p. 32) is the ordinary divergence.

2. For a V_2 referred to an orthogonal system of parametric curves

$$R_{11}\, g_{22} = R_{22}\, g_{11} = R_{1221}, \qquad R_{12} = 0,$$

$$R = g^{ij} R_{ij} = \frac{2 R_{1221}}{g_{11}\, g_{22}},$$

and consequently

$$R_{ij} = \frac{R}{2} g_{ij}.$$

3. When the fundamental form of a V_n is positive definite and θ is the angle between the vectors $\lambda_{1|}{}^{i}$ and $\lambda_{2|}{}^{i}$, then

$$\sin^2\theta = \frac{(g_{hi}\, g_{jk} - g_{hk}\, g_{ij})\, \lambda_{1|}{}^{h} \lambda_{1|}{}^{i} \lambda_{2|}{}^{j} \lambda_{2|}{}^{k}}{g_{hi}\, g_{jk}\, \lambda_{1|}{}^{h} \lambda_{1|}{}^{i} \lambda_{2|}{}^{j} \lambda_{2|}{}^{k}}.$$

4. Show that

$$\frac{\partial}{\partial x^k} \Delta_1 \theta = 2 g^{ij} \theta_{,i}\, \theta_{,jk}.$$

5. For a V_3 referred to a triply orthogonal system of surfaces

$$R_{ii} = \frac{1}{g_{jj}} R_{ijji} + \frac{1}{g_{kk}} R_{ikki} \qquad (i, j, k \neq),$$

$$R_{ij} = \frac{1}{g_{kk}} R_{ikkj} \qquad (i, j, k \neq),$$

$$R = \sum_{i,j} \frac{1}{g_{ii}} \frac{1}{g_{jj}} R_{ijji}.$$

6. Show that for a V_3 a tensor a_{ijkl} satisfying the conditions (8.10) and (8.11) has six independent components and that these can be written in the form

$$a_{ijkl} = g_{ik}\, a_{jl} - g_{il}\, a_{jk} + g_{jl}\, a_{ik} - g_{jk}\, a_{il},$$

where a_{il} is a symmetric tensor. Show also that

$$a_{jk} = \frac{1}{4} g_{jk}\, g^{il}\, g^{pq} a_{ipql} - g^{il}\, a_{ijkl}.$$

Hence if $g^{il} a_{ijkl} = 0$, then $a_{ijkl} = 0$.

7. The functions g_{ij} defined by (16.3) are invariants for V_m at points of V_n, and $a_{\alpha\beta}$ are invariants for V_n.

8. When the equations

$$a_{\alpha\beta} \frac{\partial y^\alpha}{\partial x^i} \frac{\partial y^\beta}{\partial x^j} = 0 \qquad (\alpha, \beta = 1, \cdots, m;\ i, j = 1, \cdots, n)$$

admit solutions (16.2), for the V_n thus defined there is not a metric induced by the metric of V_m. Show that in general such a V_n exists, if $m \geqq n(n+1)/2$.

17. Geodesics. Let C be a real curve defined by $x^i = f^i(t)$, t being any real parameter, and denote by A and B the points of C with the respective parametric values t_0 and t_1. The equations

$$\bar{x}^i = x^i + \varepsilon\, \omega^i,$$

where ε is an infinitesimal and ω^i are functions of the x's such that

$$\text{(17.1)} \qquad \omega^i = 0 \quad \text{for} \quad t = t_0, t_1,$$

define a curve $\bar{C}$ nearby C and passing through A and B.

Consider the integral

$$\text{(17.2)} \qquad I = \int_{t_0}^{t_1} \varphi(x^1, \cdots, x^n, \dot{x}^1, \cdots, \dot{x}^n)\, dt,$$

where $\dot{x}^i = \frac{dx^i}{dt}$ and φ is an analytic function of the $2n$ arguments. If $\bar{I}$ is the corresponding integral for $\bar{C}$, we have, on expanding φ in Taylor's series,

$$\bar{I} - I = \varepsilon \int_{t_0}^{t_1} \left[\frac{\partial \varphi}{\partial x^i} \omega^i + \frac{\partial \varphi}{\partial \dot{x}^i} \dot{\omega}^i \right] dt + \cdots,$$

where $\dot{\omega}^i = \frac{\partial \omega^i}{\partial x^j} \dot{x}^j$ and the unwritten terms are of the second and higher orders in ε. If we write

$$\delta I = \varepsilon \int_{t_0}^{t_1} \left[\frac{\partial \varphi}{\partial x^i} \omega^i + \frac{\partial \varphi}{\partial \dot{x}^i} \dot{\omega}^i \right] dt, \tag{17.3}$$

integrate the second term of the integrand by parts and make use of (17.1), we have

$$\delta I = \varepsilon \int_{t_0}^{t_1} \left[\frac{\partial \varphi}{\partial x^i} - \frac{d}{dt} \left(\frac{\partial \varphi}{\partial \dot{x}^i} \right) \right] \omega^i \, dt. \tag{17.4}$$

The integral I is said to be *stationary* and C the corresponding *extremal*, if this first variation δI is zero for every set of functions ω^i satisfying the conditions (17.1). From (17.4) it follows that a necessary and sufficient condition is that

$$\frac{d}{dt} \left(\frac{\partial \varphi}{\partial \dot{x}^i} \right) - \frac{\partial \varphi}{\partial x^i} = 0, \tag{17.5}$$

which are known as Euler's equations of condition.*

We apply this general result to the integral (12.11) for a portion of a curve C for which e is either one or minus one throughout the domain. In this case

$$\frac{\partial \varphi}{\partial \dot{x}^i} = \frac{e g_{ij} \dot{x}^j}{\sqrt{e g_{ij} \dot{x}^i \dot{x}^j}} = \frac{e g_{ij} \dot{x}^j}{\frac{ds}{dt}}, \qquad \frac{\partial \varphi}{\partial x^i} = \frac{1}{2} \frac{e \frac{\partial g_{jk}}{\partial x^i} \dot{x}^j \dot{x}^k}{\frac{ds}{dt}}.$$

Substituting in (17.5), we obtain

$$g_{ij} \ddot{x}^j + \frac{\partial g_{ij}}{\partial x^k} \dot{x}^j \dot{x}^k - \frac{1}{2} \frac{\partial g_{jk}}{\partial x^i} \dot{x}^j \dot{x}^k - g_{ij} \dot{x}^j \frac{\frac{d^2 s}{dt^2}}{\frac{ds}{dt}} = 0.$$

---

* Cf. *Bolza*, 1904, 3, p. 123; also *Bliss*, 1925, 2, p. 130.

If we make use of the Christoffel symbols formed with respect to (12.3), this equation becomes

$$(17.6) \qquad g_{ij}\frac{d^2x^j}{dt^2}+[jk,i]\frac{dx^j}{dt}\frac{dx^k}{dt}-g_{ij}\frac{dx^j}{dt}\frac{\frac{d^2s}{dt^2}}{\frac{ds}{dt}}=0.$$

Multiplying by g^{il} and summing for i, we obtain

$$(17.7) \qquad \frac{d^2x^l}{dt^2}+\left\{ {l \atop jk} \right\}\frac{dx^j}{dt}\frac{dx^k}{dt}-\frac{dx^l}{dt}\frac{\frac{d^2s}{dt^2}}{\frac{ds}{dt}}=0.$$

If in place of a general parameter t we use the arc s of the curve, equations (17.7) become

$$(17.8) \qquad \frac{d^2x^l}{ds^2}+\left\{ {l \atop jk} \right\}\frac{dx^j}{ds}\frac{dx^k}{ds}=0.$$

Thus the extremals of the integral (12.11) in which the parameter t is the arc s are integral curves of n ordinary differential equations (17.8).

These integrals satisfy the condition that along any curve

$$(17.9) \qquad g_{ij}\frac{dx^i}{ds}\frac{dx^j}{ds}=\text{const.},$$

because of (12.12). We shall show that any integral curve of (17.8) possesses this property. In fact, since the left-hand member of this equation is an invariant, its derivatives with respect to s along a curve can be obtained by taking its covariant derivative with respect to x^k, multiplying by $\frac{dx^k}{ds}$ and summing for k. Hence the condition that (17.9) shall hold along a curve, when s is a parameter, not necessarily the arc, is

See App. 2

$$(17.10) \quad g_{ij}\frac{dx^j}{ds}\frac{dx^k}{ds}\left(\frac{dx^i}{ds}\right)_{,k}\equiv g_{ij}\frac{dx^j}{ds}\left(\frac{d^2x^i}{ds^2}+\left\{ {i \atop kl} \right\}\frac{dx^k}{ds}\frac{dx^l}{ds}\right)=0.$$

It is seen that this condition is satisfied by any integral curve of (17.8), which equations may also be written in the form

$$\frac{dx^k}{ds}\left(\frac{dx^i}{ds}\right)_{,k} = 0. \tag{17.11}$$

In view of this result we have that if the constant in (17.9) is positive, negative or zero at a point of an integral curve of (17.8), it is the same all along the curve; that is, if the tangent vector at one point is non-null or null, the tangents all along the curve are of the same kind. From (17.7) it is seen that the form of (17.8) is not changed if s is replaced by $as+b$, where a and b are arbitrary constants. Hence, if the curve is not of length zero, s can be chosen so that (17.9) becomes (12.12), that is, s is the arc. On the other hand, if the constant in (17.9) is zero, the above mentioned generality of s obtains. Any integral curve of equations (17.8) is called a *geodesic*. When in particular it is a curve of length zero, we will call it a *minimal geodesic*, and we will understand that when s is used as a parameter of a minimal geodesic it is such that the differential equations of the geodesic assume the form (17.8).

Consider for example the V_4 of special relativity with the fundamental form $\varphi = (dx^1)^2 + (dx^2)^2 + (dx^3)^2 - (dx^4)^2$. Any curve of length zero in this space may be defined by equations of the form

$$x^1 = \int R\cos\theta\cos\varphi\, ds, \qquad x^2 = \int R\cos\theta\sin\varphi\, ds,$$
$$x^3 = \int R\sin\theta\, ds, \qquad x^4 = \int R\, ds,$$

where R, θ and φ are functions of s. Only in case R, θ and φ are constants are these integral curves of (17.8), which are in this case $\dfrac{d^2x^i}{ds^2} = 0$. Hence in general a curve of length zero is not a geodesic.

We return to the consideration of (17.8) in which s is the arc of the geodesic when the latter is not minimal, and is the particular parameter referred to above when the geodesic is minimal. We observe that any integral curve of (17.8) is determined by a point $P_0(x_0^1, \cdots, x_0^n)$ and a direction at P_0. Thus if we put

(17.12) $$\xi^i = \left(\frac{dx^i}{ds}\right)_0,$$

where a subscript 0 indicates the value at P_0, we have

$$x^i = x_0^i + \xi^i s + \frac{1}{2}\left(\frac{d^2 x^i}{ds^2}\right)_0 s^2 + \frac{1}{3!}\left(\frac{d^3 x^i}{ds^3}\right)_0 s^3 + \cdots$$

The coefficients of s^2 and higher powers in s are given by (17.8) and the equations resulting from (17.8) by differentiation with respect to s and replacing the second and higher derivatives of x^i by means of (17.8) and the resulting equations. Thus we have

(17.13) $$\begin{aligned} &\frac{d^3 x^i}{ds^3} + \Gamma^i_{jkl}\frac{dx^j}{ds}\frac{dx^k}{ds}\frac{dx^l}{ds} = 0, \\ &\frac{d^4 x^i}{ds^4} + \Gamma^i_{jklm}\frac{dx^j}{ds}\frac{dx^k}{ds}\frac{dx^l}{ds}\frac{dx^m}{ds} = 0, \\ &\cdots\cdots\cdots\cdots\cdots \\ &\cdots\cdots\cdots\cdots\cdots, \end{aligned}$$

where

(17.14) $$\begin{aligned} \Gamma^i_{jkl} &= \frac{1}{3}P\left(\frac{\partial}{\partial x^l}\begin{Bmatrix} i \\ j\,k \end{Bmatrix} - \begin{Bmatrix} i \\ \alpha\,k \end{Bmatrix}\begin{Bmatrix} \alpha \\ j\,l \end{Bmatrix} - \begin{Bmatrix} i \\ j\,\alpha \end{Bmatrix}\begin{Bmatrix} \alpha \\ k\,l \end{Bmatrix}\right) \\ &= \frac{1}{3}P\left(\frac{\partial}{\partial x^l}\begin{Bmatrix} i \\ j\,k \end{Bmatrix} - 2\begin{Bmatrix} i \\ \alpha\,j \end{Bmatrix}\begin{Bmatrix} \alpha \\ k\,l \end{Bmatrix}\right), \end{aligned}$$

and in general

(17.15) $$\Gamma^i_{jkl\ldots mn} = \frac{1}{N}P\left(\frac{\partial \Gamma^i_{jkl\ldots m}}{\partial x^n} - \Gamma^i_{\alpha kl\ldots m}\begin{Bmatrix} \alpha \\ j\,n \end{Bmatrix} - \cdots - \Gamma^i_{jk\ldots\alpha}\begin{Bmatrix} \alpha \\ m\,n \end{Bmatrix}\right)$$

where P before an expression indicates the sum of terms obtained by permuting the subscripts cyclically and N denotes the number of subscripts.* Hence we have

(17.16) $$x^i = x_0^i + \xi^i s - \frac{1}{2}\begin{Bmatrix} i \\ j\,k \end{Bmatrix}_0 \xi^j \xi^k s^2 - \frac{1}{3!}(\Gamma^i_{jkl})_0\, \xi^j \xi^k \xi^l s^3 - \cdots.$$

The domain of convergence of these series depends evidently upon the expressions for g_{ij} and the values of ξ^i. However for sufficiently small values of s they define an integral curve of (17.8).

* Cf. *Veblen* and *Thomas*, 1923, 4, p. 561.

18. Riemannian, normal and geodesic coördinates. In this section we introduce certain types of coördinates which have important applications. Returning to (17.16) as the equations of a particular geodesic passing through a point $P_0(x_0)$ and determined by the direction (17.12), we put

$$y^i = \xi^i s \tag{18.1}$$

and substitute it in (17.16), with the result

$$x^i = x_0^i + y^i - \frac{1}{2}\left\{{i \atop \alpha\,\beta}\right\}_0 y^\alpha y^\beta - \frac{1}{3!}(\Gamma^i_{\alpha\beta\gamma})_0 y^\alpha y^\beta y^\gamma - \cdots. \tag{18.2}$$

Since equations (18.2) do not involve the ξ's, they hold for all geodesics through P_0 and therefore constitute the equations of a transformation of coördinates. Since the Jacobian $\left|\frac{\partial x^i}{\partial y^j}\right|$ of these equations is different from zero at P_0, the series (18.2) can be inverted and we have

$$y^i = (x^i - x_0^i) + F^i(x^1 - x_0^1, \cdots, x^n - x_0^n) \qquad (i = 1, \cdots, n), \tag{18.3}$$

where F^i are series in the second and higher powers of $x^j - x_0^j$ $(j = 1, \cdots, n)$.

For a given set of values of the constants ξ^i in (18.1), these equations define a curve. When y^i in (18.2) is replaced by $\xi^i s$ we have (17.16). Consequently (18.1) are the equations of the geodesics in the new system of coördinates. These coördinates were first introduced by Riemann* and are called *Riemannian coördinates*. In these coördinates the equations of the geodesics through P_0 are of the same form as the equations for straight lines through the origin in euclidean geometry.

From the form of equations (18.1) it is seen that these coördinates are valid only for a domain about P_0 such that no two geodesics through P_0 meet again in the domain, and from (18.3) it follows that this domain is that for which the series (18.2) may be inverted into (18.3).

If we write the fundamental form in the y's thus

$$\varphi = \bar{g}_{ij}\, dy^i\, dy^j, \tag{18.4}$$

* 1854, 1, p. 261.

and indicate by $\overline{\left\{ {i \atop jk} \right\}}$ and $\overline{[ij, k]}$ the Christoffel symbols formed with respect to (18.4), the equations of the geodesics are

$$\frac{d^2 y^i}{ds^2} + \overline{\left\{ {i \atop jk} \right\}} \frac{dy^j}{ds} \frac{dy^k}{ds} = 0. \tag{18.5}$$

Since the expression (18.1) must satisfy these equations, we have

$$\overline{\left\{ {i \atop jk} \right\}} \xi^j \xi^k = 0, \tag{18.6}$$

and on multiplication by s^2

$$\overline{\left\{ {i \atop jk} \right\}} y^j y^k = 0, \tag{18.7}$$

which equations hold throughout the domain. Conversely, if these conditions are satisfied, equations (18.5) are satisfied by (18.1) and the y's are Riemannian coördinates.

By applying to (18.5) considerations similar to those applied to (17.8) we obtain similarly to (17.16)

$$y^i = \xi^i s - \frac{1}{2} \overline{\left\{ {i \atop \alpha\beta} \right\}}_0 \xi^\alpha \xi^\beta s^2 - \cdots.$$

Since this must reduce to (18.1) for arbitrary values of ξ^i it follows that

$$\overline{\left\{ {i \atop \alpha\beta} \right\}}_0 = 0. \tag{18.8}$$

Since the functions $\overline{\Gamma}$ defined by equations analogous to (17.14) and (17.15) are symmetric, we have also

$$(\overline{\Gamma}^i_{\alpha\beta\gamma})_0 = 0, \cdots, (\overline{\Gamma}^i_{\alpha\beta\cdots\lambda\mu})_0 = 0. \tag{18.9}$$

From (7.3) and (7.4) it follows that equations (18.8) are equivalent to

$$\left(\frac{\partial \overline{g}_{ij}}{\partial y^k}\right)_0 = 0 \qquad (i, j, k = 1, \cdots, n). \tag{18.10}$$

Hence:

At the origin of Riemannian coördinates the first derivatives of the components of the fundamental tensor in these coördinates are zero.

It follows also from (18.8) and the general formula for covariant differentiation that at the origin of Riemannian coördinates first covariant derivatives reduce to ordinary derivatives. Evidently (18.10) is a special case of this result, since $\bar{g}_{ij,k} = 0$.

If another general system of coördinates x'^i are used, we have a set of equations (17.16) in the primed quantities from which we obtain another set of Riemannian coördinates y'^i by equations analogous to (18.3), and the equations of the geodesics in this coördinate system are

$$y'^i = \left(\frac{dx'^i}{ds}\right)_0 s = \xi'^i s.$$

Since

$$\xi'^i = \left(\frac{dx'^i}{ds}\right)_0 = \left(\frac{\partial x'^i}{\partial x^j}\frac{dx^j}{ds}\right)_0 = a^i_j \xi^j, \tag{18.11}$$

where the a's are constants, we have:

When the coördinates x^i of a space are subjected to an arbitrary analytic transformation, the Riemannian coördinates determined by the x's and a point undergo a linear transformation with constant coefficients.

Since the a's in (18.11) are the values of $\dfrac{\partial x'^i}{\partial x^j}$ at the point, it is evident that conversely when a linear transformation of the Riemannian coördinates is given, corresponding analytic transformations of the x's exist but are not uniquely defined.

At the point P_0 the coefficients $\bar{g}_{ij}$ in (18.4) are constants. From § 9 it follows that real linear transformations of the y's with constant coefficients can be found for which (18.4) reduces to a form at P_0 involving only squares of the differentials and the signs of these terms depend upon the signature of the differential form. These particular Riemannian coördinates have been called *normal coördinates* by Birkhoff.* See App. 3

The transformation defined by (18.2) belongs to the class of transformations of the type

---

* 1923, 2, p. 124.

$$(18.12)\quad x^i = x_0^i + x'^i + \frac{1}{2} c^i_{\alpha\beta} x'^\alpha x'^\beta + \frac{1}{3!} c^i_{\alpha\beta\gamma} x'^\alpha x'^\beta x'^\gamma + \cdots$$

where the c's are symmetric in the subscripts. From (18.12) we have at P_0 of coördinates x_0^i and $x'^i = 0$ in the respective systems

$$\left(\frac{\partial x^i}{\partial x'^\alpha}\right)_0 = \delta^i_\alpha, \qquad \left(\frac{\partial^2 x^i}{\partial x'^\alpha \partial x'^\beta}\right)_0 = c^i_{jk}\,\delta^j_\alpha\,\delta^k_\beta = c^i_{jk}\left(\frac{\partial x^j}{\partial x'^\alpha}\right)_0 \left(\frac{\partial x^k}{\partial x'^\beta}\right)_0.$$

Hence if $\left\{ {i \atop jk} \right\}'$ indicates the Christoffel symbols in the x''s, we have from (7.14)

$$\left\{ {i \atop jk} \right\}'_0 = \left\{ {i \atop jk} \right\}_0 + c^i_{jk}.$$

Therefore a necessary and sufficient condition that $\left\{ {i \atop jk} \right\}'_0 = 0$ is that $c^i_{jk} = -\left\{ {i \atop jk} \right\}_0$. Accordingly the equations

$$(18.13)\quad \begin{aligned} x^i = x_0^i + x'^i - \frac{1}{2}\left\{ {i \atop \alpha\beta} \right\}_0 x'^\alpha x'^\beta + \frac{1}{3!} c^i_{\alpha\beta\gamma} x'^\alpha x'^\beta x'^\gamma + \cdots \\ \cdots + \frac{1}{m!} c^i_{\alpha_1\cdots\alpha_m} x'^{\alpha_1} \cdots x'^{\alpha_m} + \cdots, \end{aligned}$$

where the c's are arbitrary constants symmetric in the subscripts,* define a transformation of coördinates such that

$$(18.14)\quad \left(\frac{\partial g'_{ij}}{\partial x'^k}\right)_0 = 0.$$

The x''s so defined are called *geodesic coördinates.* Hence:

At the origin of a geodesic coördinate system first covariant derivatives are ordinary derivatives.

The equations in geodesic coördinates of the geodesic through the origin determined by $\xi^i = \left(\frac{dx^i}{ds}\right)_0$ are

$$(18.15)\quad x^i = \xi^i s - \frac{1}{3!}(\Gamma^i_{\alpha\beta\gamma})_0\, \xi^\alpha \xi^\beta \xi^\gamma s^3 - \cdots$$

* This assumption is no restriction as to generality.

Comparing these expressions with (18.1) we see that Riemannian coördinates are the geodesic coördinates for which the Γ's vanish for $x^i = 0$.

19. Geodesic form of the linear element. Finite equations of geodesics. If $f(x^1, \cdots, x^n)$ is any real function such that $\Delta_1 f \neq 0$, the normals to the hypersurface $f = 0$ are not null vectors (§ 14), and consequently the geodesics determined at each point of $f = 0$ by the direction of the normal are not curves of length zero. If we change coördinates taking this hypersurface for $x^1 = 0$, and the geodesics for the curves of parameter x^1, and take for the coördinate x^1 the length of arc of these geodesics measured from $x^1 = 0$, from (12.5) it follows that in this coördinate system

$$g_{11} = e_1, \tag{19.1}$$

where e_1 is plus or minus one. From the equations of the geodesics which result from (17.6) when we take $t = s = x^1$ we have

$$\frac{\partial g_{1i}}{\partial x^1} = 0.$$

For $i \neq 1$ by hypothesis $g_{1i} = 0$ for $x^1 = 0$, it follows that $g_{1i} = 0$ identically. Hence the linear element is

$$ds^2 = e(e_1 \, dx_1^2 + g_{\alpha\beta} \, dx^\alpha \, dx^\beta) \quad (\alpha, \beta = 2, \cdots, n). \tag{19.2}$$

We call this the *geodesic form* of the linear element. As a result we have the theorem:

*If f is any real function of the x's such that $\Delta_1 f \neq 0$ and geodesics be drawn normal to the hypersurface $f = 0$ and on each geodesic the same length be laid off from $f = 0$, the locus of the end points is a hypersurface orthogonal to the geodesics.**

These hypersurfaces are said to be *geodesically parallel* to the hypersurface $f = 0$.

Incidentally we have the theorem:

---

* This is the generalization of a theorem of Gauss for surfaces in euclidean 3-space, cf. 1909, 1, p. 206. Also, we remark that the first assumption of the theorem is satisfied, if (12.3) is definite.

A necessary and sufficient condition that the curves of parameter x^1 *be geodesic and the coördinate* x^1 *be the arc is that* g_{11} *be constant* e_1 *and* g_{1i} *for* $i = 2, \cdots, n$ *be independent of* x^1.

For the quadratic form (19.2) we have

$$\Delta_1 x^1 = e_1. \tag{19.3}$$

Conversely, if f is any solution of the differential equation

$$\Delta_1 f = e_1, \tag{19.4}$$

where e_1 is plus or minus one, the surfaces $f =$ const. are orthogonal to a congruence of geodesics, and the length of any geodesic between two hypersurfaces $f = c_1$, and $f = c_2$ is $c_2 - c_1$. In fact, if we give f the significance of f^1 in (14.10) and proceed as in § 14, we get the fundamental form (14.13). With respect to this form equation (19.4) reduces to $g'^{11} = e_1$. Since $g'^{11} = \frac{1}{g'_{11}}$, the form (14.13) reduces to (19.2).

A *complete* solution of either of the equations (19.4), that is, for $e_1 = 1$ or -1, is a function f involving $n-1$ arbitrary constants $a_1, \cdots, a_{n-1}$ in addition to an additive constant c.* The covariant components of the normals to the corresponding hypersurfaces

$$f(x^1, \cdots, x^n, a_1, \cdots, a_{n-1}) = c \tag{19.5}$$

are $\frac{\partial f}{\partial x^i}$, each hypersurface being determined by a value of c. Consider now any point P and a non-null vector at the point whose covariant components are λ_i. According as $g^{ij} \lambda_i \lambda_j$ is positive or negative, we take the solution of (19.4) for $e_1 = 1$ or -1. Then the n equations

$$\frac{\partial f}{\partial x^i} = \varrho \lambda_i$$

determine the a's and the factor ϱ, and equation (19.5) the value of c so that one of the hypersurfaces (19.5) shall have the given direction λ_i for its normal at P.

---

* *Goursat*, 1891, 1, p. 98.

If we imagine the expression (19.5) substituted in (19.4) and differentiate with respect to a_i, we obtain

$$\Delta_1\left(f, \frac{\partial f}{\partial a_i}\right) = 0.$$

Consequently the hypersurfaces

$$\frac{\partial f}{\partial a_i} = b_i, \tag{19.6}$$

where the b's are constants, are orthogonal to the hypersurfaces (19.5) and meet in the geodesics orthogonal to the latter hypersurfaces. Since we have shown that one of the hypersurfaces (19.5) can be chosen so that a given direction at a point is normal to it, we have the theorem:

When a complete solution (19.5) *of* (19.4) *is known, equations* (19.6) *for arbitrary values of the b's are the equations of the non-minimal geodesics, and the arc of the geodesics is given by the value of f.**

Exercises.

1. If the coördinates at points of a geodesic are expressed in terms of s [cf. (17.8)] and φ is any function of the x's, then

$$\frac{d^m \varphi}{ds^m} = \varphi_{,r_1 r_2 \cdots r_m} \frac{dx^{r_1}}{ds} \frac{dx^{r_2}}{ds} \cdots \frac{dx^{r_m}}{ds}.$$

Levy, 1925, 1.

2. If for every point in space and for a special coördinate system associated with each point a tensor equation is satisfied, the tensor equation holds throughout the space for any coördinate system.

3. Show that at the origin of a system of geodesic coördinates defined by (18.13) any component of a tensor in the x's is equal to the component with the same indices in the x''s; in particular this applies to the fundamental tensor.

4. If x^i are geodesic coördinates with a point P for origin, and they are subjected to the transformation

$$x^i = x'^i + \frac{1}{6} c^i_{\alpha\beta\gamma} x'^\alpha x'^\beta x'^\gamma,$$

where the c's are constants symmetric in α, β and γ, the x''s are geodesic with P for origin and at P

$$\frac{\partial}{\partial x'^\gamma} \begin{Bmatrix} i \\ \alpha\beta \end{Bmatrix}' - \frac{\partial}{\partial x^\gamma} \begin{Bmatrix} i \\ \alpha\beta \end{Bmatrix} = c^i_{\alpha\beta\gamma}.$$

---

* This is the generalization of a theorem in the theory of surfaces. Cf. 1909, 1, p. 217; also *Bianchi*, 1902, 1, p. 338.

5. If in the transformations of Ex. 4

$$c^i_{\alpha\beta\gamma} = -\frac{1}{3}\left[\frac{\partial}{\partial x^\gamma}\left\{{i \atop \alpha\beta}\right\} + \frac{\partial}{\partial x^\alpha}\left\{{i \atop \beta\gamma}\right\} + \frac{\partial}{\partial x^\beta}\left\{{i \atop \gamma\alpha}\right\}\right]_P,$$

then at P in the x''s

$$\frac{\partial}{\partial x'^\gamma}\left\{{i \atop \alpha\beta}\right\}' + \frac{\partial}{\partial x'^\alpha}\left\{{i \atop \beta\gamma}\right\}' + \frac{\partial}{\partial x'^\beta}\left\{{i \atop \gamma\alpha}\right\}' = 0.$$

There are $\frac{1}{6}\, n^2(n+1)(n+2)$ of these equations. Show also that for a V_n the second derivatives of g'_{ij} at P are uniquely determined by these equations and (8.3) as linear functions of R'^h_{ijk}. *Eddington*, 1923, 1, p. 79.

6. Show, with the aid of Exs. 3 and 5, that for a V_n the components of any tensor involving only g_{ij} and their first and second derivatives are functions of g_{ij} and R^h_{ijk}. *Eddington*, 1923, 1, p. 79.

7. Show that for a V_n the only covariant symmetric tensor of the second order, whose components are linear in the second derivatives of g_{ij} and involve also g_{ij} and their first derivatives, are of the form

$$R_{ij} + g_{ij}(aR + b),$$

where a and b are invariants.

8. For the generalized Liouville form of the fundamental form, namely

$$(X_1 + X_2 + \cdots + X_n)\sum_i e_i (dx^i)^2,$$

where X_i is a function of x^i alone, a complete integral of $\Delta_1\theta = 1$ is

$$\theta = c + \int \sum_i \sqrt{e_i(X_i + a_i)}\, dx^i,$$

where c and the a's are constants, the latter being subject to the condition $a_1 + \cdots + a_n = 0$. *Bianchi*, 1902, 1, p. 338.

20. Curvature of a curve. Given any non-minimal curve in a V_n which is not a geodesic and let the coördinates be expressed in terms of its arc. If we write

$$\frac{d^2x^i}{ds^2} + \left\{{i \atop jk}\right\}\frac{dx^j}{ds}\frac{dx^k}{ds} = \mu^i, \tag{20.1}$$

See App. 4 it is evident from the form (17.11) of the left-hand member of this equation that μ^i are the contravariant components of a vector. Moreover, in consequence of (17.10) we have

$$g_{ij}\,\mu^i\frac{dx^j}{ds} = 0, \tag{20.2}$$

that is, the vector μ^i is orthogonal to the curve at each point.

An invariant ϱ is defined by the equation

$$\frac{1}{\varrho} = \sqrt{|g_{ij}\,\mu^i\,\mu^j|}. \tag{20.3}$$

At the origin of Riemannian coördinates equations (20.1) are

$$\frac{d^2 x^i}{d s^2} = \mu^i. \tag{20.4}$$

Thus $1/\varrho$ is the generalization of the first curvature in euclidean 3-space and $\mu^i\varrho$ of the direction-cosines of the principal normal of the curve. Accordingly we call ϱ, defined by (20.3), the *radius of first curvature* of the curve and the vector of components μ^i the *principal normal.* We have at once:

*When the first curvature of a curve is zero at all its points, either it is a geodesic and its principal normal is indeterminate or it is a curve for which the principal normal is a null vector.**

By means of (20.4) the equations of the curve are expressible in the form

$$x^i = \left(\frac{d x^i}{d s}\right)_0 s + \frac{1}{2}(\mu^i)_0\, s^2 + \cdots. \tag{20.5}$$

The equations of the geodesic through the origin which has the same direction as the given curve at the point are

$$\overline{x}^i = \left(\frac{d x^i}{d s}\right)_0 s.$$

Hence the distance d between points of the curve and the geodesic for the same value of s, to within terms of the third and higher order, is given by

$$d = \sqrt{|g_{ij}(x^i - \overline{x}^i)(x^j - \overline{x}^j)|} = \frac{1}{2}s^2\sqrt{|g_{ij}\,\mu^i\,\mu^j|} = \frac{1}{2}\frac{s^2}{\varrho}, \tag{20.6}$$

as in the case of euclidean 3-space.†

* When the fundamental form is definite, the second possibility does not arise.

† Cf. 1909, 1, p. 18.

In consequence of the remark following (17.11) it follows that when a curve is minimal but not a geodesic, the preceding developments apply with the understanding that s in (20.6) is the parameter in terms of which the equations of the minimal geodesics tangent to the curve are expressible in the form (17.8).

We have from (20.6):

A necessary and sufficient condition that a curve and its tangent geodesic at a point have contact of the second or higher order is that the curvature be zero.

In terms of Riemannian coördinates with a given point as origin, the surface consisting of the geodesics through the origin in the pencil of directions determined by the tangent and the principal normal of a curve at the origin is given by the equations

$$\bar{x}^i = \left[a\left(\frac{dx^i}{ds}\right)_0 + b\,(\mu^i)_0\right]s,$$

where a and b are parameters. If we take $a = 1$, $b = \frac{1}{2}s$, we have from (20.5) that the curve so determined coincides with the curve to within terms of the third and higher orders. Hence:

The surface formed by the geodesics through a point of a curve in the pencil of directions determined by the tangent and principal normal to the curve at the point osculates the curve.

We call this surface the *osculating geodesic surface* of the curve. It is an evident generalization of the osculating plane of a curve in euclidean 3-space.

If in the right-hand members of equations (20.1) the functions μ^i are arbitrary, we have a system of differential equations admitting a solution for each point determined by a direction at the point, as in the case of equations (17.8).

21. Parallelism. In this section we define parallelism of vectors. As the basis of this definition we take a property of parallelism in the euclidean plane, namely that all vectors parallel to one another make the same angle with a straight line, that is, with a geodesic.

Consider now any V_2 and in it a non-minimal geodesic C at points of which the coördinates $x^i\,(i = 1, 2)$ are expressed in terms of the arc s, let $\lambda^i\,(x)$ be the components of unit vectors at points of C and not tangent to C. The cosine of the angle between the vector at a point and the tangent to C at the point is $g_{ij}\lambda^i \frac{dx^j}{ds}$. The condition that this angle be constant along C is

App. 5 replaces lines 3-9

$$\frac{dx^k}{ds}\left(g_{ij}\lambda^i\frac{dx^j}{ds}\right)_{,k} = g_{ij}\frac{dx^k}{ds}\left[\lambda^i_{,k}\frac{dx^j}{ds}+\lambda^i\left(\frac{dx^j}{ds}\right)_{,k}\right] = 0,$$

which reduces in consequence of (17.11) to

$$g_{ij}\frac{dx^j}{ds}\lambda^i_{,k}\frac{dx^k}{ds} = 0.$$

Since λ^i are the components of a unit vector, we have $\lambda_i\lambda^i = e$, from which it follows that

$$g_{ij}\lambda^j\lambda^i_{,k}\frac{dx^k}{ds} = 0.$$

By hypothesis $g = |g_{ij}| \neq 0$ and $\begin{vmatrix} \lambda^1 & \lambda^2 \\ \frac{dx^1}{ds} & \frac{dx^2}{ds} \end{vmatrix} \neq 0$. Consequently from the preceding equations we have

$$\lambda^i_{,k}\frac{dx^k}{ds} = \left(\frac{\partial\lambda^i}{\partial x^k}+\lambda^l\left\{{i \atop kl}\right\}\right)\frac{dx^k}{ds} = 0. \tag{21.1}$$

For the euclidean plane, and indeed for a euclidean space of any order, referred to cartesian coördinates the condition that a vector-field be a parallel field is that λ^i be constants. In this case the expression in parenthesis in (21.1) vanishes, since the Christoffel symbols are zero; consequently in any coördinate system the condition for parallelism is

$$\lambda^i_{,k} = \frac{\partial\lambda^i}{\partial x^k}+\lambda^l\left\{{i \atop kl}\right\} = 0. \tag{21.2}$$

From (11.17) we have

$$\lambda^i_{,jk}-\lambda^i_{,kj} = -\lambda^l R^i_{ljk},$$

and consequently the condition of integrability of (21.2) is

$$\lambda^l R^i_{ljk} = 0. \tag{21.3}$$

When the fundamental form of a space is such that a coördinate system can be chosen in terms of which the coefficients g_{ij} are constant and only then, the components R^i_{ljk} of the Riemann tensor vanish (§ 10). In this case equations (21.3) are satisfied identically, and consequently equations (21.2) are completely integrable; that is, a solution of (21.2) is determined by arbitrary initial values of the λ's. In this case we have a field of vectors parallel to an arbitrary vector. If equations (21.2) and (21.3) are consistent, we will have one or more fields of parallel vectors; this question will be considered in § 23. However, in a space with a general fundamental form this is not possible. Consequently we introduce the idea of vectors parallel at points of a curve, and take (21.1) as the definition of *parallelism along any curve, not necessarily a geodesic, with respect to the metric of the space, whatever be the order of the space.* Thus if we take a curve C defined analytically
See by the x's as functions of s, equations (21.1) admit a solution
App. 6 determined by an arbitrary direction at an initial point of the curve. Not only the curve but also the metric of the space are involved in these equations, and consequently we speak of such a solution as defining a set of vectors parallel along the curve with respect to the metric of the space, or for brevity with respect to V_n. This is the *parallelism of Levi-Civita,** who first proposed this definition, but from another point of view (cf. § 24).

As a first consequence of this definition, we have that, if in (21.1) we put $\lambda^i = \frac{dx^i}{ds}$, we get the equation (17.8) of the geodesics. Hence:

Geodesics are characterized by the property that the tangents are parallel with respect to the curve.

This is an evident generalization of the property of constancy of direction of a straight line in euclidean space.

Again if $\lambda_{1|}{}^i$ and $\lambda_{2|}{}^i$ are two sets of solutions of (21.1) we have that $g_{ij}\lambda_{1|}{}^i\lambda_{2|}{}^j$ is constant along the curve. Hence:

At every point of a curve the two directions parallel with respect to the curve to two directions at a given point P of the curve make a constant angle.

In particular, when the curve is a geodesic and its tangents are

---

* 1917, 1; cf. also, *Severi*, 1917, 2, p. 230.

one set of directions we have the property in a V_2 which served as the basis for the definition of parallelism.*

Equations (21.1) are equivalent to

$$g_{ij}\frac{dx^k}{ds}\left(\frac{\partial \lambda^i}{\partial x^k}+\lambda^l\left\{{i \atop k\,l}\right\}\right) = 0, \tag{21.4}$$

since by hypothesis the determinant g of the g_{ij}'s is different from zero, and consequently the covariant components satisfy

$$\lambda_{j,k}\frac{dx^k}{ds} = 0. \tag{21.5}$$

22. Parallel displacement and the Riemann tensor. For a general parameter t equation (21.1) becomes

$$\frac{d\lambda^i}{dt}+\lambda^l\left\{{i \atop k\,l}\right\}\frac{dx^k}{dt} = 0. \tag{22.1}$$

Instead of speaking of the solution determined by an initial direction as a set of parallel vectors, we may speak of the vectors arising from a given vector by parallel displacement along a curve. In particular, it is interesting to consider the effect of parallel displacement of a vector about a small closed circuit.†

Take a surface defined by $x^i = f^i(u, v)$, where the functions f^i and their derivatives up to the third exist and are continuous at P, and consider the circuit consisting of $P(u, v)$, $Q(u+\Delta u, v)$, $R(u+\Delta u, v+\Delta v)$, $S(u, v+\Delta v)$ and P. If a vector λ^i be transported parallel to itself about this circuit, we have

$$(\lambda^i)_Q = (\lambda^i)_P+\left(\frac{d\lambda^i}{du}\right)_P\Delta u+\frac{1}{2}\left(\frac{d^2\lambda^i}{du^2}\right)_P(\Delta u)^2+\cdots,$$

$$(\lambda^i)_R = (\lambda^i)_Q+\left(\frac{d\lambda^i}{dv}\right)_Q\Delta v+\frac{1}{2}\left(\frac{d^2\lambda^i}{dv^2}\right)_Q(\Delta v)^2+\cdots,$$

* *Levi-Civita*, 1917, 1, p. 184.

† This question was considered by *Schouten*, 1918, 1, p. 64 and by *Pérès*, 1919, 1; it was considered for the general case of an affine connection by *Weyl*, 1921, 1, p. 106; see also *Dienes*, 1922, 2, and *Synge*, 1923, 3; the method followed in the text is similar to that of Synge.

$$(\lambda^i)_S = (\lambda^i)_R - \left(\frac{d\lambda^i}{du}\right)_R \Delta u + \frac{1}{2}\left(\frac{d^2\lambda^i}{du^2}\right)_R (\Delta u)^2 + \cdots,$$

$$(\bar{\lambda}^i)_P = (\lambda^i)_S - \left(\frac{d\lambda^i}{dv}\right)_S \Delta v + \frac{1}{2}\left(\frac{d^2\lambda^i}{dv^2}\right)_S (\Delta v)^2 + \cdots,$$

where the terms not written are of the third and higher orders, and the quantities such as $\left(\frac{d\lambda^i}{dv}\right)_Q$, $\left(\frac{d^2\lambda^i}{dv^2}\right)_Q$ are given by (22.1) and the equations resulting from the differentiation of this equation. If all of the above equations be added, we have

$$\Delta(\lambda^i)_P = (\bar{\lambda}^i)_P - (\lambda^i)_P = \Delta u\left[\left(\frac{d\lambda^i}{du}\right)_P - \left(\frac{d\lambda^i}{du}\right)_R\right] + \Delta v\left[\left(\frac{d\lambda^i}{dv}\right)_Q - \left(\frac{d\lambda^i}{dv}\right)_S\right]$$
(22.2)
$$+ \frac{1}{2}(\Delta u)^2\left[\left(\frac{d^2\lambda^i}{du^2}\right)_P + \left(\frac{d^2\lambda^i}{du^2}\right)_R\right] + \frac{1}{2}(\Delta v)^2\left[\left(\frac{d^2\lambda^i}{dv^2}\right)_Q + \left(\frac{d^2\lambda^i}{dv^2}\right)_S\right] + \cdots.$$

If we assume that the x's are geodesic with P as origin, so that $\left\{ {i \atop jk} \right\}_P = 0$, we have from (22.1), in which the Christoffel symbols are evaluated by means of their expansions about P,

$$\left(\frac{d\lambda^i}{du}\right)_P = 0, \qquad \left(\frac{d\lambda^i}{dv}\right)_Q = -\left(\frac{\partial}{\partial x^m}\left\{ {i \atop jk} \right\}\frac{\partial x^m}{\partial u}\frac{\partial x^j}{\partial v}\lambda^k\right)_P \Delta u + \cdots,$$

$$\left(\frac{d\lambda^i}{du}\right)_R = -\left(\frac{\partial}{\partial x^m}\left\{ {i \atop jk} \right\}\frac{\partial x^m}{\partial u}\frac{\partial x^j}{\partial u}\lambda^k\right)_P \Delta u$$
$$- \left(\frac{\partial}{\partial x^m}\left\{ {i \atop jk} \right\}\frac{\partial x^m}{\partial v}\frac{\partial x^j}{\partial u}\lambda^k\right)_P \Delta v + \cdots,$$

$$\left(\frac{d\lambda^i}{dv}\right)_S = -\left(\frac{\partial}{\partial x^m}\left\{ {i \atop jk} \right\}\frac{\partial x^m}{\partial v}\frac{\partial x^j}{\partial v}\lambda^k\right)_P \Delta v + \cdots,$$

$$\left(\frac{d^2\lambda^i}{du^2}\right)_P + \left(\frac{d^2\lambda^i}{du^2}\right)_R = -2\left(\frac{\partial}{\partial x^m}\left\{ {i \atop jk} \right\}\frac{\partial x^m}{\partial u}\frac{\partial x^j}{\partial u}\lambda^k\right)_P + \cdots,$$

$$\left(\frac{d^2\lambda^i}{dv^2}\right)_Q + \left(\frac{d^2\lambda^i}{dv^2}\right)_S = -2\left(\frac{\partial}{\partial x^m}\left\{ {i \atop jk} \right\}\frac{\partial x^m}{\partial v}\frac{\partial x^j}{\partial v}\lambda^k\right)_P + \cdots.$$

When these expressions are substituted in (22.2), we obtain

$$\Delta(\lambda^i)_P = \left[\left(\frac{\partial}{\partial x^m}\left\{ {i \atop jk} \right\} - \frac{\partial}{\partial x^j}\left\{ {i \atop mk} \right\}\right)\frac{\partial x^j}{\partial u}\frac{\partial x^m}{\partial v}\lambda^k\right]_P \Delta u\,\Delta v + \cdots.$$

Since the left-hand member is a contravariant vector in V_n, and $\frac{\partial x^j}{\partial u}$, $\frac{\partial x^m}{\partial v}$, λ^k are the components of contravariant vectors, it follows that in a general coördinate system this equation is

$$(22.3) \qquad \Delta(\lambda^i)_P = \left(R^i_{kmj}\frac{\partial x^j}{\partial u}\frac{\partial x^m}{\partial v}\lambda^k\right)_P \Delta u \Delta v + \cdots.$$

From the considerations of § 21 it follows that $\Delta(\lambda^i)_P = 0$ when $R^i_{kmj} = 0$. When this condition is not satisfied, it follows from (22.3) that when a general vector is displaced around an infinitesimal circuit the difference between its final and original direction is of the second order and depends upon the value of the components R^i_{kmj} at the starting point and upon the circuit. An exception to this case is treated in the next section.

23. Fields of parallel vectors. From (21.1) it follows that when a set of functions λ^i satisfy the equations

$$(23.1) \qquad \lambda^i_{,k} = \frac{\partial \lambda^i}{\partial x^k} + \lambda^l \left\{ {i \atop kl} \right\} = 0,$$

any two vectors of the vector-field are parallel with respect to any curve joining points of these vectors.* The conditions of integrability of these equations are (21.3), that is

$$(23.2) \qquad \lambda^l R^i_{ljk} = 0.$$

Unless $R^i_{ljk} = 0$, which is assumed not to be the case, the components of such vector-fields must satisfy (23.2) as well as (23.1). Differentiating (23.2) covariantly with respect to x^{m_1} and expressing the condition that (23.1) is satisfied, we get

$$(23.3) \qquad \lambda^l R^i_{ljk,m_1} = 0.$$

Continuing this process, we get a sequence of necessary conditions

$$(23.4) \qquad \begin{aligned} \lambda^l R^i_{ljk,m_1m_2} &= 0, \\ \cdots & \cdots \\ \cdots & \cdots \\ \lambda^l R^i_{ljk,m_1m_2\cdots m_s} &= 0, \\ \cdots & \cdots \end{aligned}$$

* in the region of V_n in which (23.1) apply.

If the equations (23.2), (23.3), (23.4) are algebraically inconsistent, there is no field of parallel vectors. To be consistent it is necessary that equations (23.2) and the first $q (\geqq 0)$ sets of equations (23.3) and (23.4) admit a complete system of p sets of linearly independent solutions $\lambda_{1|}{}^i, \cdots, \lambda_{p|}{}^i$, for $p \geqq 1$, $i = 1, \cdots, n$, in terms of which all other solutions are linearly expressible, such that these p sets of solutions satisfy also the $(q+1)$th set of equations (23.4). Thus any set of solutions λ^i is given by

$$\lambda^i = \varphi^{(1)} \lambda_{1|}{}^i + \cdots + \varphi^{(p)} \lambda_{p|}{}^i, \tag{23.5}$$

where the φ's are functions of the x's, which we seek to determine so that λ^i is a set of solutions of (23.1).

In the first place we remark that if $\lambda_{\sigma|}{}^i$ is any one of the p sets of solutions and we substitute it in (23.2) and the first q sets of (23.3) and (23.4), and differentiate these equations covariantly, then since $\lambda_{\sigma|}{}^i$ satisfies the $(q+1)$th set also it follows that $\lambda_{\sigma|}{}^i{}_{,m}$ is a solution of (23.2) and the first q sets of (23.3) and (23.4). Consequently it is expressible in the form

$$\lambda_{\sigma|}{}^i{}_{,m} = \mu^{(1)}_{\sigma|m} \lambda_{1|}{}^i + \cdots + \mu^{(p)}_{\sigma|m} \lambda_{p|}{}^i, \tag{23.6}$$

where the p^2 covariant vectors $\mu^{(\alpha)}_{\sigma|m}$ $(\alpha, \sigma = 1, \cdots, p;\ m = 1, \cdots, n)$ are to be determined; here α and σ indicate the vector and m the component. They are determined by the condition that (§ 11)

$$\lambda_{\sigma|}{}^i{}_{,ml} - \lambda_{\sigma|}{}^i{}_{,lm} = -\lambda_{\sigma|}{}^j R^i{}_{jml} = 0,$$

in consequence of (23.2). Substituting from (23.6) in this equation and making use of (23.6) in the reduction, we obtain

$$\left[\mu^{(\beta)}_{\sigma|m,l} - \mu^{(\beta)}_{\sigma|l,m} + \left(\mu^{(\alpha)}_{\sigma|m}\mu^{(\beta)}_{\alpha|l} - \mu^{(\alpha)}_{\sigma|l}\mu^{(\beta)}_{\alpha|m}\right)\right] \lambda_{\beta|}{}^i = 0. \quad (\alpha, \beta = 1, \cdots, p).*$$

Since the rank of the matrix $\| \lambda_{\beta|}{}^i \|$ is p, these equations are equivalent to the system

$$\frac{\partial \mu^{(\beta)}_{\sigma|m}}{\partial x^l} - \frac{\partial \mu^{(\beta)}_{\sigma|l}}{\partial x^m} + \left(\mu^{(\alpha)}_{\sigma|m}\mu^{(\beta)}_{\alpha|l} - \mu^{(\alpha)}_{\sigma|l}\mu^{(\beta)}_{\alpha|m}\right) = 0 \quad \begin{pmatrix} \alpha, \beta, \sigma = 1, \cdots, p; \\ l, m = 1, \cdots n \end{pmatrix}. \tag{23.7}$$

---

* In this equation α and β are summed from 1 to p; the same is true of a repeated index of this sort in the following equations.

When now we require that λ^i as given by (23.5) shall satisfy (23.1) we obtain, in consequence of (23.6), since the rank of $\|\lambda^{i}_{\sigma|}\|$ is p,

$$\frac{\partial \varphi^{(\beta)}}{\partial x^k} + \varphi^{(\alpha)} \mu^{(\beta)}_{\alpha|k} = 0. \tag{23.8}$$

Because of (23.7) this system of equations is completely integrable, See App. 7
and consequently the solution involves p arbitrary constants. In view of the above results we have the theorem:

If the system of equations (23.2), (23.3), (23.4) *is algebraically consistent, there exists one or more fields of parallel vectors; more specifically, if* (23.2) *and the first* $q\,(\geqq 0)$ *sets of* (23.3) *and* (23.4) *admit a complete system* of p sets *of solutions which also satisfy the* $(q+1)$*th set of these equations, there exist fields of parallel vectors depending on* p *arbitrary constants.*

Since equations (23.8) admit a solution determined by an arbitrary set of initial values, we see that when the conditions of the theorem are satisfied, any vector at any point P in space in the p-fold bundle determined by the p vectors $\lambda_{\sigma|}{}^{i}$ is parallel to a vector in the bundle at any other point.*

We have just obtained the conditions for fields of parallel vectors in invariantive form. Now we shall show how such fields may be obtained by making a suitable choice of coördinates. Using the preceding notation and indicating by $\lambda'_{\sigma|}{}^{i}$ the See
components of p independent fields in coördinates x'^i, we have App. 8

$$\lambda'_{\sigma|}{}^{i} = \lambda_{\sigma|}{}^{j} \frac{\partial x'^i}{\partial x^j}. \tag{23.9}$$

Consider the system of p linear partial differential equations

$$X_\sigma(\theta) \equiv \lambda_{\sigma|}{}^{j} \frac{\partial \theta}{\partial x^j} = 0, \quad (\sigma = 1, \cdots, p;\ j = 1, \cdots, n), \tag{23.10}$$

where $X_\sigma(\theta)$ is an abbreviation. If $X_\tau X_\sigma(\theta)$ has the significance

$$X_\tau X_\sigma(\theta) = \lambda_{\tau|}{}^{k} \frac{\partial}{\partial x^k}\left(\lambda_{\sigma|}{}^{j} \frac{\partial \theta}{\partial x^j}\right),$$

---

* This problem for a single field of parallel vectors was treated by *Levi-Civita*, 1917, 1, p. 194; cf. *Eisenhart*, 1922, 3, p. 209; also *Veblen* and *Thomas*, 1923, 4, pp. 589–591.

the operator

$$(X_\tau,\ X_\sigma)\,\theta \equiv X_\tau X_\sigma(\theta) - X_\sigma X_\tau(\theta)$$

is called the *Poisson operator*. A fundamental theorem of systems of linear partial differential equations is: A necessary and sufficient condition that a system (23.10) be completely integrable, that is, that it admit $n-p$ independent solutions, is that (X_τ, X_σ) be linearly expressible in terms of the X's.*

When now we apply this general theory to the case where $\lambda_{\sigma|}{}^i$ satisfy (23.1), we find that $(X_\sigma, X_\tau)\,\theta \equiv 0$ and consequently equations (23.10) admit $n-p$ independent solutions. If we take them for the coördinates $x'^{p+1}, \cdots, x'^n$, it follows from (23.9) that $\lambda'_{\sigma|}{}^t = 0$ for $t = p+1, \cdots, n$. Again if we omit one of the equations from (23.10), say $X_r(\theta) = 0$, the remaining system is complete and admits in addition to $x'^{p+1}, \cdots, x'^n$ another independent solution x'^r. In this way the x''s are defined so that all of the components of the λ's are zero except those with the same subscript and superscript. If it is assumed that these vectors are unit vectors, we have accordingly in the new coördinate system

App. 9 replaces the remainder of this section

$$(23.11)\quad \lambda_{\sigma|}{}^\sigma = \frac{1}{\sqrt{e_\sigma g_{\sigma\sigma}}},\quad \lambda_{\sigma|}{}^t = 0 \quad (\sigma = 1, \cdots, p;\ t = 1, \cdots, n;\ t \neq \sigma).$$

If these expressions are substituted in (23.1), we get

$$\frac{\partial}{\partial x^k} \log \sqrt{g_{\sigma\sigma}} - \left\{ {\sigma \atop k\sigma} \right\} = 0,$$

$$\left\{ {j \atop k\sigma} \right\} = 0 \quad \left({\sigma = 1, \cdots, p; \atop j,\ k = 1, \cdots, n;\ j \neq \sigma} \right),$$

where $\left\{ {\sigma \atop k\sigma} \right\}$ is not summed for σ, but consists of a single term. If we multiply the first of these equations by $g_{\sigma l}$ and subtract from it the second multiplied by g_{jl} and summed for j, we get the equivalent set of equations

$$g_{\sigma l} \frac{\partial}{\partial x^k} \log \sqrt{g_{\sigma\sigma}} - [k\sigma,\ l] = 0,$$

* *Goursat*, 1891, 1, p. 52.

that is,

$$(23.12)\qquad g_{\sigma l}\frac{\partial}{\partial x^k}\log g_{\sigma\sigma}-\frac{\partial g_{kl}}{\partial x^\sigma}-\frac{\partial g_{\sigma l}}{\partial x^k}+\frac{\partial g_{k\sigma}}{\partial x^l}=0.$$

For the case $k=\sigma$, these equations reduce to

$$\frac{\partial}{\partial x^\sigma}\left(\frac{g_{\sigma l}}{\sqrt{e_\sigma g_{\sigma\sigma}}}\right)=e_\sigma\frac{\partial}{\partial x^l}\sqrt{e_\sigma g_{\sigma\sigma}}.$$

In accordance with these equations we define p functions ψ_σ by

$$\sqrt{e_\sigma g_{\sigma\sigma}}=e_\sigma\frac{\partial\psi_\sigma}{\partial x^\sigma},\qquad \frac{g_{\sigma l}}{\sqrt{e_\sigma g_{\sigma\sigma}}}=\frac{\partial\psi_\sigma}{\partial x^l},$$

from which have

$$(23.13)\qquad g_{\sigma l}=e_\sigma\frac{\partial\psi_\sigma}{\partial x^\sigma}\frac{\partial\psi_\sigma}{\partial x^l}\quad(\sigma=1,\cdots,p;\ l=1,\cdots,n).$$

From these expressions it follows that ψ_σ must involve x^σ, otherwise the space is of less than n dimensions.

Again if neither k nor l in (23.12) is σ, we have

$$(23.14)\quad g_{kl}=e_\sigma\frac{\partial\psi_\sigma}{\partial x^k}\frac{\partial\psi_\sigma}{\partial x^l}+\varphi_{kl\sigma}\quad\left(\begin{matrix}\sigma=1,\cdots,p;k,l=1,\cdots n;\\ k\neq\sigma,\ l\neq\sigma\end{matrix}\right),$$

where $\varphi_{kl\sigma}$ is a function independent of x^σ.

From (23.13) and 23.14) it follows that for each value of σ the fundamental form can be written

$$\varphi=e_\sigma(d\psi_\sigma)^2+g_{rs}\,dx^r\,dx^s\qquad(r,s=1,\cdots,n;r\neq\sigma,s\neq\sigma),$$

where g_{rs} are independent of x^σ.

If then we put $x'^\sigma=\psi_\sigma$, $x'^j=x^j\ (j\neq\sigma)$, the curves of parameter x'^σ are the same as those of parameter x^σ, and these curves are geodesics (cf. § 19). Hence we have:

When a V_n admits p independent fields of parallel unit vectors, the vectors of each field are the tangent vectors to a congruence of geodesics.

Conversely, if the fundamental form of a space is reducible to the form

$$\varphi = e_1 (dx^1)^2 + g_{rs}\, dx^r\, dx^s \qquad (r, s = 2, \cdots, n), \tag{23.15}$$

it is found from (23.12) that a necessary and sufficient condition that the tangents to the curves of parameter x^1 form a parallel field is that g_{rs} be independent of x^1. In this case all the spaces $x^1 =$ const. have the same fundamental form and consequently any one of them can be brought into coincidence with any other by a *translation*, that is, by a motion in which each point describes the same distance along the geodesic normal to the sub-space. In the case $p > 1$ the space admits p independent translations; thus any one of the subspaces of each of the family of subspaces $\psi_\sigma =$ const. can be brought into coincidence with any other of the family by a translation.

If, in particular, we take $\psi_\sigma = x^\sigma + \varphi_\sigma (x^{p+1}, \cdots, x^n)$ for $\sigma = 1, \cdots, p$, it follows from (23.13) and (23.14) that for a V_n with the fundamental form

$$\varphi = e_1 (dx^1)^2 + \cdots + e_p (dx^p)^2 + g_{\alpha\beta}\, dx^\alpha\, dx^\beta \quad (\alpha, \beta = p+1, \cdots n),$$

where $g_{\alpha\beta}$ are arbitrary functions of $x^{p+1}, \cdots, x^n$ the tangents to curves of parameters $x^1, x^2, \cdots, x^p$ form fields of parallel vectors.*

24. Associate directions. Parallelism in a sub-space. Let C be any non-minimal curve in a V_n at points of which the coördinates x^i are expressed in terms of the arc,† and let λ^i be the components of a unit or null vector-field; in either case we have

$$\lambda_i \lambda^i{}_{,k} = 0. \tag{24.1}$$

If we put

$$\frac{dx^k}{ds} \lambda^i{}_{,k} = \mu^i, \tag{24.2}$$

it is seen from (21.1) that $\mu^i = 0$, if the vectors at points of C are parallel with respect to the curve; otherwise, as follows from

* Cf. *Eisenhart*, 1925, 3, for the complete solution of the problem.

† If the curve is minimal, we take for s the parameter in terms of which the equations of the tangent geodesics are of the form (17.8); note the remark following equation (17.11).

the form of the left-hand member of (24.2), the functions μ^i are the contravariant components of a vector, which Bianchi* has called the *associate direction* for the vector λ^i along the curve. From (24.1) and (24.2) we have

$$\lambda_i \mu^i = 0, \tag{24.3}$$

and consequently:

If a set of vectors at points of a curve are not parallel with respect to the curve, there is determined at each point of the curve an associate direction and it is orthogonal to the given vector at the point.

The invariant $1/r$ defined by

$$\frac{1}{r} = \sqrt{|g_{ij} \mu^i \mu^j|} \tag{24.4}$$

we call, with Bianchi, the *associate curvature* of the vector λ^i with respect to the curve. When, in particular, the vectors λ^i are tangent to the curve, equations (24.2) and (24.4) reduce to (20.1) and (20.3), and consequently the associate direction and curvature are the principal normal and first curvature of the curve.

Consider the space V_n as immersed in a space V_m of coördinates y^α, the equations of V_n being (16.2).† Let ξ^α be the components in the y's of the vector-field whose components in the x's are λ^i, that is [cf. (16.8)],

$$\xi^\beta = \lambda^j \frac{\partial y^\beta}{\partial x^j}. \tag{24.5}$$

Differentiating these equations with respect to s, we have

$$\frac{d\xi^\beta}{ds} = \frac{d\lambda^j}{ds}\frac{\partial y^\beta}{\partial x^j} + \lambda^j \frac{\partial^2 y^\beta}{\partial x^j \partial x^i}\frac{dx^i}{ds}. \tag{24.6}$$

If η^α denote the components of the associate direction of ξ^α in V_m (which is not necessarily the same as the associate direction of λ^i in V_n), we have analogously to (24.2)

---

* 1922, 4, p. 161.

† Throughout the remainder of this section Greek indices take the values $1, \cdots, m$ and Latin $1, \cdots, n$, unless stated otherwise.

$$\eta^\beta = \frac{d y^\alpha}{d s}\left(\frac{\partial \xi^\beta}{\partial y^\alpha} + \xi^\gamma \left\{{\beta \atop \alpha\, \gamma}\right\}_a\right), \tag{24.7}$$

where the Christoffel symbols $\left\{{\beta \atop \alpha\, \gamma}\right\}_a$ are formed with respect to the fundamental tensor $a_{\alpha\beta}$ of V_m. Because of (24.5) and (24.6) this may be written

$$\eta^\beta = \frac{d\lambda^j}{d s}\frac{\partial y^\beta}{\partial x^j} + \lambda^j \frac{d x^i}{d s}\left(\frac{\partial^2 y^\beta}{\partial x^i\, \partial x^j} + \left\{{\beta \atop \alpha\, \gamma}\right\}_a \frac{\partial y^\alpha}{\partial x^i}\frac{\partial y^\gamma}{\partial x^j}\right). \tag{24.8}$$

If we denote by $[ij, k]_g$ the Christoffel symbols of the first kind formed with respect to (12.3); we have from (16.3) by direct calculation

$$[ij, k]_g = a_{\beta\delta}\frac{\partial y^\delta}{\partial x^k}\left(\frac{\partial^2 y^\beta}{\partial x^i\, \partial x^j} + \left\{{\beta \atop \alpha\, \gamma}\right\}_a \frac{\partial y^\alpha}{\partial x^i}\frac{\partial y^\gamma}{\partial x^j}\right). \tag{24.9}$$

When (24.8) is multiplied by $a_{\beta\delta}\dfrac{\partial y^\delta}{\partial x^k}$ and summed for β, the resulting equation is reducible by means of (16.3) and (24.9) to

$$\begin{aligned} a_{\beta\delta}\frac{\partial y^\delta}{\partial x^k}\eta^\beta &= g_{jk}\frac{d\lambda^j}{d s} + \lambda^j \frac{d x^i}{d s}[ij, k]_g \\ &= g_{ik}\frac{d x^j}{d s}\left(\frac{\partial \lambda^i}{\partial x^j} + \lambda^l \left\{{i \atop l\, j}\right\}_g\right). \end{aligned} \tag{24.10}$$

If the vectors ξ^α are parallel with respect to the curve in V_m, then $\eta^\beta = 0$, and from (24.10) and (21.4) we have that the vectors are parallel in V_n. Hence:

If a curve C lies in a V_n which is immersed in a V_m and vectors are parallel along C with respect to V_m, they are parallel with respect to V_n.

As previously remarked (§ 16), if the fundamental form of V_n is definite, it is possible to find a euclidean V_m enveloping it and the requirement that vectors in V_n be parallel with respect to V_m leads to parallelism with respect to V_n. This was the point of departure for Levi-Civita's definition of parallelism in any space.*

---

* 1917, 1.

As a consequence of the above theorem and the first theorem of § 21 we have:

If a curve is a geodesic of a space, it is a geodesic of any sub-space in which it lies.

If vectors along a curve are parallel with respect to V_n but not with respect to V_m, we have from (24.10)

$$a_{\alpha\beta}\frac{\partial y^\alpha}{\partial x^k}\eta^\beta = 0, \tag{24.11}$$

that is, the associate vector is normal to V_n, and conversely. Hence:

A necessary and sufficient condition that vectors along a curve in V_n be parallel with respect to V_n, when they are not parallel with respect to an enveloping space V_m, is that the vectors in V_m associate to these vectors be normal to V_n.

*When a geodesic in a space V_n is not a geodesic in an enveloping space V_m, its principal normals as a curve in V_m are normal to V_n.**

Consider two spaces V_n and V_n' immersed in a V_m such that at each point of a curve C every normal to one is normal to the other; in this case the spaces V_n and V_n' are said to be *tangent* to one another along C. From the next to the last theorem we have:

If two spaces V_n and V_n' in a V_m are tangent along a curve C, vectors parallel to one another along C with respect to V_n are parallel with respect to V_n' and vice-versa.

Two spaces V_n and V_q for $q < n$ in a V_m are said to be *tangent* along a curve C, if every normal to V_n at each point of C is normal to V_q. Hence:

If in a V_m two spaces V_n and V_q for $q < n$ are tangent along a curve C, vectors parallel to one another along C with respect to V_n are parallel with respect to V_q.

Two subspaces V_n and V_n' immersed in a V_m are said to be *applicable*, if there exists a transformation of the coördinates x^i and x'^i of these spaces such that the fundamental forms are transformable into one another. Since the equations of parallelism involve only the components of the fundamental tensor and their first derivatives, we have:

---

* This a generalization of a characteristic property of geodesics on a surface in euclidean space, 1909, 1, p. 204; cf. *Bianchi*, 1922, 4.

If two spaces V_n and V'_n in a V_m are applicable, to vectors parallel along a curve with respect to V_n there correspond vectors parallel along the corresponding curve in V'_n.

As a simple example of several of these theorems, we consider a sphere in euclidean space and a circular cone tangent to the sphere along a small circle C. If we have a set of vectors parallel along C with respect to the sphere, they are parallel with respect to the cone, and when the cone is rolled out upon a plane the vectors are parallel in the euclidean sense.

We consider the converse problem: Given a curve C and at each point of it a vector μ^i, to find all sets of vectors λ^i such that the vectors μ^i are associate to λ^i. We denote by $\lambda_{\sigma|}{}^i$ ($\sigma = 1, \cdots, n-1$) the components of $n-1$ unit vectors orthogonal to μ^i. Then λ^i, if they exist, are given by

$$\lambda^i = t^1 \lambda_{1|}{}^i + t^2 \lambda_{2|}{}^i + \cdots + t^{n-1} \lambda_{n-1|}{}^i \equiv t^\sigma \lambda_{\sigma|}{}^i \tag{24.12}$$
$$(\sigma = 1, \cdots, n-1),$$

in accordance with the first theorem of this section. Substituting in (24.2), we have

$$\mu^i = \lambda_{\sigma|}{}^i \frac{d t^\sigma}{d s} + t^\sigma \mu_{\sigma|}{}^i, \tag{24.13}$$

where $\mu_{\sigma|}{}^i$ are the components of the associate vector of $\lambda_{\sigma|}{}^i$.

Multiplying (24.13) by $\lambda_{\tau|i}$ and summing for i, we have

$$\alpha_{\tau\sigma} \frac{d t^\sigma}{d s} + t^\sigma \mu_{\sigma|}{}^i \lambda_{\tau|i} = 0, \tag{24.14}$$

where

$$\alpha_{\tau\sigma} = g_{ij} \lambda_{\tau|}{}^i \lambda_{\sigma|}{}^j. \qquad (\sigma, \tau = 1, \cdots, n-1). \tag{24.15}$$

We assume that the t's in (24.12) are chosen so that λ^i are the components of a unit vector, if it is not a null vector. Hence we have

$$\alpha_{\tau\sigma} t^\sigma t^\tau = e \text{ or } 0 \qquad (\tau, \sigma = 1, \cdots, n-1). \tag{24.16}$$

We consider first the case when μ^i is not a null vector, in which case the $n-1$ vectors $\lambda_{\sigma|}{}^i$ can be chosen mutually orthogonal (§ 13). Then

(24.17) $$\alpha_{\sigma\sigma} = e_\sigma, \qquad \alpha_{\sigma\tau} = 0 \qquad (\sigma \neq \tau),$$

and equations (24.14) become

(24.18) $$\frac{dt^\tau}{ds} + e_\tau t^\sigma b_{\sigma\tau} = 0,$$

where $b_{\sigma\tau} = \mu_{\sigma|}{}^i \lambda_{\tau|i}$. Differentiating $\lambda_{\sigma|}{}^i \lambda_{\tau|i} = 0$ $(\sigma \neq \tau)$ with respect to s and applying (24.2), we have $b_{\sigma\tau} + b_{\tau\sigma} = 0$. In consequence of this relation any set of solutions of (24.18) satisfy the condition $\sum_\sigma e_\sigma (t^\sigma)^2 = \text{const.}$; consequently if (24.16) and (24.17) are satisfied by the initial values, they are satisfied for all values of s. Hence equations (24.18) admit ∞^{n-2} sets of solutions satisfying (24.16), where $\alpha_{\tau\sigma}$ are given by (24.17). Hence:

*Given a set of non-null vectors along a curve C, there exist ∞^{n-2} sets of vectors λ^i along C with respect to which the given vectors are associate; each set is determined by choosing the components λ^i at a point of C.**

When μ^i are the components of a null vector, we have

(24.19) $$\mu^i = c^\sigma \lambda_{\sigma|}{}^i \qquad (\sigma = 1, \cdots, n-1),$$

in accordance with the considerations at the close of § 13. Moreover, we have

(24.20) $$\mu_{\sigma|}{}^i = \varrho_\sigma \xi^i + c^\tau{}_\sigma \lambda_{\tau|}{}^i \quad (\sigma, \tau = 1, \cdots, n-1),$$

where ξ^i are the components of a vector linearly independent of the $n-1$ vectors $\lambda_{\sigma|}{}^i$. Since the n vectors ξ^i and $\lambda_{\sigma|}{}^i$ are all independent, equations (24.13) are equivalent to

(24.21) $$\frac{dt^\sigma}{ds} = c^\sigma - c^\sigma{}_\tau t^\tau,$$
$$t^\sigma \varrho_\sigma = 0.$$

Differentiating (24.19) covariantly with respect to x^k and multiplying by $\frac{dx^k}{ds}$, we have, in consequence of equations of the form (24.2),

$$\mu^i{}_{,k} \frac{dx^k}{ds} = c^\sigma (\xi^i \varrho_\sigma + c^\tau{}_\sigma \lambda_{\tau|}{}^i) + \frac{dc^\sigma}{ds} \lambda_{\sigma|}{}^i.$$

* Cf. *Bianchi*, 1922, 4, p. 166, where this theorem is established for spaces with a definite fundamental form.

Multiplying by μ_i and summing for i, we have, since $\mu_i \xi^i \neq 0$,

$$c^\sigma \varrho_\sigma = 0. \tag{24.22}$$

Differentiating the second of (24.21) and making use of the first and of (24.22), we obtain $t^\sigma\left(\frac{d\varrho_\sigma}{ds} - c^\tau_{\ \sigma}\varrho_\tau\right) = 0$. Proceeding in like manner with this equation, we find

$$t^\sigma\left(\frac{d^2\varrho_\sigma}{ds^2} - 2c^\tau_{\ \sigma}\frac{d\varrho_\tau}{ds} - \frac{dc^\tau_{\ \sigma}}{ds}\varrho_\tau + c^\tau_{\ \sigma}c^\alpha_{\ \tau}\varrho_\alpha\right) + c^\sigma\left(\frac{d\varrho_\sigma}{ds} - c^\tau_{\ \sigma}\varrho_\tau\right) = 0$$
$$(\alpha, \sigma, \tau = 1, \cdots, n-1).$$

From this process it is seen that the determination of vectors λ^i for which a given null vector μ^i is the associate depends upon the character of the latter, that is, whether sooner or later we obtain an equation by this process which is satisfied in consequence of its predecessors.

We will not proceed further with this general case, but will establish the theorem:

If a set of null vectors are parallel with respect to a curve C, they are the associates with respect to this curve of ∞^{n-1} sets of vectors.

In fact, if $\frac{dx^k}{ds}\mu^i_{\ ,k} = 0$, any set of solutions of the equations

$$\frac{dx^k}{ds}\lambda^i_{\ ,k} = \mu^i$$

satisfy the condition $\mu^i\lambda_i = \text{const}$. Hence any set of solutions whose initial values are such that $\mu^i\lambda_i = 0$ satisfy the conditions of the theorem.*

Exercises.

1. When in (20.1) $\mu^i = a^i_{\ j}\frac{dx^j}{ds}$, either the associate tensor a_{ij} is skew-symmetric, or $a_{jk}\frac{dx^j}{ds}\frac{dx^k}{ds} = 0$ is a first integral of (20.1).

---

* The existence of solutions λ^i of the above equations is the problem of the existence of solutions of a system of ordinary linear differential equations of the first order (cf. § 21).

2. Let P_1, P_2, P_3 be the vertices of a geodesic triangle in a V_2 and $\varphi_1, \varphi_2, \varphi_3$ the interior angles of the triangle at these respective points; show that when the tangent vector at P_1 to the geodesic $P_1 P_2$ is transported parallel to itself around the triangle in the direction $P_1 P_2 P_3$, it makes the angle $\pi - \varphi_1 - \varphi_2 - \varphi_3$ with its original direction at P_1. *Levi-Civita*, 1925, 4, p. 224.

3. A necessary and sufficient condition that the tangents to the curves $x^2 =$ const. on a V_2 be parallel with respect to a curve C is that C be an integral curve of

$$\left\{ {2 \atop 1i} \right\} dx^i = 0 \qquad (i = 1, 2).$$

Bianchi, 1922, 4, p. 167.

4. When the coördinates of a V_2 are chosen so that the fundamental form is $e_1 (dx^1)^2 + 2 g_{12}\, dx^1 dx^2 + e_2 (dx^2)^2$, and only in this case, the tangents to the parametric curves of either family are parallel with respect to the curves of the other family. *Bianchi*, 1922, 4, p. 170.

5. When the fundamental form of the surface considered in § 22 is definite at the point P, equations (22.3) can be written

$$(\Delta \lambda^i)_P = (R^i_{kmj} \lambda^k)_P \, \xi_{1|}^{\,j} \xi_{2|}^{\,m} \frac{\Delta \Sigma}{\sin \theta} + \cdots,$$

where $\Delta \Sigma$ is the area enclosed by the circuit, θ is the angle between the parametric curves at P and $\xi_{1|}^{\,i}$ and $\xi_{2|}^{\,i}$ are the components in V_n of the tangents to these curves at P.

6. If μ_i are the components of any vector field and $\mu_i \lambda^i = \cos \alpha$, the change in α at a point P when the vector λ^i is transported around a small circuit as in § 22 is given by (cf. Ex. 5)

$$(\Delta \alpha)_P = -(R^i_{kmj} \lambda^k)_P \, \xi_{1|}^{\,j} \xi_{2|}^{\,m} \mu_i \frac{\Delta \Sigma}{\sin \theta \sin \alpha}.$$

Pérès, 1919, 1, p. 427.

7. When in equations (23.13) and (23.14) for $\sigma = 1, 2$

$$\psi_1 = e_1 f_1 + a f_2 + A_1, \qquad \psi_2 = e_2 f_2 + a f_1 + A_2,$$

where f_1 and f_2 are independent of x^2 and x^1 respectively, a is an arbitrary constant and A_1 and A_2 are arbitrary functions of $x^3, \cdots, x^n$, the tangents to the curves of parameters x^1 and x^2 constitute fields of parallel vectors.

25. Curvature of V_n at a point. Let $\lambda_{1|}^{\,i}$ and $\lambda_{2|}^{\,i}$ be the components of two contravariant vector-fields. The vectors at a point P determine a pencil of directions defined by

$$\xi^i = \alpha \lambda_{1|}^{\,i} + \beta \lambda_{2|}^{\,i}, \tag{25.1}$$

where α and β are parameters. The geodesics through P in this pencil of directions constitute a *geodesic surface S*. The Gaussian curvature of S at P was taken by Riemann* to be the definition of the *curvature of V_n at P for the given orientation*, that is, the orientation determined by $\lambda_{1|}^{i}$ and $\lambda_{2|}^{i}$.

We assume that the coördinates x^i of V_n are Riemannian with P as origin (§ 18). Then the surface S is defined by

$$x^i = \lambda_{1|}^{i} u^1 + \lambda_{2|}^{i} u^2, \tag{25.2}$$

where $u^1 = \alpha s$ and $u^2 = \beta s$ for any geodesic through P, and $\lambda_{1|}^{i}$ and $\lambda_{2|}^{i}$ are constants.†

In terms of u^1 and u^2 the fundamental form of S is

$$\varphi = b_{\alpha\beta}\, du^\alpha\, du^\beta, \tag{25.3}$$

where (cf. § 16)

$$b_{\alpha\beta} = g_{ij} \frac{\partial x^i}{\partial u^\alpha} \frac{\partial x^j}{\partial u^\beta}.\ddagger \tag{25.4}$$

From a formula analogous to (24.9) we have in this case, as a consequence of (25.2),

$$[\alpha\beta, \gamma]_b = g_{lk}\, \lambda_{\gamma|}^{k}\, \lambda_{\alpha|}^{i}\, \lambda_{\beta|}^{j} \left\{ {l \atop ij} \right\}_g. \tag{25.5}$$

For $n = 2$ all the Riemann symbols of the first kind (§ 8) are zero or differ from $\overline{R}_{1212}$ at most in sign, because of the identities (8.10).§

In this case we have for two coördinate systems, u^i and u'^i,

$$\overline{R}'_{1212} = \overline{R}_{1212} \left(\frac{\partial u^1}{\partial u'^1} \frac{\partial u^2}{\partial u'^2} - \frac{\partial u^1}{\partial u'^2} \frac{\partial u^2}{\partial u'^1} \right)^2,$$

as follows from the general equations (4.6), and also for the determinant $b = |b_{\alpha\beta}|$ from (9.3)

$$b' = b \left(\frac{\partial u^1}{\partial u'^1} \frac{\partial u^2}{\partial u'^2} - \frac{\partial u^1}{\partial u'^2} \frac{\partial u^2}{\partial u'^1} \right)^2.$$

---

* 1854, 1, p. 261.

† We observe that s is not uniquely determined when the geodesic is of length zero [cf. the remarks following equation (17.11)].

‡ Throughout this section it is understood that Greek indices take the values 1 and 2.

§ We indicate by $\overline{R}_{\alpha\beta\gamma\delta}$ these symbols formed with respect to (25.3).

Hence

(25.6) $$K = \frac{\overline{R}_{1212}}{b} = \frac{\overline{R}_{1212}}{b_{11}\, b_{22} - b_{12}^2}$$

is an invariant. Since

$$b^{11} = \frac{b_{22}}{b}, \qquad b^{12} = -\frac{b_{12}}{b}, \qquad b^{22} = \frac{b_{11}}{b},$$

we have

(25.7) $$K b_{11} = \overline{R}^2{}_{121}, \quad K b_{12} = \overline{R}^2{}_{221} = \overline{R}^1{}_{112}, \quad K b_{22} = \overline{R}^1{}_{212}.$$

From these equations it follows that K as defined by (25.6) is the Gaussian curvature of S.*

From (25.5) it follows that at P the origin of Riemannian coördinates all the symbols $[\alpha\beta, \gamma]_b$ are zero, and from (8.8)

$$\overline{R}_{1212} = \frac{\partial}{\partial u^1}[22, 1]_b - \frac{\partial}{\partial u^2}[12, 1]_b.$$

When the expressions from (25.5) are substituted, we obtain, because of (18.8) and (8.3),

$$\begin{aligned}\overline{R}_{1212} &= g_{lk}\, \lambda_{1|}{}^k \lambda_{2|}{}^i \lambda_{2|}{}^j \lambda_{1|}{}^m \left(\frac{\partial}{\partial x^m}\left\{{l \atop ij}\right\} - \frac{\partial}{\partial x^i}\left\{{l \atop mj}\right\}\right) \\ &= g_{lk}\, \lambda_{1|}{}^k \lambda_{2|}{}^i \lambda_{2|}{}^j \lambda_{1|}{}^m R^l{}_{imj} = R_{kimj}\, \lambda_{1|}{}^k \lambda_{2|}{}^i \lambda_{1|}{}^m \lambda_{2|}{}^j.\end{aligned}$$

Since the expression on the right is an invariant, it holds in any coördinate system.

We have from (25.4) and (25.2)

(25.8) $$b_{11}\, b_{22} - b_{12}^2 = (g_{hj}\, g_{ik} - g_{hk}\, g_{ij})\, \lambda_{1|}{}^h \lambda_{2|}{}^i \lambda_{1|}{}^j \lambda_{2|}{}^k.$$

Hence (25.6) may be written in the form

(25.9) $$K = \frac{R_{hijk}\, \lambda_{1|}{}^h \lambda_{2|}{}^i \lambda_{1|}{}^j \lambda_{2|}{}^k}{(g_{hj}\, g_{ik} - g_{hk}\, g_{ij})\, \lambda_{1|}{}^h \lambda_{2|}{}^i \lambda_{1|}{}^j \lambda_{2|}{}^k},$$

which is the expression in any coördinate system for the curvature at a point P for the orientation determined by $\lambda_{1|}{}^i$ and $\lambda_{2|}{}^i$.

---

* 1909, 1, p. 155.

26. The Bianchi identity. The theorem of Schur. We recall from (8.3) that the components $R^h{}_{ijk}$ of the Riemann tensor are defined by

$$(26.1)\quad R^h{}_{ijk} = \frac{\partial}{\partial x^j}\left\{ {h \atop ik} \right\} - \frac{\partial}{\partial x^k}\left\{ {h \atop ij} \right\} + \left\{ {h \atop mj} \right\}\left\{ {m \atop ik} \right\} - \left\{ {h \atop mk} \right\}\left\{ {m \atop ij} \right\}.$$

If we choose geodesic coördinates at a point P, then at P

$$R^h{}_{ijk,l} = \frac{\partial^2}{\partial x^j \partial x^l}\left\{ {h \atop ik} \right\} - \frac{\partial^2}{\partial x^k \partial x^l}\left\{ {h \atop ij} \right\}.$$

From this and similar expressions for the other terms in the left-hand member of the following equation it follows that

$$(26.2)\qquad R^h{}_{ijk,l} + R^h{}_{ikl,j} + R^h{}_{ilj,k} = 0$$

at P. Since the terms of this equation are components of a tensor, this equation holds for any coördinate system and at each point. Hence (26.2) is an identity throughout the space for $h, i, j, k, l = 1, \cdots, n$. It is known as the *identity of Bianchi* who was the first to discover it.* Since g_{ij} and g^{ij} behave like constants in covariant differentiation, we have from (26.2)

$$(26.3)\qquad R_{hijk,l} + R_{hikl,j} + R_{hilj,k} = 0.$$

Because of the identities (8.10) equation (26.2) can be written

$$R^h{}_{ijk,l} - R^h{}_{ilk,j} + g^{hm} R_{milj,k} = 0.$$

If we contract for h and k, we obtain

$$R_{ij,l} - R_{il,j} + g^{hm} R_{milj,h} = 0,$$

where R_{ij} are the components of the Ricci tensor (§ 8). If this equation be multiplied by g^{il}, and i and l be summed, we get

$$(26.4)\qquad R^l{}_{j,l} = \frac{1}{2}\frac{\partial R}{\partial x^j},$$

* *Bianchi*, 1902, 1, p. 351.

where

(26.5) $$R = g^{il} R_{il}$$

is called the *curvature invariant*, or *scalar curvature*, of the space.* Equations (26.4) are important in the general theory of relativity.

From (25.9) it follows that a necessary and sufficient condition that the curvature at every point of space be independent of the orientation is that (cf. Ex. 14, p. 32)

(26.6) $$R_{hijk} = b\,(g_{hj} g_{ik} - g_{hk} g_{ij}),$$

where b is at most a function of the x's. Since we have from (26.6)

$$R_{hijk,l} = \frac{\partial b}{\partial x^l}(g_{hj} g_{ik} - g_{hk} g_{ij}),$$

it follows from (26.3) that

$$\frac{\partial b}{\partial x^l}(g_{hj} g_{ik} - g_{hk} g_{ij}) + \frac{\partial b}{\partial x^j}(g_{hk} g_{il} - g_{hl} g_{ik}) + \frac{\partial b}{\partial x^k}(g_{hl} g_{ij} - g_{hj} g_{il}) = 0.$$

If we assume that j, k and l are different, on multiplying this equation by g^{hj} and summing for h, we obtain $g_{ik}\frac{\partial b}{\partial x^l} - g_{il}\frac{\partial b}{\partial x^k} = 0$. If i is allowed to take values from 1 to n, it follows that $\frac{\partial b}{\partial x^l} = \frac{\partial b}{\partial x^k} = 0$, since the determinant g is not zero by hypothesis. Hence b is constant and we have the following theorem due to Schur:†

If the Riemannian curvature of a space at each point is the same for every orientation, it does not vary from point to point.

A space of this kind is said to be of *constant Riemannian curvature.* Equations (26.6), where b is constant, are the necessary and sufficient conditions for such a space.

In § 10 it was shown that a necessary and sufficient condition that there exist a coördinate system for a V_n for which the components

* Cf. *Levi-Civita*, 1917, 3, p. 388.

† 1886, 1, p. 563.

g_{ij} of the fundamental tensor are constants is that $R^h{}_{ijk} = 0$ for $h, i, j, k = 1, \cdots, n$. In this case as follows from (25.9) $K = 0$ for every orientation at every point of V_n, and is a special case of (26.6) with $b = 0$. When the fundamental form is definite, V_n is a euclidean space of n dimensions and the special coördinate system is cartesian. We denote by S_n a space for which $R^h{}_{ijk} = 0$ for $h, i, j, k = 1, \cdots, n$ and call it a *flat* space.

27. Isometric correspondence of spaces of constant curvature. Motions in a V_n. When the fundamental forms of any two spaces of the same order are transformable into one another, we say that the spaces are *isometric* and that the equations of the transformation define the *isometric correspondence.* In § 24 we have applied the term applicable to two isometric sub-spaces of a space V_m; some writers use this term as synonymous with isometric, but we prefer the term isometric when the two spaces are not looked upon as sub-spaces of an enveloping space, since applicable has the connotation of applicability.

Returning to the consideration of equations (10.5) and their interpretation in § 26, we give the third theorem § 10 the form:

*Any two spaces of n dimensions of the same constant curvature are isometric, and the equations of the isometric correspondence involve $n(n+1)/2$ arbitrary constants.**

The geometrical properties of a surface in euclidean 3-space which depend upon the fundamental form alone as distinguished from its properties as a sub-space of the enveloping euclidean space are called *intrinsic.* We apply this term to the properties of any V_n depending only upon its fundamental form. As a result of the above theorem we have:

Two spaces of n dimensions of the same constant curvature whose fundamental forms have the same signatures have the same intrinsic properties.

We have seen in § 26 that a necessary and sufficient condition that a space V_n be of constant curvature K_0 is that the components of the fundamental tensor satisfy the conditions

$$(27.1) \qquad R_{hijk} = K_0 (g_{hj} g_{ik} - g_{hk} g_{ij}).$$

* In order that the correspondence be real, the signatures of the fundamental forms of the two spaces must be the same.

We inquire whether there exists a system of coördinates x^i in such a space for which the fundamental form is

$$\varphi = \sum_i^{1,\dots,n} \frac{e_i (dx^i)^2}{U^2}, \tag{27.2}$$

where U is a function of the x's and the e's are plus or minus one. Making use of (15.8), we find that the conditions (27.1) applied to (27.2) reduce to

$$\begin{aligned} &\frac{\partial^2 U}{\partial x^i \partial x^j} = 0, \\ &U\left(e_i \frac{\partial^2 U}{\partial x^{j2}} + e_j \frac{\partial^2 U}{\partial x^{i2}}\right) = e_i e_j \left[K_0 + \sum_k^{1,\dots,n} e_k \left(\frac{\partial U}{\partial x^k}\right)^2\right] \quad (i \neq j). \end{aligned} \tag{27.3}$$

From the first of these equations it follows that

$$U = X_1 + \dots + X_n,$$

where X_i is a function of x^i alone. From the second of (27.3) and the equation obtained therefrom by replacing j by l, we get $X_j'' e_j = X_l'' e_l$, where the primes denote differentiation with respect to the argument. Since the first and second terms involve x^j and x^l at most, it follows from this equation that $X_i'' e_i = 2a$, where a is an arbitrary constant, and therefore that

$$X_i = e_i (a x^{i2} + 2 b_i x^i + c_i),$$

where the b's and c's are arbitrary constants. If we substitute these expressions in the second of (27.3), we obtain the following conditions upon these constants:

$$K_0 = 4 \sum_i e_i (a c_i - b_i^2). \tag{27.4}$$

When, in particular, we take all of the b's equal to zero and choose the c's so that $\sum_i e_i c_i = 1$, then (27.2) becomes

$$\varphi = \frac{e_1 (dx^1)^2 + \dots + e_n (dx^n)^2}{\left[1 + \frac{K_0}{4}(e_1 x^{12} + e_2 x^{22} + \dots + e_n x^{n2})\right]^2}. \tag{27.5}$$

This is known as the *Riemannian form* for a space of constant curvature*. From the first theorem of this section we have:

The coördinates of any space of constant curvature can be chosen so that its fundamental form assumes the Riemannian form (27.5).

In order to give a geometric interpretation to the first theorem of this section, we consider two points P and P' of two spaces V_n and V'_n of the same constant curvature. As we are concerned primarily with real isometric correspondences, we assume that the signatures (§ 9) of the fundamental forms at P and P' are the same. We take any ennuple of mutually orthogonal non-null vectors at P for the directions of the parametric curves at P and similarly at P', and choose the coördinates so that at P and P' the fundamental forms are respectively

$$\text{(27.6)}\qquad \begin{aligned} \varphi &= (dx^1)^2 + \cdots + (dx^p)^2 - (dx^{p+1})^2 - \cdots - (dx^n)^2, \\ \varphi' &= (dx'^1)^2 + \cdots + (dx'^p)^2 - (dx'^{p+1})^2 - \cdots - (dx'^n)^2. \end{aligned}$$

Returning to the considerations of § 10, we observe that if we take

$$\text{(27.7)}\qquad p^i_j = \delta^i_j$$

for the values of x^i at P, the conditions (10.3) are satisfied and also (10.4) in consequence of (27.1). By the arguments of § 10 there exists a solution of (10.1) and (10.2), determined by the initial values (27.7), which satisfies (10.3) and (10.4) for all values of x^i. We remark that (27.7) is the condition that the direction of the curve of parameter x^i at P corresponds to the direction of the curve of parameter x'^i at P'. From the first of (27.6) it follows that the components λ^i of the directions of the curves of parameter x^i for $i = 1, \cdots, p$ at P are such that the invariant $g_{ij}\lambda^i\lambda^j$ is positive, and for $i = p+1, \cdots. n$ this invariant is negative; similarly for the directions of the parametric curves at P'. According as this invariant is positive or negative we say that the corresponding vector is *positive* or *negative*. Accordingly we have the theorem:

If V_n and V'_n are two spaces of the same constant curvature,

---

* *Riemann*, 1854, 1, p. 264.

*and P and P′ are two points of these spaces at which the signatures of the fundamental forms are the same, a real isometric correspondence can be established between V_n and V'_n such that P and any orthogonal ennuple at P corresponds to P′ and any orthogonal ennuple at P′, subject to the restriction that positive and negative vectors at P correspond to vectors of the same kind at P′.**

When, in particular, we apply the preceding considerations to one space instead of two, we have an isometric correspondence of V_n with itself such that P and an arbitrary orthogonal ennuple at P correspond to a point P' and an arbitrary orthogonal ennuple at P'. Thus we interpret the equations between the x's and x''s as an isometric point transformation of the space into itself. This is evidently a generalization of a point transformation of a euclidean space into itself; when the equations of such a transformation involve parameters, they may be interpreted as defining a motion of a portion of the space into another portion.

In order to consider more fully the question of a motion of a portion of a space into another portion, we recall that when a euclidean space is refered to cartesian coördinates x^i, the equations of a general motion are defined by

$$\bar{x}^i = a^i_{\ j} x^j + b^i, \tag{27.8}$$

where the a's and b's are constants subject to the conditions

$$\sum_i (a^i_{\ j})^2 = 1, \qquad \sum_i a^i_{\ j} a^i_{\ k} = 0 \qquad (j \neq k). \tag{27.9}$$

From (27.8) and (27.9) we have

$$\sum_i (dx^i)^2 = \sum_i (d\bar{x}^i)^2. \tag{27.10}$$

If now the x^is are replaced by functions of any coördinates x'^i and and $\bar{x}^i$ by the same functions of $\bar{x}'^i$, equation (27.10) becomes

$$g'_{ij}\, dx'^i\, dx'^j = \bar{g}'_{ij}\, d\bar{x}'^i\, d\bar{x}'^j,$$

* Evidently there is no such restriction when the fundamental forms of V_n and V'_n are definite.

where g'_{ij} and $\overline{g}'_{ij}$ are the same functions of the x's and $\overline{x}$'s respectively. Dropping the primes we have the result that the equations of a motion in euclidean space referred to general coördinates satisfy the differential equations

$$(27.11) \qquad g_{ij} = \overline{g}_{kl} \frac{\partial \overline{x}^k}{\partial x^i} \frac{\partial \overline{x}^l}{\partial x^j},$$

where g_{ij} and $\overline{g}_{ij}$ are the same functions of the x's and $\overline{x}$'s respectively.

We generalize this result and say that when the fundamental tensor of a V_n is such that equations (27.11) admit a solution

$$(27.12) \qquad \overline{x}^i = \varphi^i(x', \cdots, x^n)$$

involving one or more parameters, these equations define a *motion* of V_n into itself; when, in particular, (27.12) do not involve a parameter these equations define merely an isometric correspondence of the space with itself. In order to determine whether a space V_n admits motions into itself, we have only to apply the processes of § 10 to the case where g_{ij} and $\overline{g}_{ij}$ are the same functions of the x's and $\overline{x}$'s. This general problem will be considered in Chapter 6. For the present we remark that the third theorem of § 10 may be given the form:

*A space V_n of constant curvature admits a group of motions of $n(n+1)/2$ parameters; conversely, when a V_n admits a group of motions of $n(n+1)/2$ parameters, its curvature is constant.**

From the fourth theorem of this section and the above considerations we have also:

If the signature of the fundamental form of a space of constant curvature is the same at all points, there exists a motion of the portion of the space in the neighborhood of a point P into the portion in the neighborhood of any other point P' such that an orthogonal ennuple at P goes into an arbitrary ennuple at P', with the restriction that a positive or negative vector of the former goes into one of the same kind at P'.

* Cf. *Bianchi*, 1902, 1, p. 348.

28. Conformal spaces. Spaces conformal to a flat space. If the fundamental tensors g_{ij} and $\overline{g}_{ij}$ of two spaces V_n and $\overline{V}_n$ are in the relation

$$\overline{g}_{ij} = e^{2\sigma} g_{ij}, \tag{28.1}$$

where σ is any function of the x's, from (12.5) it follows that the magnitudes of the vectors of components dx^i at points of V_n and $\overline{V}_n$ with the same coördinates are proportional and from (13.4) that the angles between two corresponding directions at corresponding points are equal. Accordingly we say that the correspondence between V_n and $\overline{V}_n$ is *conformal*, and that V_n and $\overline{V}_n$ are *conformal spaces*. The condition (28.1) is necessary as well as sufficient.

From (28.1) we have

$$\overline{g}^{ij} = e^{-2\sigma} g^{ij}, \tag{28.2}$$

and from (7.1) and (7.2) we derive the following relations between the Christoffel symbols formed with respect to the two tensors:

$$\begin{aligned} \overline{[ij,k]} &= e^{2\sigma}([ij,k] + g_{ik}\sigma_{,j} + g_{jk}\sigma_{,i} - g_{ij}\sigma_{,k}), \\ \overline{\left\{ {l \atop ij} \right\}} &= \left\{ {l \atop ij} \right\} + \delta_i^l \sigma_{,j} + \delta_j^l \sigma_{,i} - g_{ij} g^{lm} \sigma_{,m}, \end{aligned} \tag{28.3}$$

where $\sigma_{,i} = \dfrac{\partial \sigma}{\partial x^i}$. If $\sigma_{,ij}$ denote the second covariant derivatives of σ with respect to the g's and we write

$$\sigma_{ij} = \sigma_{,ij} - \sigma_{,i}\sigma_{,j}, \tag{28.4}$$

when we substitute these expressions in equations analogous to (8.8), we have

$$\begin{aligned} e^{-2\sigma} \overline{R}_{hijk} = R_{hijk} &+ g_{hk}\sigma_{ij} + g_{ij}\sigma_{hk} - g_{hj}\sigma_{ik} - g_{ik}\sigma_{hj} \\ &+ (g_{hk}g_{ij} - g_{hj}g_{ik}) \Delta_1 \sigma, \end{aligned} \tag{28.5}$$

where $\Delta_1 \sigma$ is defined by (14.1).

By means of (28.2) and (28.5) we have for the expressions for the components of the Ricci tensor (§ 8) for $\overline{V}_n$

$$(28.6)\quad \overline{R}_{ij} = \overline{g}^{hk}\overline{R}_{hijk} = R_{ij} + (n-2)\sigma_{ij} + g_{ij}[\Delta_2\sigma + (n-2)\Delta_1\sigma],$$

where $\Delta_2\sigma$ is defined by (14.3), and the invariant curvature is given by

$$(28.7)\quad \overline{R} = \overline{g}^{ij}\overline{R}_{ij} = e^{-2\sigma}[R + 2(n-1)\Delta_2\sigma + (n-1)(n-2)\Delta_1\sigma].$$

The case $n = 1$ evidently is of no interest. Since any quadratic differential form in two variables is reducible to the form $\lambda[(dx^1)^2 \pm (dx^2)^2]$ in an infinity of ways*, any V_2 is conformal to any other. In what follows we understand that $n > 2$.

In consequence of (28.1) equation (28.7) can be written

$$(28.8)\quad \overline{g}_{ij}\overline{R} = g_{ij}[R + 2(n-1)\Delta_2\sigma + (n-1)(n-2)\Delta_1\sigma].$$

Eliminating $\Delta_2\sigma$ from this equation and (28.6), we obtain

$$(28.9)\quad \begin{aligned}\sigma_{ij} = \frac{1}{n-2}(\overline{R}_{ij} - R_{ij}) - \frac{1}{2(n-1)(n-2)}(\overline{g}_{ij}\overline{R} - g_{ij}R)\\ - \frac{1}{2}g_{ij}\Delta_1\sigma.\end{aligned}$$

Because of (28.2) equations (28.5) can be written

$$(28.10)\quad \begin{aligned}\overline{R}^h{}_{ijk} = R^h{}_{ijk} + \delta^h_k\sigma_{ij} - \delta^h_j\sigma_{ik} + g^{hl}(g_{ij}\sigma_{lk} - g_{ik}\sigma_{lj})\\ + (\delta^h_k g_{ij} - \delta^h_j g_{ik})\Delta_1\sigma.\end{aligned}$$

If the expression (28.9) for σ_{ij} and analogous expressions for σ_{ik}, σ_{lk} and σ_{lj} be substituted in (28.10), the resulting equations are reducible to

$$(28.11)\quad \overline{C}^h{}_{ijk} = C^h{}_{ijk},$$

where

$$(28.12)\quad \begin{aligned}C^h{}_{ijk} = R^h{}_{ijk} + \frac{1}{n-2}(\delta^h_j R_{ik} - \delta^h_k R_{ij} + g_{ik}R^h{}_j - g_{ij}R^h{}_k)\\ + \frac{R}{(n-1)(n-2)}(\delta^h_k g_{ij} - \delta^h_j g_{ik}).\end{aligned}$$

---

* 1909, 1, pp. 93, 102.

Evidently $C^h{}_{ijk}$ are the components of a tensor, and as follows from (28.11) this tensor is the same for V_n and $\overline{V}_n$ in conformal correspondence. It was called the *conformal curvature tensor* by Weyl*, who was the first to consider it.

When $n = 3$ and the coördinates are chosen so that $g_{ij} = 0$ $(i \neq j)$ (§ 15), it is readily shown that (cf. Ex. 15, p. 32).

The conformal curvature tensor is a zero tensor in a V_3†.

In consequence of (26.2) we have from (28.12)

$$(28.13)\quad \begin{aligned} C^h{}_{ijk,l} + C^h{}_{ikl,j} + C^h{}_{ilj,k} = \frac{1}{n-2}(\delta^h_j R_{ikl} + \delta^h_k R_{ilj} \\ + \delta^h_l R_{ijk} + g_{ik} R^h{}_{jl} + g_{il} R^h{}_{kj} + g_{ij} R^h{}_{lk}), \end{aligned}$$

where we have put

$$(28.14)\quad \begin{aligned} R_{ijk} &= R_{ij,k} - R_{ik,j} + \frac{1}{2(n-1)}(g_{ik} R_{,j} - g_{ij} R_{,k}), \\ R^h{}_{jk} &= g^{hi} R_{ijk}. \end{aligned}$$

Raising the index i and contracting for i and j, we have in consequence of (26.4)

$$(28.15)\quad R^i{}_{ik} = 0.$$

Contracting (28.12) for h and k, we have $C_{ij} = 0$. When we make use of this result and (28.15) in contracting (28.13) for h and k, we obtain

$$(28.16)\quad C^h{}_{ilj,h} = \frac{n-3}{n-2} R_{ilj}.$$

From (27.5) it is seen that any space of constant curvature is conformal to a flat-space S_n (§ 26). We seek the necessary and sufficient conditions that a V_n be conformal to an S_n.

In order that $\overline{V}_n$ in the preceding discussion be an S_n, it is necessary and sufficient that $\overline{R}_{hijk} = 0$ (§ 26). From (28.11) and (28.12) it follows at once that $C_{hijk} = 0$, that is,

---

* 1918, 2, p. 404.

† *Weyl*, 1918, 2., p. 404.

$$(28.17)\quad \begin{aligned} R_{hijk} &+ \frac{1}{n-2}(g_{jh}R_{ik} - g_{hk}R_{ij} + g_{ik}R_{hj} - g_{ij}R_{hk}) \\ &+ \frac{R}{(n-1)(n-2)}(g_{hk}g_{ij} - g_{hj}g_{ik}) = 0. \end{aligned}$$

Since $\bar{R}_{ij} = 0$ also, we have from (28.9)

$$(28.18)\quad \sigma_{,ij} = \sigma_{,i}\sigma_{,j} + \frac{1}{n-2}\left(\frac{R g_{ij}}{2(n-1)} - R_{ij}\right) - \frac{1}{2}g_{ij}\Delta_1\sigma.$$

Moreover, when σ satisfies these equations, equations (28.8) for $\bar{R} = 0$ are satisfied. The conditions of integrability of (28.18) are [cf. (11.14)]

$$\sigma_{,ijk} - \sigma_{,ikj} = \sigma_{,l} R^{l}_{\ ijk}.$$

Substituting from (28.18), we find as the conditions

$$(28.19)\quad R_{ij,k} - R_{ik,j} + \frac{1}{2(n-1)}(g_{ik}R_{,j} - g_{ij}R_{,k}) = 0.$$

For $n \neq 3$ this condition is a consequence of (28.17) as follows from (28.16). Hence we have the theorem:

*Any V_2 can be mapped conformally on an S_2; a necessary and sufficient condition that a V_n for $n > 2$ can be mapped conformally on an S_n is that the tensor R_{ijk} be a zero tensor when $n = 3$ and when $n > 3$ that C_{hijk} be a zero tensor.**

Exercises.

1. A coördinate system can be chosen so that $\dfrac{\partial g_{ij}}{\partial x^k} = 0$ along a given curve.

Fermi, 1922, 5; *Levi-Civita*, 1925, 4, p. 190; 1926, 8, p. 298.

2. A space for which

$$R_{ij} = \frac{R}{n} g_{ij}$$

is called an *Einstein space*. Every V_2 is an Einstein space (cf. Ex. 2, p. 47). Show that an Einstein space V_3 has constant curvature.

Schouten and *Struik*, 1921, 3, p. 214.

---

* *Weyl*, 1918, 2, p. 404, showed that the vanishing of C_{hijk} is a necessary condition. *Schouten*, 1921, 2, p. 80, that it is sufficient when $n > 3$; he also derived the above conditions for a V_3.

3. Show that a space of constant curvature K_0 is an Einstein space, and that $R = K_0(1-n)n$.

4. If an Einstein space is conformal to a flat space, it is a space of constant curvature. *Schouten* and *Struik*, 1921, 3, p. 214.

5. Show by means of (26.4) that when $n > 2$ the scalar curvature of an Einstein space is constant. *Herglotz*, 1916, 2, p. 203.

6. A V_4 for which

$$g_{11} = -\left(1 - \frac{2a}{x^1}\right)^{-1}, \qquad g_{22} = -(x^1)^2, \qquad g_{33} = -(x^1 \sin x^2)^2,$$

$$g_{44} = 1 - \frac{2a}{x^1}, \qquad g_{ij} = 0 \qquad (i \neq j),$$

where a is an arbitrary constant, is an Einstein space for which $R = 0$. *Schwarzschild*, 1916, 3, p. 195.

7. A V_4 for which

$$g_{11} = -A^{-1}, \qquad g_{22} = -(x^1)^2, \qquad g_{33} = -(x^1 \sin x^2)^2,$$

$$g_{44} = A, \qquad A = 1 + \frac{a(x^1)^2}{3} + \frac{c}{x^1}, \qquad g_{ij} = 0 \qquad (i \neq j),$$

where a and c are arbitrary constants is an Einstein space. Show that when $c = 0$ the V_4 has constant Riemannian curvature. *Kottler*, 1918, 3, p. 443.

8. In order that the tensor

$$a^i_{\,j} = R^i_{\,j} + \delta^i_j (aR + b),$$

where $R^i_{\,j} = g^{ik} R_{kj}$ and where a and b are invariants, shall satisfy the conditions $a^i_{\,j,i} = 0$, it is necessary and sufficient that it be of the form

$$a^i_{\,j} = R^i_{\,j} + \delta^i_j \left(-\frac{1}{2} R + c\right),$$

where c is an arbitrary constant.

9. Let K be the curvature at a point P of a V_n determined by the vectors $\lambda_{1|}{}^i$ and $\lambda_{2|}{}^i$; when $\lambda_{1|}{}^i$ is displaced parallel to itself around a small circuit and returns to P, the change in the angle α with the vector $\lambda_{2|}{}^i$ is given by $\Delta\alpha = -K\Delta\Sigma$, where $\Delta\Sigma$ is the area enclosed by the circuit (cf. Ex. 6, p. 79). *Pérès*, 1919, 1, p. 428.

10. If $\lambda_{1|}{}^i$ and $\lambda_{2|}{}^i$ are the components of two families of unit vectors,* the vectors of each family being parallel with respect to a curve C, the curvature K determined at each point by the vectors $\lambda_{1|}{}^i$ and $\lambda_{2|}{}^i$ at the point satisfies the equation

$$\frac{dK}{ds} = e_1 e_2 R_{ijkl,m}\, \lambda_{1|}{}^i \lambda_{2|}{}^j \lambda_{1|}{}^k \lambda_{2|}{}^l \frac{dx^m}{ds}.$$

In order that K be constant along C for all sets of parallel vectors $\lambda_{1|}{}^i$ and $\lambda_{2|}{}^i$, it is necessaıy and sufficient that

$$R_{ijkl,m} \frac{dx^m}{ds} = 0.$$

* orthogonal to one another.

In order that this property hold for any curve, it is necessary and sufficient that $R_{ijkl,m} = 0$. *Levy*, 1925, 1.

11. If σ is any function of the x's such that $\Delta_1 \sigma \neq 0$, and $g^{hi} \sigma_{,h} C_{ijkl} = 0$ for $h, i, j, k, l = 1, \cdots, 4$, then $C_{ijkl} = 0$. (Cf. Ex. 12, p. 32.)

Brinkmann, 1924, 2, p. 277.

12. If $\sigma = -\frac{1}{2n} \log g$ in (28.1), then $\overline{g} = \text{const.}$ for $\overline{V}_n$ in this coördinate system and $\overline{\left\{ {i \atop ij} \right\}} = 0$.

13. Show that the quantities

$$K^i_{jk} = \left\{ {i \atop jk} \right\} - \frac{1}{n}\left(\delta^i_j \left\{ {l \atop lk} \right\} + \frac{1}{n} \delta^i_k \left\{ {l \atop lj} \right\} - g^{il} g_{jk} \left\{ {h \atop hl} \right\} \right)$$

have the same values at corresponding points of two spaces whose fundamental tensors are connected by (28.2). *Thomas*, 1925, 5, p. 257.

14. By expressing integrability conditions of the equations of transformation of the quantities K^i_{jk} of Ex. 13 under a change of coördinate systems, show that the following quantities are the components of a tensor:

$$(n-2)\ F^i_{jkl} + \delta^i_k F_{jl} - \delta^i_l F_{jk} + g_{jl} F^i_k - g_{jk} F^i_l + \frac{F}{n-1} (\delta^i_l g_{jk} - \delta^i_k g_{jl}),$$

where F^i_{jkl} is formed from the K's in the same way that R^i_{jkl} is formed from the Christoffel symbols of the second kind, and where $F_{jl} = F^i_{jli}$. Show also that the above expression is equal to $(n-2)\, C^i_{jkl}$. *Thomas*, 1925, 5, p. 258.

15. Show that, if each Christoffel symbol in the covariant derivative of $g^{ij} g_{kl}$ is replaced by the corresponding K^i_{jk} (cf. Ex. 13), the result is identically zero. Hence show that in the system of coördinates y^i, defined by

$$x^i = x^i_0 + y^i - \frac{1}{2} (K^i_{jk})_0\, y^j y^k,$$

the components of the conformal tensor $g^{ij} g_{kl}$ are stationary at the origin.

Thomas, 1925, 5, p. 259.

16. Show by means of (27.4) that the most general conformal map of a euclidean space upon itself for $n > 2$ is obtained as the product of inversions with respect to a hypersphere, motions and transformations of similitude.

Bianchi, 1902, 1, p. 375, 376.

17. Obtain the theorem for any flat space analogous to that of Ex. 16.

18. A necessary and sufficient condition that a V_n for $n > 2$ can be mapped conformally on an Einstein space $\overline{V}_n$ is that there exist a function σ satisfying the equations

$$\sigma_{,ij} - \sigma_{,i}\, \sigma_{,j} + \Lambda g_{ij} = L_{ij},$$

where

$$A = \frac{1}{2} \Delta_1 \sigma - \frac{\overline{R} e^{2\sigma}}{2n(n-1)}, \quad (n-2) L_{ij} = \frac{R}{2(n-1)} g_{ij} - R_{ij},$$

$\overline{R}$ being the constant scalar curvature of $\overline{V}_n$; then $\overline{g}_{ij} = e^{2\sigma} g_{ij}$.

Brinkmann, 1924, 2, p. 271.

19. Show that the conditions of integrability of the equations of Ex. 18 are

$$\sigma_{,h} C^h_{\ ijk} = -\frac{1}{n-2} R_{ijk},$$

where R_{ijk} is defined by (28.14), and that consequently the equations are completely integrable only in case V_n can be mapped on an S_n.

Brinkmann, 1924, 2, p. 272.

20. In order that an Einstein space can be mapped conformally on an Einstein space, it is necessary that the function σ in § 28 satisfy the equations

$$\sigma_{,ij} = \sigma_{,i} \sigma_{,j} + \frac{g_{ij}}{2n(n-1)} [\overline{R} e^{2\sigma} - R - n(n-1) \Delta_1 \sigma]$$

where $\overline{R}$ and R are the constant scalar curvatures of the two spaces.

Brinkmann, 1925, 6, p. 121.

21. Show by means of Ex. 4, p. 47 that for any solution of the equations of Ex. 20

$$\Delta_1 \sigma = \frac{1}{n(n-1)} (\overline{R} e^{2\sigma} + 2 c e^{\sigma} + R),$$

where c is a constant; and consequently, if $\Delta_1 \sigma = 0$, the scalar curvatures of the two spaces must be zero. *Brinkmann*, 1925, 6, p. 122.

22. An Einstein space V_n can be mapped conformally on another Einstein space by means of a function σ for which $\Delta_1 \sigma \neq 0$, if, and only if, its fundamental form is reducible to

$$\varphi = f g_{\alpha\beta} dx^\alpha dx^\beta + \frac{1}{f} (dx^n)^2 \quad (\alpha, \beta = 1, \cdots, n-1),$$

where

$$f = \frac{1}{n(n-1)} [R(x^n)^2 + 2 a x^n + b],$$

a and b being constants, and the functions $g_{\alpha\beta}$ are independent of x^n and such that $g_{\alpha\beta} dx^\alpha dx^\beta$ is the fundamental form of an Einstein V_{n-1}.

Brinkmann, 1925, 6, p. 125.

CHAPTER III

Orthogonal ennuples

29. Determination of tensors by means of the components of an orthogonal ennuple and invariants. If the equations (13.12) of an orthogonal ennuple are written in the form

$$(29.1)\qquad \lambda_{h|i}\,\lambda_{k|}{}^{i} = 0 \quad (h \neq k), \qquad \lambda_{h|i}\,\lambda_{h|}{}^{i} = e_h \quad (h, k = 1, \cdots, n),$$

and we solve the $n-1$ equations of the first set for $\lambda_{h|i}$, we get

$$\frac{\lambda_{h|1}}{\Lambda^{h|}{}_1} = \frac{\lambda_{h|2}}{\Lambda^{h|}{}_2} = \cdots = \frac{\lambda_{h|n}}{\Lambda^{h|}{}_n},$$

where $\Lambda^{h|}{}_r$ denotes the cofactor of $\lambda_{h|}{}^{r}$ in the determinant $|\lambda_{h|}{}^{r}|$ divided by this determinant; hence $\lambda_{h|}{}^{r}\Lambda^{h|}{}_s = \delta^r_s$. From the second of (29.1) it follows that the value of these ratios is e_h, and consequently

$$(29.2)\qquad \Lambda^{h|}{}_i = e_h\,\lambda_{h|i}.$$

If we solve the equations

$$g_{ij}\,\lambda_{h|}{}^{i} = \lambda_{h|j} \qquad (h = 1, \cdots, n)$$

for g_{ij} and make use of (29.2), we obtain

$$(29.3)\qquad g_{ij} = \sum_h^{1,\cdots,n} e_h\,\lambda_{h|i}\,\lambda_{h|j}.$$

From these equations follow

$$(29.4)\qquad \sum_h^{1,\cdots,n} e_h\,\lambda_{h|i}\,\lambda_{h|}{}^{j} = \delta^j_i$$

and

$$(29.5)\qquad \sum_h^{1,\cdots,n} e_h\,\lambda_{h|}{}^{i}\,\lambda_{h|}{}^{j} = g^{ij}$$

Consider now any covariant tensor of the mth order of components $a_{r_1 \cdots r_m}$. The quantities $c_{h_1 \cdots h_m}$, defined by

$$c_{h_1 \cdots h_m} = a_{r_1 \cdots r_m} \lambda_{h_1|}{}^{r_1} \cdots \lambda_{h_m|}{}^{r_m}, \tag{29.6}$$

are scalars. If these expressions for $c_{h_1 \cdots h_m}$ are substituted in the right-hand member of the equation

$$a_{s_1 \cdots s_m} = \sum_{h_1, \cdots, h_m}^{1, \cdots, n} c_{h_1 \cdots h_m} e_{h_1} \cdots e_{h_m} \lambda_{h_1|s_1} \cdots \lambda_{h_m|s_m}, \tag{29.7}$$

this equation reduces to an identity because of (29.4). Hence:

The components of any tensor are expressible in terms of invariants and the components of an orthogonal ennuple.*

30. Coefficients of rotation. Geodesic congruences. In conformity with (29.6) we define a set of invariants γ_{lhk} by the equations

$$\gamma_{lhk} = \lambda_{l|i,j} \lambda_{h|}{}^{i} \lambda_{k|}{}^{j}, \tag{30.1}$$

where $\lambda_{l|i,j}$ $(i,j = 1, \cdots, n)$ are the components of the covariant derivative of $\lambda_{l|i}$ with respect to the fundamental form of the space. Equations (30.1) are equivalent by (29.7) to

$$\lambda_{l|i,j} = \sum_{h,k}^{1, \cdots, n} e_h e_k \gamma_{lhk} \lambda_{h|i} \lambda_{k|j}. \tag{30.2}$$

From the first of equations (29.1) we have by covariant differentiation [cf. (11.11)]

$$\lambda_{h|i,j} \lambda_{k|}{}^{i} + \lambda_{k|i,j} \lambda_{h|}{}^{i} = 0.$$

Substituting from equations of the form (30.2), multiplying by $\lambda_{l|}{}^{j}$ and summing for j, we obtain

$$\gamma_{hkl} + \gamma_{khl} = 0 \qquad (h \neq k); \tag{30.3}$$

in particular we have

$$\gamma_{hhl} = 0. \tag{30.4}$$

---

* Cf. *Ricci* and *Levi-Civita*, 1901, 1, p. 147.

So far as these identities go there are $n^2(n-1)/2$ independent invariants γ_{hkl}. However, they are not arbitrary but are subject to the conditions arising from the conditions of integrability of equations (30.2).

The conditions of integrability of (30.2) are of the form (cf. § 11)

$$\lambda_{l|i,jk} - \lambda_{l|i,kj} = \lambda_{l|h} R^h{}_{ijk}. \tag{30.5}$$

If the expressions obtained by differentiating (30.2) covariantly and a similar equation in $\lambda_{l|i,k}$ be substituted in (30.5) and the resulting equation be multiplied by $\lambda_{p|}{}^i \lambda_{q|}{}^j \lambda_{r|}{}^k$ and summed for i, j and k, this equation is reducible by means of (30.1) to

$$\gamma_{lpqr} = R_{hijk} \lambda_{l|}{}^h \lambda_{p|}{}^i \lambda_{q|}{}^j \lambda_{r|}{}^k, \tag{30.6}$$

where by definition

$$\begin{aligned}\gamma_{lpqr} = \frac{\partial \gamma_{lpq}}{\partial s_r} - \frac{\partial \gamma_{lpr}}{\partial s_q} + \sum_m^{1,\dots,n} e_m \, [\gamma_{lpm}(\gamma_{mqr} - \gamma_{mrq}) \\ + \gamma_{mlr}\gamma_{mpq} - \gamma_{mlq}\gamma_{mpr}],\end{aligned} \tag{30.7}$$

and where for any invariant function we write

$$\frac{\partial f}{\partial s_r} = \lambda_{r|}{}^i \frac{\partial f}{\partial x^i}. \tag{30.8}$$

As thus defined $\frac{\partial f}{\partial s_r}$ is the ratio of two differentials. We call it an *intrinsic derivative.*

From (8.10) and (30.6) it follows that

$$\gamma_{lpqr} = -\gamma_{plqr} = -\gamma_{lprq} = \gamma_{qrlp}. \tag{30.9}$$

From (30.8) we have

$$\begin{aligned}\frac{\partial}{\partial s_k}\frac{\partial f}{\partial s_h} &= \lambda_{k|}{}^i \frac{\partial}{\partial x^i}\left(\lambda_{h|}{}^j \frac{\partial f}{\partial x^j}\right) = \lambda_{k|}{}^i (\lambda_{h|}{}^j{}_{,i} f_{,j} + \lambda_{h|}{}^j f_{,ji}) \\ &= \sum_l^{1,\dots,n} e_l \gamma_{hlk} \lambda_{l|}{}^j f_{,j} + \lambda_{h|}{}^j \lambda_{k|}{}^i f_{,ji}.\end{aligned}$$

Since $f_{,ji} = f_{,ij}$, it follows that

$$(30.10)\qquad \frac{\partial}{\partial s_k}\frac{\partial f}{\partial s_h} - \frac{\partial}{\partial s_h}\frac{\partial f}{\partial s_k} = \sum_{l}^{1,\ldots,n} e_l(\gamma_{lkh} - \gamma_{lhk})\frac{\partial f}{\partial s_l}.$$

This is the form which the condition of integrability of intrinsic derivatives assumes.

In order to give a geometric interpretation to the invariants γ_{lhk}, we consider a point P_0 of V_n and the curve C_m of the congruence $\lambda_{m|}{}^i$ through P_0; along C_m we have

$$(30.11)\qquad \frac{\partial x^j}{\partial s_m} = \lambda_{m|}{}^j.$$

Denote by $\theta_{h\bar{l}}$ the angle at any point P of C_m between the vector $\lambda_{h|i}$ at P and the vector $\bar{\lambda}_{l|i}$ at P parallel to $\lambda_{l|}{}^i$ at P_0 with respect to a displacement from P_0 to P along C_m; then

$$\cos\theta_{h\bar{l}} = \bar{\lambda}_{l|}{}^i\,\lambda_{h|i}.$$

By hypothesis $\lambda_{m|}{}^j\,\bar{\lambda}_{l|i,j} = 0$ and consequently (§ 11)

$$(30.12)\qquad \begin{aligned}\frac{\partial}{\partial s_m}\cos\theta_{h\bar{l}} &= \bar{\lambda}_{l|}{}^i\,\lambda_{m|}{}^j\,\lambda_{h|i,j} = \bar{\lambda}_{l|}{}^i\,\lambda_{m|}{}^j\sum_{p,q}^{1,\ldots,n} e_p\,e_q\,\gamma_{hpq}\,\lambda_{p|i}\,\lambda_{q|j}\\ &= \bar{\lambda}_{l|}{}^i\sum_{p}^{1,\ldots,n} e_p\,\gamma_{hpm}\,\lambda_{p|i}.\end{aligned}$$

At P_0 $\bar{\lambda}_{l|}{}^i = \lambda_{l|}{}^i$ and consequently at P_0

$$(30.13)\qquad \frac{\partial}{\partial s_m}\cos\theta_{hl} = \gamma_{hlm}.$$

Hence we have:

If P_0 is any point of V_n and P is a nearby point on the curve C_m of the congruence $\lambda_{m|}{}^i$ through P_0, then $\gamma_{hlm}\,ds_m$ is equal, to within terms of higher order, to minus the difference of the cosine of the angle between the vectors $\lambda_{h|}{}^i$ and $\lambda_{l|}{}^i$ at P_0 and the cosine of the angle between the vector $\lambda_{h|}{}^i$ at P and the vector at P parallel to $\lambda_{l|}{}^i$ at P_0 with respect to C_m.

When the space is euclidean, $\gamma_{hlm}\,ds_m$ is the component in the direction $\lambda_{l|}{}^i$ of the rotation of the vector $\lambda_{h|}{}^i$ as P_0 moves to P.

Consequently we speak of γ_{hlm} in the general case as the *coefficients of rotation* of the ennuple.*

From (30.2) we have

$$\lambda_{l|}{}^{j}\,\lambda_{l|i,j} = \sum_h e_h\,\gamma_{lhl}\,\lambda_{h|i}. \tag{30.14}$$

From (17.11) it follows that the right-hand member is zero, when, and only when, the curves of the congruence $\lambda_{l|}{}^{i}$ are geodesics. If this expression equated to zero be multiplied by $\lambda_{k|}{}^{i}$ and summed for i, we obtain the theorem:

A necessary and sufficient condition that the curves of the congruence $\lambda_{l|}{}^{i}$ *be geodesics is that*

$$\gamma_{hll} = 0 \qquad (h = 1, \cdots, n). \tag{30.15}$$

In the general case we have from (30.14) and (20.1)

$$\mu_{l|}{}^{i} = \sum_h e_h\,\gamma_{lhl}\,\lambda_{h|}{}^{i}, \tag{30.16}$$

where $\mu_{l|}{}^{i}$ are the components of the principal normal of a curve of direction $\lambda_{l|}{}^{i}$. From (30.16) and (20.3) we have

$$\frac{1}{\varrho_l^2} = g_{ij}\,\mu_{l|}{}^{i}\,\mu_{l|}{}^{j} = \sum_h e_h\,\gamma_{hll}^2. \tag{30.17}$$

Hence when the principal normals are not null vectors, the first curvature is given by

$$\frac{1}{\varrho_l} = \sqrt{\left|\sum_h e_h\,\gamma_{hll}^2\right|}, \tag{30.18}$$

and the principal normals are positive or negative vectors (§ 27) according to the sign of the right-hand member of (30.17). Also from (30.17) we have that the principal normals to the curves $\lambda_{l|}{}^{i}$ are null vectors, when, and only when,

$$\sum_h e_h\,\gamma_{hll}^2 = 0 \qquad (h = 1, \cdots, n), \tag{30.19}$$

and (30.15) is not satisfied.

---

* *Levi-Civita*, 1917, 1, p. 192.

31. Determinants and matrices. Certain theorems concerning determinants and matrices can be given simple form by the use of quantities $\varepsilon_{i_1 i_2 \cdots i_n} = \varepsilon^{i_1 i_2 \cdots i_n}$ which are defined to be zero, when two or more of the indices are the same, and 1 or -1 according as the indices are obtainable from the natural sequence $1, \cdots, n$ by an even or odd number of transpositions.* Thus the determinant $a = |a^i_j|$, in which i indicates the column and j the row for $i, j = 1, \cdots, n$, may be written in either of the forms

$$a = \varepsilon_{i_1 i_2 \cdots i_n} a^{i_1}_1 a^{i_2}_2 \cdots a^{i_n}_n \tag{31.1}$$

or

$$a = \varepsilon^{i_1 i_2 \cdots i_n} a^1_{i_1} a^2_{i_2} \cdots a^n_{i_n}. \tag{31.2}$$

From these equations it is seen at once that a determinant changes sign, if the elements of two rows (or columns) are interchanged, and that a determinant is zero, if corresponding elements of two rows (or columns) are the same. These properties are put in evidence also by the following identities which are consequences of (31.1) and (31.2):

$$\varepsilon_{j_1 j_2 \cdots j_n} a = \varepsilon_{i_1 i_2 \cdots i_n} a^{i_1}_{j_1} a^{i_2}_{j_2} \cdots a^{i_n}_{j_n}; \quad \varepsilon^{j_1 j_2 \cdots j_n} a = \varepsilon^{i_1 \cdots i_n} a^{j_1}_{i_1} a^{j_2}_{i_2} \cdots a^{j_n}_{i_n}. \tag{31.3}$$

As an example of the use of the ε's we establish the law for multiplication of determinants. Let a and $b = |b^i_j|$ be two determinants of the nth order. By (31.1) and (31.3) we have

$$\begin{aligned} a \cdot b &= a \varepsilon_{j_1 \cdots j_n} b^{j_1}_1 \cdots b^{j_n}_n \\ &= \varepsilon_{i_1 i_2 \cdots i_n} a^{i_1}_{j_1} a^{i_2}_{j_2} \cdots a^{i_n}_{j_n} b^{j_1}_1 b^{j_2}_2 \cdots b^{j_n}_n \\ &= \varepsilon_{i_1 i_2 \cdots i_n} c^{i_1}_1 c^{i_2}_2 \cdots c^{i_n}_n, \end{aligned}$$

where $c^i_k = a^i_j b^j_k$.

As defined the ε's have n indices when the indices take the values $1, \cdots, n$. We define also a set of quantities $\delta^{i_1 i_2 \cdots i_p}_{\alpha_1 \alpha_2 \cdots \alpha_p}$ for $p \leqq n$. By definition these quantities are zero, when two or more

* Cf. *Eddington*, 1923, 1, p. 107.

superscripts (or subscripts) are the same, or when the superscripts do not have the same set of p values as the subscripts; also any δ is $+1$ or -1 according as the superscripts and the subscripts differ from one another by an even or odd number of permutations.* As an immediate consequence of the definitions we have

$$\varepsilon_{j_1\cdots j_n} a = \delta^{i_1\cdots i_n}_{j_1\cdots j_n} a^1_{i_1} a^2_{i_2} \cdots a^n_{i_n}, \tag{31.4}$$

$$\varepsilon^{j_1\cdots j_n} a = \delta^{j_1\cdots j_n}_{i_1\cdots i_n} a^{i_1}_1 a^{i_2}_2 \cdots a^{i_n}_n. \tag{31.5}$$

Also we have the identity

$$\delta^{i_1\cdots i_n}_{j_1\cdots j_n} \delta^{j_1\cdots j_n}_{k_1\cdots k_n} = n!\, \delta^{i_1\cdots i_n}_{k_1\cdots k_n}. \tag{31.6}$$

Moreover, from (31.3) and (31.4) we have

$$\delta^{k_1\cdots k_n}_{j_1\cdots j_n} a^1_{k_1} a^2_{k_2} \cdots a^n_{k_n} = \varepsilon_{i_1\cdots i_n} a^{i_1}_{j_1} a^{i_2}_{j_2} \cdots a^{i_n}_{j_n}. \tag{31.7}$$

Consider now two matrices

$$\| c^l_\beta \|, \qquad \| d^\beta_m \|, \tag{31.8}$$

See App. 10 where the Greek letters take the values $1, \cdots, n$ and determine the column, and the Latin $1, \cdots, p\,(< n)$ and determine the row. We put

$$b^l_m = c^l_\gamma d^\gamma_m \tag{31.9}$$

and establish the following theorem which we shall use later:

The determinant of the quantities b^l_m defined by (31.9) *is the sum of the products of corresponding determinants of the pth order of the matrices* (31.8).

From (31.9) and (31.1)

$$b = |b^l_m| = \varepsilon_{i_1 i_2\cdots i_p} c^{i_1}_{\gamma_1} c^{i_2}_{\gamma_2} \cdots c^{i_p}_{\gamma_p} d^{\gamma_1}_1 d^{\gamma_2}_2 \cdots d^{\gamma_p}_p,$$

which by (31.7) may be written

$$b = \delta^{\beta_1\beta_2\cdots\beta_p}_{\gamma_1\gamma_2\cdots\gamma_p} c^1_{\beta_1} c^2_{\beta_2} \cdots c^p_{\beta_p} d^{\gamma_1}_1 d^{\gamma_2}_2 \cdots d^{\gamma_p}_p$$

* Cf. *Murnaghan*, 1925, 7.

and by (31.6)

$$b = \frac{1}{p!} \delta^{\beta_1 \cdots \beta_p}_{\alpha_1 \cdots \alpha_p} \delta^{\alpha_1 \cdots \alpha_p}_{\gamma_1 \cdots \gamma_p} c^1_{\beta_1} \cdots c^p_{\beta_p} d^{\gamma_1}_1 \cdots d^{\gamma_p}_p. \tag{31.10}$$

For any term of this sum to be different from zero, the β's and γ's must take on the same set of values and each permutation of the α's over these values gives a term; there are consequently $p!$ terms for a given set of β's and γ's each of which is obtained by multiplying together

$$\delta^{\beta_1 \cdots \beta_p}_{\alpha_1 \cdots \alpha_p} c^1_{\beta_1} c^2_{\beta_2} \cdots c^p_{\beta_p}, \qquad \delta^{\alpha_1 \cdots \alpha_p}_{\gamma_1 \cdots \gamma_p} d^{\gamma_1}_1 d^{\gamma_2}_2 \cdots d^{\gamma_p}_p$$

for the α's in the same order. But from (31.4) und (31.5) these expressions for a given set of α's are seen to be corresponding determinants of the matrices (31.8) to within the equal multipliers $\varepsilon_{\alpha_1 \cdots \alpha_p}$ and $\varepsilon^{\alpha_1 \cdots \alpha_p}$, whose product is 1. Hence the expression on the right in (31.10) reduces to the sum of the products of corresponding determinants of (31.8), as was to be proved.*

32. The orthogonal ennuple of Schmidt. Associate directions of higher orders. The Frenet formulas for a curve in a V_n. Let $\xi_{1|}^{\;i}$ be the components of a unit vector, that is,

$$g_{ij} \xi_{1|}^{\;i} \xi_{1|}^{\;j} = e_1, \tag{32.1}$$

and let $\xi_{\sigma|}^{\;i}$ for $\sigma = 2, \cdots, n$ be the components of any $n-1$ other vectors such that these n vectors are linearly independent. We put

$$g_{ij} \xi_{l|}^{\;i} \xi_{m|}^{\;j} = b^l_m = b^m_l \qquad (l, m = 1, \cdots, n),\dagger \tag{32.2}$$

and we denote by b_p the determinant of b^α_β for $\alpha, \beta = 1, \cdots, p$, thus,

$$b_p = |b^\alpha_\beta| \qquad (\alpha, \beta = 1, \cdots, p). \tag{32.3}$$

From (32.2), (32.3) and the results of § 31 we have that b_p is the sum of the products of corresponding p row determinants of the

---

* For another proof of this theorem, see *Kowalewski*, 1909, 2, p. 77.

† Normally one would use b_{lm} but the notation used makes for simplicity in what follows.

matrices $\|g_{ij}\,\xi_{\alpha|}{}^{i}\|$ and $\|\xi_{\beta}^{j}\|$. Consequently when the fundamental form of V_n is positive definite, all of the determinants b_p for $p = 1, \cdots, n$ are positive;* when the fundamental form is indefinite, we assume that the vectors $\xi_{\sigma|}{}^{i}$ are such that $b_p \neq 0$ for $p = 1, \cdots, n$.

Consider now the vector of components $\lambda_{p|}{}^{i}$ which are expressed linearly in terms of the components $\xi_{\sigma|}{}^{i}$ for $\sigma = 1, \cdots, p$, as follows

$$\lambda_{p|}{}^{i} = e_p \sqrt{\frac{e_p\, b_p}{b_{p-1}}}\, \xi_{\alpha|}{}^{i}\, B_p^{\alpha} \qquad (\alpha = 1, \cdots, p), \tag{32.4}$$

where e_p is chosen so that the radical is real and B_p^{α} is the cofactor of b_p^{α} in b_p divided by b_p. From (32.1) and (32.3) it follows that $b_1 = e_1$. In order that (32.4) may hold for $p = 1$ and that $\lambda_{1|}{}^{i} = \xi_{1|}{}^{i}$, we define b_0 as 1.

From (32.4) we have

$$g_{ij}\,\lambda_{p|}{}^{i}\,\xi_{q|}{}^{j} = e_p \sqrt{\frac{e_p\, b_p}{b_{p-1}}}\, \delta_p^q \qquad (q \leqq p). \tag{32.5}$$

Assuming that $q < p$, we have from the definition of $\lambda_{q|}{}^{i}$ similar to (32.4) and from (32.5)

$$g_{ij}\,\lambda_{p|}{}^{i}\,\lambda_{q|}{}^{j} = 0 \qquad (p \neq q). \tag{32.6}$$

If both sides of (32.4) be multiplied by $g_{ij}\,\lambda_{p|}{}^{j}$ and summed for i, we have in consequence of (32.5)

$$g_{ij}\,\lambda_{p|}{}^{i}\,\lambda_{p|}{}^{j} = e_p. \tag{32.7}$$

Thus the vectors defined by (32.4) for $p = 1, \cdots, n$ form an orthogonal ennuple, as first shown by E. Schmidt†.

Consider now any curve C in V_n and unit vectors of a field $\lambda_{1|}{}^{i}$ at points of C which are assumed not to be parallel along C. If we put

---

* This is seen by considering any point P and choosing the coördinate system so that at P $g_{ii} = 1$, $g_{ij} = 0$ $(i \neq j)$, in which case any b_p is the sum of squares.

† 1908, 1, p. 61; cf. also *Kowalewski*, 1909, 2, pp. 423–426.

$$\frac{dx^j}{ds}\lambda_{1|}{}^i{}_{,j} = e_1\,\xi_{2|}{}^i, \tag{32.8}$$

then $\xi_{2|}{}^i$ are the components of the vector associate to $\lambda_{1|}{}^i$ (§ 24). Since $b_2^1 = b_1^2 = 0$ for this case, we must assume that this vector is not a null vector, if we desire b_2 as defined by (32.3) to be different from zero. We define $n-2$ other vectors along C by the equations

$$\frac{dx^j}{ds}\xi_{r|}{}^i{}_{,j} = \xi_{r+1|}{}^i \qquad (r = 2,\cdots, n-1).* \tag{32.9}$$

We assume that these n vectors are linearly independent and that $b_p \neq 0$ for $p = 1,\cdots, n$. Then equations (32.4) define an orthogonal ennuple of directions at points of C which we call the *associate directions of* $\lambda_{1|}{}^i$ *of orders* $1,\cdots, n-1$.

At points of C the components $\frac{dx^j}{ds}$ of the tangent vector to C are expressible in the form

$$\frac{dx^j}{ds} = a^r\lambda_{r|}{}^j \qquad (j, r = 1,\cdots, n), \tag{32.10}$$

where the a's are invariants. From (32.10) and

$$\lambda_{p|}{}^i{}_{,j} = \sum_{k,l}^{1,\cdots,n} e_k\, e_l\, \gamma_{pkl}\,\lambda_{k|}{}^i\,\lambda_{l|j} \tag{32.11}$$

we have

$$\frac{dx^j}{ds}\lambda_{p|}{}^i{}_{,j} = \sum_{k}^{1,\cdots,n} e_k\,\alpha_{pk}\,\lambda_{k|}{}^i, \tag{32.12}$$

where

$$\alpha_{pk} = a^r\gamma_{pkr}. \tag{32.13}$$

Because of (30.3) we have also

$$\alpha_{pk} + \alpha_{kp} = 0. \tag{32.14}$$

From (32.4) and (32.9) it follows that $\frac{dx^j}{ds}\lambda_{p|}{}^i{}_{,j}$ is at most a

* For the development of this section to apply we assume that none of the vectors $\xi_{r|}{}^i$ are parallel with respect to C.

linear expression in $\xi_{1|}^{\ i}, \cdots, \xi_{p+1|}^{\ i}$ and therefore in $\lambda_{1|}^{\ i}, \cdots, \lambda_{p+1|}^{\ i}$. Consequently $\alpha_{pk} = 0$ for $k > p+1$. Combining this result with (32.14), we have

See App. 11

$$\alpha_{p\,p+1} = -\alpha_{p+1\,p} = \frac{1}{\varrho_p}, \tag{32.15}$$
$$\alpha_{pk} = 0 \qquad [k \neq (p \pm 1)],$$

where ϱ_p is defined by the first of these equations. Accordingly equations (32.12) reduce to

$$\frac{dx^j}{ds} \lambda_{p|}{}^{i}{}_{,j} = \frac{-e_{p-1}}{\varrho_{p-1}} \lambda_{p-1|}^{\ i} + \frac{e_{p+1}}{\varrho_p} \lambda_{p+1|}^{\ i} \quad (p = 2, \cdots, n-1), \tag{32.16}$$

from which we have

$$\lambda_{p+1|i} \frac{dx^j}{ds} \lambda_{p|}{}^{i}{}_{,j} = \frac{1}{\varrho_p} \qquad (p = 2, \cdots, n-1). \tag{32.17}$$

From (32.8) and § 24 it follows that (32.16) apply also to the case $p = 1$ with the understanding that $1/\varrho_0 = 0$. Also from (32.12) and (32.15) for $p = n$ we have (32.16) for $p = n$ with the understanding that $1/\varrho_n = 0$.

We call $1/\varrho_p$ for $p = 1, \cdots, n-1$ the *associate curvatures of order* $1, \cdots, n-1$ of the vector $\xi_{1|}^{\ i} (= \lambda_{1|}^{\ i})$ for the curve C. We can find their expressions in terms of the determinants b_p by differentiating covariantly equations (32.4) with respect to x^j and substituting in (32.17). This gives, in consequence of (32.9),

$$\frac{1}{\varrho_p} = \lambda_{p+1|i} \left[\frac{\partial}{\partial x^j} \left(\sqrt{\frac{e_p b_p}{b_{p-1}}} B_p^\alpha \right) \xi_{\alpha|}^{\ i} \frac{\partial x^j}{\partial s} + \sqrt{\frac{e_p b_p}{b_{p-1}}} B_p^\alpha \xi_{\alpha+1|}^{\ i} \right],$$
$$(\alpha = 1, \cdots, p),$$

which is reducible by means of (32.5) to

$$\frac{1}{\varrho_p} = \sqrt{\frac{e_p e_{p+1} b_{p-1} b_{p+1}}{b_p^2}} \qquad (p = 1, \cdots, n-1). \tag{32.18}$$

When, in particular, the vector $\lambda_{1|}^{\ i}$ is the tangent vector to C, we have in (32.10) $a^1 = 1$, $a^\sigma = 0$ for $\sigma \neq 1$ and from (32.13) $\alpha_{pk} = \gamma_{pk1}$. From (32.17), (20.1) and (20.3) it follows that $1/\varrho_1$

is the first curvature of C. In this case we say that $1/\varrho_p$ are the *first, second,* $\cdots$, $n-1$th *curvatures of C*. Moreover, equations (32.16) for $p = 1, \cdots, n$ are a generalization of the Frenet formulas for a curve in euclidean space in cartesian coördinates, as is readily seen by replacing covariant derivatives by ordinary derivatives.* Hence we follow Blaschke in calling (32.16) the *formulas of Frenet* for a curve in a Riemannian space.†

Exercises.

1. If $\bar{\gamma}_{ijk}$ denote the coefficients of rotation for the orthogonal ennuple defined by (13.14), show that

$$\bar{\gamma}_{ijk} = \gamma_{pqr} t_i^p t_j^q t_k^r + \sum_r^{1,\ldots,n} e_r t_{i,p}^r t_j^r t_k^s \lambda_{s|}{}^p,$$

and that

$$\bar{\gamma}_{ijkl} = t_i^p t_j^q t_k^r t_l^s \gamma_{pqrs}.$$

2. Show that $\varepsilon_{i_1 i_2 \cdots i_n} \sqrt{g}$ are the components of a covariant tensor (§ 31).
Ricci and *Levi-Civita*, 1901, 1, p, 135.

3. Show that the components of the contravariant tensor of order n associate to the tensor of Ex. 2 by means of g_{ij} are $\varepsilon^{i_1 \cdots i_n}/\sqrt{g}$.
Ricci and *Levi-Civita*, 1901, 1, p. 138.

4. Show that the first covariant derivatives of the tensors of Exs. 2 and 3 are zero. *Ricci* and *Levi-Civita*, 1901, 1, p. 138.

5. Show that

$$\delta_{j_1 \cdots j_m}^{i_1 \cdots i_m} = \begin{vmatrix} \delta_{j_1}^{i_1} & \cdots\cdots & \delta_{j_m}^{i_1} \\ \cdot & \cdot\;\cdot\;\cdot & \cdot \\ \cdot & \cdot\;\cdot\;\cdot & \cdot \\ \delta_{j_1}^{i_m} & \cdots\cdots & \delta_{j_m}^{i_m} \end{vmatrix},$$

and consequently that the δ's are the components of a tensor of order $2m$.
Murnaghan, 1925, 7, p. 238.

33. Principal directions determined by a symmetric covariant tensor of the second order.

Let a_{ij} be the components of a symmetric covariant tensor of the second order and consider the determinant equation

$$|a_{ij} - \varrho g_{ij}| = 0. \tag{33.1}$$

---

* Cf. 1909, 1, p. 17.

† *Blaschke*, 1920, 1, p. 97, considered the case when the fundamental form is definite, in which case the only restriction is that $\lambda_{1|}^{i}$, $\xi_{2|}^{i}$ and the vectors $\xi_{r|}^{i}$ defined by (32.9) be linearly independent. When the form is indefinite, it must be assumed also that the determinants b_p defined by (32.3) be different from zero; in particular, this requires that the curve C be not minimal.

In another coördinate system x'^i we have

$$(33.2) \qquad a_{ij} = a'_{lm} \frac{\partial x'^l}{\partial x^i} \frac{\partial x'^m}{\partial x^j}, \qquad g_{ij} = g'_{lm} \frac{\partial x'^l}{\partial x^i} \frac{\partial x'^m}{\partial x^j},$$

so that (33.1) becomes

$$|a'_{lm} - \varrho g'_{lm}| \cdot \left| \frac{\partial x'^k}{\partial x^i} \right|^2 = 0.$$

Since by hypothesis the Jacobian is not zero, this equation is of the same form as (33.1) and thus the roots ϱ of (33.1) are invariants.

If ϱ_h is a real simple root of (33.1), the equations

$$(33.3) \qquad (a_{ij} - \varrho_h g_{ij}) \lambda_{h|}{}^i = 0$$

define, to within a factor, n quantities $\lambda_{h|}{}^i$, which are the contravariant components of a real vector-field, as is seen by changing the coördinates and making use of (33.2). If ϱ_k is another real simple root of (33.1), we have a second vector-field defined by

$$(33.4) \qquad (a_{ij} - \varrho_k g_{ij}) \lambda_{k|}{}^i = 0.$$

Multiplying (33.3) by $\lambda_{k|}{}^j$ and (33.4) by $\lambda_{h|}{}^j$, summing for j in each case and subtracting, we have, since $\varrho_h \neq \varrho_k$ by hypothesis,

$$(33.5) \qquad g_{ij} \lambda_{h|}{}^i \lambda_{k|}{}^j = 0,$$

that is, the two vector-fields are orthogonal.

From the algebraic theory* it follows that if the roots of (33.1) are real and the elementary divisors are simple, there exists a real transformation of the variables x^i such that at a point P the forms

$$(33.6) \qquad \varphi = g_{ij}\, dx^i\, dx^j, \qquad \psi = a_{ij}\, dx^i\, dx^j$$

are reducible to

$$(33.7) \qquad \begin{aligned} \varphi &= c_1 (dx^1)^2 + \cdots + c_n (dx^n)^2, \\ \psi &= c_1 \varrho_1 (dx^1)^2 + \cdots + c_n \varrho_n (dx^n)^2, \end{aligned}$$

* Cf. *Bromwich*, 1906, 1, pp. 30, 50.

where the c's are constants none of which is zero and $\varrho_1, \cdots, \varrho_n$ are the roots of (33.1), which are not necessarily different. In particular, if φ is a definite form, the roots of (33.1) are real, and the c's have the same signs.*

If ϱ_1 is a simple root, then at P the solutions of equations (33.3) are $\lambda_{1|}{}^1 = 1$, $\lambda_{1|}{}^\alpha = 0$ $(\alpha = 2, \cdots, n)$, to within a multiplier. Hence the vector is not a null vector. Accordingly if all the roots of (33.1) are real and simple, equations (33.3) define n mutually orthogonal non-null vectors, that is, an orthogonal ennuple (§ 13).

When p of the roots are equal, say $\varrho_1 = \cdots = \varrho_p$, then for $h = 1, \cdots, p$, equations (33.3) reduce to $(\varrho_{p+\sigma} - \varrho_h)\lambda_{h|}{}^{p+\sigma} = 0$ for $\sigma = 1, \cdots, n-p$, ($p + \sigma$ being not summed). These equations are satisfied by the p linearly independent vectors whose components are

$$\lambda_{\alpha|}{}^i = \delta^i_\alpha \quad (\alpha = 1, \cdots, p;\ i = 1, \cdots, n),$$

which evidently are non-null vectors. Moreover, any other solution is a linear combination of these vectors. Consequently for a multiple root of order p the rank of (33.1) is $n-p$, and there are ∞^{p-1} sets of solutions.

If the coördinates are any whatever and $\lambda_{\alpha|}{}^i$ for $\alpha = 1, \cdots, p$ are the components of p independent solutions, then

$$\xi_{\alpha|}{}^i = \mu_\alpha{}^\beta \lambda_{\beta|}{}^i \quad (\alpha, \beta = 1, \cdots, p;\ i = 1, \cdots, n) \tag{33.8}$$

are another set of solutions. If we choose the functions $\mu_\alpha{}^\beta$ so that

$$\mu_\alpha{}^\beta \mu_\gamma{}^\delta g_{ij} \lambda_{\beta|}{}^i \lambda_{\delta|}{}^j = 0 \qquad \mu_\alpha{}^\beta \mu_\alpha{}^\delta g_{ij} \lambda_{\beta|}{}^i \lambda_{\delta|}{}^j \neq 0 \quad (\alpha \neq \gamma),$$

the p vectors of components $\xi_{\alpha|}{}^i$ are mutually orthogonal and are not null vectors. The determination of the μ's is equivalent to finding an orthogonal ennuple in a space of p dimensions whose fundamental tensor $\bar{g}_{\alpha\beta}$ is defined by $\bar{g}_{\alpha\beta} = g_{ij}\lambda_{\alpha|}{}^i \lambda_{\beta|}{}^j$. At a point P in the coördinate system giving (33.7), we have $\lambda_{\alpha|}{}^i = 0$ for $i = p+1, \cdots, n$, and consequently

$$\bar{g} = |\bar{g}_{\alpha\beta}| = c_1 \cdots c_p |\lambda_{\alpha|}{}^\beta|^2 \neq 0 \quad (\alpha, \beta = 1, \cdots, p).$$

---

* *Bôcher*, 1907, 1, pp. 171, 305.

Hence functions $\mu_\alpha{}^\beta$ satisfying these conditions can be obtained in accordance with the results of § 13.

Gathering the foregoing results together we have the theorem:

If a_{ij} are the components of a symmetric covariant tensor such that the elementary divisors of equation (33.1) *are simple and the roots are real, equations* (33.3) *define a real orthogonal ennuple; this is unique when the roots are simple; when a root is of order p, there are $\infty^{p(p-1)/2}$ sets of mutually orthogonal non-null vectors corresponding to this root.*

The directions at each point defined by these vectors are called the *principal directions* determined by the tensor a_{ij}; the n congruences defined by the ennuple the *principal congruences* and $\varrho_1, \cdots, \varrho_n$ the *principal invariants.*

Since the vectors are not null vectors, the components can be chosen so that

$$g_{ij}\,\lambda_{h|}{}^i\,\lambda_{h|}{}^j = e_h \qquad (h = 1, \cdots, n), \tag{33.9}$$

and we have from (33.3)

$$a_{ij}\,\lambda_{h|}{}^i\,\lambda_{k|}{}^j = 0, \qquad (h \neq k), \tag{33.10}$$

$$\varrho_h = e_h\,a_{ij}\,\lambda_{h|}{}^i\,\lambda_{h|}{}^j.$$

Hence if none of the roots of (33.1) is zero, that is, if the determinant $|a_{ij}| \neq 0$, we have

$$a_{ij}\,\lambda_{h|}{}^i\,\lambda_{h|}{}^j \neq 0 \qquad (h = 1, \cdots, n). \tag{33.11}$$

Conversely, if $\lambda_{h|}{}^i$ are the components of n mutually orthogonal unit vectors, and a_{ij} are the components of a symmetric tensor such that the first of (33.10) is satisfied, then these vectors define the principal directions determined by a_{ij}. For, if we define n invariants ϱ_h by (33.10), we have as a consequence of (33.5), (33.9) and (33.10)

$$(a_{ij} - \varrho_h\,g_{ij})\,\lambda_{h|}{}^i\,\lambda_{k|}{}^j = 0 \qquad (h, k = 1, \cdots, n).$$

Since the determinant of the λ's is different from zero, these equations are equivalent to (33.3), which establishes the theorem.

If we write equations (33.3) in the form

$$a_{kj}\,\lambda_{h|}{}^k = \varrho_h\,\lambda_{h|j},$$

multiply by $e_h \lambda_{h|i}$, sum for h and make use of (29.4), we obtain

$$a_{ij} = \sum_h e_h \varrho_h \lambda_{h|i} \lambda_{h|j}. \tag{33.12}$$

When both of the forms (33.6) are indefinite, there is a possibility that the elementary divisors are not simple. We consider this case for 4-spaces and it can be shown that the results are general. If one, or more, of the elementary divisors are multiple and real at a point P, a real coördinate system can be chosen for which at P the coefficients of the forms are of one of the following types.*

Type 1.

$$g_{12} = 1, \quad g_{33} = k_3, \quad g_{44} = k_4,$$
$$a_{11} = k_1, \quad a_{12} = \varrho_1, \quad a_{33} = \varrho_3 k_3, \quad a_{44} = \varrho_4 k_4,$$

where the k's are constants, all the other g's and a's being zero. The elementary divisors are $(\varrho - \varrho_1)^2$, $(\varrho - \varrho_3)$, $(\varrho - \varrho_4)$.

1°. $\varrho_1, \varrho_3, \varrho_4 \neq$. The vectors given by (33.3) are

$$(0, 1, 0, 0), \quad (0, 0, 1, 0), \quad (0, 0, 0, 1),$$

of which the first is a null vector and the others are not.

2°. $\varrho_3 = \varrho_4$. The vectors are the first of the above, and any vector of the pencil determined by the last two.

3°. $\varrho_1 = \varrho_3$. The vectors are the last of the above and any vector determined by the first two. Any vector of the pencil is orthogonal to (0, 1, 0, 0).

4°. $\varrho_1 = \varrho_3 = \varrho_4$. Any vector for which the first component is zero.

Type 2.

$$g_{12} = 1, \quad g_{34} = 1,$$
$$a_{11} = k_1, \quad a_{12} = \varrho_1, \quad a_{33} = k_3, \quad a_{34} = \varrho_3.$$

The elementary divisors are $(\varrho - \varrho_1)^2$, $(\varrho - \varrho_3)^2$.

1°. $\varrho_1 \neq \varrho_3$. The vectors are (0, 1, 0, 0) and (0, 0, 0, 1), and both are null vectors.

2°. $\varrho_1 = \varrho_3$. Any vector of the pencil determined by the vectors of the preceeding case.

---

* Cf. *Bromwich*, 1906, 1, p. 46.

Type 3.

$$g_{12} = 1, \quad g_{33} = k_3, \quad g_{44} = k_4,$$
$$a_{12} = \varrho_1, \quad a_{23} = 1, \quad a_{33} = \varrho_1 k_3, \quad a_{44} = \varrho_4 k_4.$$

The elementary divisors are $(\varrho - \varrho_1)^3$, $(\varrho - \varrho_4)$.

1°. $\varrho_1 \neq \varrho_4$. The vectors are (1, 0, 0, 0) and (0, 0, 0, 1), of which the first is a null vector.

2°. $\varrho_1 = \varrho_4$. Any vector of the pencil determined by the preceeding two.

Type 4.

$$g_{12} = 1, \quad g_{34} = 1,$$
$$a_{12} = \varrho_1, \quad a_{23} = 1, \quad a_{34} = \varrho_1, \quad a_{44} = k_4.$$

There is one elementary divisor $(\varrho - \varrho_1)^4$ and one vector (1, 0, 0, 0), which is a null vector.

When two or more of the ϱ's are equal, the corresponding elementary divisors are said to have the same *base*.

Combining the results of this section and recalling that when the elementary divisors are simple there are n of them, although some may have the same base, we have:

The number of principal directions defined by (33.3) *is equal to the number of elementary divisors; when* $p\,(>1)$ *of the divisors have the same base, the vectors corresponding to this base are any linear combination of* p *independent vectors; to a divisor which is not simple there corresponds a null vector when the base is not the same as any other, and when it is the same as another base one or more of the* p *vectors is a null vector, according as it is the base of one or more divisors which are not simple.*

Thus in case the divisors are simple there are n principal directions, and only in this case.

If we write

$$\varrho = \frac{a_{ij} \lambda^i \lambda^j}{g_{ij} \lambda^i \lambda^j}, \tag{33.13}$$

the finite maxima and minima values of ϱ at a point are given by the directions for which $\frac{\partial \varrho}{\partial \lambda^j} = 0$, for $j = 1, \cdots, n$, that is,

$$(a_{ij} - \varrho g_{ij}) \lambda^i = 0.$$

Hence we have:

At a point the finite maxima and minima of ϱ *defined by* (33.13) *are given by the principal directions at the point.*

If the fundamental form is definite, ϱ is finite for all directions. If it is indefinite, ϱ is infinite for all null directions, except those which are principal directions; this exception arises when the elementary divisors of (33.1) are not simple.

34. Geometrical interpretation of the Ricci tensor. The Ricci principal directions. Let $\lambda_{h|}{}^i$ be the components of any unit vector, and $\lambda_{k|}{}^i$ for $k = 1, \cdots, n; k \neq h$, the components of $n-1$ unit vectors forming an orthogonal ennuple with the given vector. The Riemannian curvature at a point for the orientation determined by $\lambda_{h|}{}^i$ and any vector $\lambda_{k|}{}^i$, denoted by r_{hk}, is given by [cf. (25.9)]

$$r_{hk} = e_h e_k R_{pqrs} \lambda_{h|}{}^p \lambda_{k|}{}^q \lambda_{h|}{}^r \lambda_{k|}{}^s. \tag{34.1}$$

Since the right-hand member of this equation is zero for $k = h$, we assume that $r_{hh} = 0$.

In consequence of (29.5) we have

$$\sum_k^{1,\cdots,n} r_{hk} = e_h R_{pqrs} \lambda_{h|}{}^p \lambda_{h|}{}^r g^{qs} = -e_h R_{ij} \lambda_{h|}{}^i \lambda_{h|}{}^j. \tag{34.2}$$

Hence $\sum_k r_{hk}$ is the sum of the Riemannian curvatures determined by the vector $\lambda_{h|}{}^i$ and $n-1$ mutually orthogonal non-null vectors orthogonal to it; moreover, from (34.2) it is seen that it is independent of the choice of these $n-1$ vectors. We denote it by ϱ_h and call it the *mean curvature* of the space for the direction $\lambda_{h|}{}^i$. This result is due to Ricci,* who gave this geometrical interpretation of the tensor which Einstein chose later as the basis of the general theory of relativity.

If we write (34.2) in the form

$$\varrho_h = -\frac{R_{ij} \lambda_{h|}{}^i \lambda_{h|}{}^j}{g_{ij} \lambda_{h|}{}^i \lambda_{h|}{}^j}, \tag{34.3}$$

* 1904, 2, p. 1234.

we see (§ 33) that the finite maximum and minimum values of the mean curvature correspond to the principal directions determined by the Ricci tensor, that is, the directions given by

$$(R_{ij} + \varrho g_{ij}) \lambda^i = 0. \tag{34.4}$$

From (33.12) it follows that for these principal directions

$$R_{ij} = -\sum_h e_h \varrho_h \lambda_{h|i} \lambda_{h|j}. \tag{34.5}$$

We call these the *Ricci principal directions* of the space.

A necessary and sufficient condition that the principal directions for a tensor a_{ij} be indeterminate is that $a_{ij} = \varrho g_{ij}$. In this case we say that the space is *homogeneous with respect to the tensor* a_{ij}. We have at once:

A necessary and sufficient condition that a space be homogeneous with respect to the Ricci tensor is that

$$R_{ij} = \frac{1}{n} R g_{ij}, \tag{34.6}$$

that is, that it be an Einstein space (cf. Ex. 2, p. 92).

35. Condition that a congruence of an orthogonal ennuple be normal. By definition a congruence of curves in a V_n is *normal* when they are the orthogonal trajectories of a family of hypersurfaces $f(x^1, \cdots, x^n) = \text{const}$. If dx^i are the components of any displacement in one of these hypersurfaces, then

$$\frac{\partial f}{\partial x^i} dx^i = 0. \tag{35.1}$$

Consequently if $\lambda_{n|}{}^i$ are the components of a normal congruence of an ennuple, we must have

$$\frac{\partial f}{\partial x^i} \equiv f_{,i} = \mu \lambda_{n|i}, \tag{35.2}$$

where μ is an invariant (§ 14), and from (35.1) it follows that f must be such that we have

$$X_h(f) \equiv \lambda_{h|}{}^i \frac{\partial f}{\partial x^i} = 0 \quad (h = 1, \cdots, n-1). \tag{35.3}$$

In order that these $n-1$ equations may admit a solution which is not a constant, they must constitute a complete system. A necessary and sufficient condition is that

$$(X_h, X_k)f \equiv X_h X_k(f) - X_k X_h(f)$$

be a linear function of $X_h(f)$ for $h, k = 1, \cdots, n-1$ (§ 23). From (35.3) we have, in consequence of (30.2),

$$\begin{aligned} X_h X_k(f) &= \lambda_{h|}{}^j (\lambda_{k|}{}^i f_{,ij} + f_{,i} \lambda_{k|}{}^i{}_{,j}) \\ &= \lambda_{h|}{}^j \lambda_{k|}{}^i f_{,ij} + f_{,i} \sum_l e_l \gamma_{klh} \lambda_{l|}{}^i \\ &= \lambda_{h|}{}^j \lambda_{k|}{}^i f_{,ij} - \sum_\alpha^{1,\cdots,n-1} e_\alpha \gamma_{\alpha kh} X_\alpha(f) - e_n \gamma_{nkh} \frac{\partial f}{\partial s_n}. \end{aligned}$$

Hence

$$(X_h, X_k)f = \sum_\alpha^{1,\cdots,n-1} e_\alpha (\gamma_{\alpha hk} - \gamma_{\alpha kh}) X_\alpha(f) + e_n (\gamma_{nhk} - \gamma_{nkh}) \frac{\partial f}{\partial s_n}.$$

Since $\lambda_{n|}{}^i$ is not expressible linearly in terms of $\lambda_{h|}{}^i$ for $h = 1, \cdots, n-1$, $\frac{\partial f}{\partial s_n}$ is not expressible in terms of the $X(f)$'s. Hence:

A necessary and sufficient condition that the congruence $\lambda_{n|}{}^i$ of an orthogonal ennuple be normal is that

$$\gamma_{nhk} = \gamma_{nkh} \qquad (h, k = 1, \cdots, n-1). \tag{35.4}$$

From (35.4), (30.2) and (30.15) we have:

A necessary and sufficient condition that a geodesic congruence $\lambda_{n|i}$ be normal is that $\lambda_{n|i,j}$ be a symmetric tensor.

Suppose that the conditions (35.4) are satisfied. Equating the expressions for $f_{,ij}$ obtained from (35.2) and for $f_{,ji}$ from $f_{,j} = \mu \lambda_{n|j}$, we get

$$\mu_{,j} \lambda_{n|i} + \mu \lambda_{n|i,j} = \mu_{,i} \lambda_{n|j} + \mu \lambda_{n|j,i}.$$

Multiplying by $\lambda_{n|}{}^j$ and summing for j, we have, in consequence of (30.2) and (30.3),

$$e_n \frac{\partial \log \mu}{\partial x^i} = \nu \lambda_{n|i} - \sum_l e_l \gamma_{lnn} \lambda_{l|i}, \qquad \nu = \lambda_{n|}{}^j \frac{\partial}{\partial x^j} \log \mu. \tag{35.5}$$

Expressing the condition of integrability of these equations, we obtain

$$\nu_{,j}\lambda_{n|i}-\nu_{,i}\lambda_{n|j}+\nu(\lambda_{n|i,j}-\lambda_{n|j,i})$$
$$+\sum_l e_l\left[\frac{\partial\gamma_{lnn}}{\partial x^i}\lambda_{l|j}-\frac{\partial\gamma_{lnn}}{\partial x^j}\lambda_{l|i}+\gamma_{lnn}(\lambda_{l|j,i}-\lambda_{l|i,j})\right]=0.$$

Multiplying by $\lambda_{h|}{}^j\lambda_{n|}{}^i$ and summing for i and j, we have for the determination of ν the equations

$$(35.6)\qquad e_n\frac{\partial\nu}{\partial s_h}+\frac{\partial\gamma_{hnn}}{\partial s_n}+\nu\gamma_{hnn}+\sum_l e_l\gamma_{lnn}(\gamma_{lhn}-\gamma_{lnh})=0 \qquad (h,l=1,\cdots,n-1).$$

Multiplying the above equation by $\lambda_{h|}{}^i\lambda_{k|}{}^j$ and summing for i and j we have, in consequence of (35.4), the identities

$$(35.7)\qquad \frac{\partial\gamma_{hnn}}{\partial s_k}-\frac{\partial\gamma_{knn}}{\partial s_h}+\sum_l e_l\gamma_{lnn}(\gamma_{lhk}-\gamma_{lkh})=0 \qquad (h,k,l=1,\cdots,n-1).$$

We consider, in particular, the case when the congruence $\lambda_{n|}{}^i$ is normal to a family of hypersurfaces $f=$ const., where f is a solution of the differential equation

$$(35.8)\qquad g^{ij}f_{,ij}=0.$$

These have been called *isothermic* hypersurfaces by Ricci and Levi-Civita* and are an immediate generalization of isothermic surfaces as defined by Lamé.†

From (35.2) and (35.8) we have

$$g^{ij}f_{,ij}=g^{ij}(\mu_{,j}\lambda_{n|i}+\mu\sum_{h,k}e_h e_k\gamma_{nhk}\lambda_{h|i}\lambda_{k|j})$$
$$=\mu_{,j}\lambda_{n|}{}^j+\mu\sum_h e_h\gamma_{nhh}=0.$$

From this equation it follows that ν in (35.5) has the value $-\sum_h e_h\gamma_{nhh}$ in this case, and consequently

$$(35.9)\qquad e_n\frac{\partial\log\mu}{\partial x^i}=-\sum_h e_h\gamma_{nhh}\lambda_{n|i}-\sum_h e_h\gamma_{hnn}\lambda_{h|i}\quad(h=1,\cdots,n-1).$$

* 1901, 1, p. 152.

† 1857, 1, p. 1.

Conversely, if the expression on the right is the component of a gradient, the function f defined by (35.2) satisfies (35.8). Hence:

A necessary and sufficient condition that a congruence $\lambda_{n|i}$ *be normal to a family of isothermic hypersurfaces is that* (35.4) *be satisfied and the right-hand member of* (35.9) *be the component of a gradient.*

36. N-tuply orthogonal systems of hypersurfaces. From the definition of an n-tuply orthogonal system of hypersurfaces in § 15 it follows that the curves of intersection of these hypersurfaces form an ennuple of mutually orthogonal normal congruences. As there considered the coördinates x^i are such that the congruences are the parametric curves. When the coördinates are general, we are able to find the condition that all the congruences of an orthogonal ennuple be normal by remarking that in this case, as follows from (35.4), we must have

$$\gamma_{hkl} = \gamma_{hlk} \qquad (h, k, l = 1, \cdots, n;\ h, k, l \neq).$$

By means of equations of this form and the identities (30.3) we have

$$\gamma_{hkl} = \gamma_{hlk} = -\gamma_{lhk} = -\gamma_{lkh} = \gamma_{klh} = \gamma_{khl} = -\gamma_{hkl},$$

that is, $\gamma_{hkl} = 0$. Hence:

A necessary and sufficient condition that the congruences of an orthogonal ennuple be normal is

$$\gamma_{hkl} = 0 \qquad (h, k, l = 1, \cdots, n;\ h, k, l \neq).^* \tag{36.1}$$

As remarked in § 15 such an ennuple does not exist in a general V_n. The conditions, in general form, which a V_n must satisfy in order that such an ennuple exist are to be found by a consideration of the equations which the components $\lambda_{h|}{}^i$ of the ennuple and the invariants γ_{hkk} must satisfy in this case. From (30.6) and (30.7), when (36.1) hold, we have

$$R_{hijk}\,\lambda_{l|}{}^h\,\lambda_{p|}{}^i\,\lambda_{q|}{}^j\,\lambda_{r|}{}^k = 0 \qquad (l, p, q, r \neq), \tag{36.2}$$

$$R_{hijk}\,\lambda_{l|}{}^h\,\lambda_{p|}{}^i\,\lambda_{p|}{}^j\,\lambda_{r|}{}^k = \frac{\partial \gamma_{lpp}}{\partial s_r} + e_p\,\gamma_{lpp}\,\gamma_{rpp} - e_r\,\gamma_{lrr}\,\gamma_{rpp}, \tag{36.3}$$

* Cf. *Ricci* and *Levi-Civita*, 1901, 1, p. 151.

$$
\begin{aligned}
(36.4)\qquad & R_{hijk}\,\lambda_{l|}{}^{h}\,\lambda_{p|}{}^{i}\,\lambda_{p|}{}^{j}\,\lambda_{l|}{}^{k} \\
& = \frac{\partial \gamma_{lpp}}{\partial s_l} + \frac{\partial \gamma_{pll}}{\partial s_p} + e_l\,\gamma_{pll}^2 + e_p\,\gamma_{lpp}^2 + \sum_m e_m\,\gamma_{mll}\,\gamma_{mpp}.
\end{aligned}
$$

Since the left-hand member of (36.3) is unaltered when l and r are interchanged, we must have

$$
(36.5)\qquad \frac{\partial \gamma_{lpp}}{\partial s_r} - \frac{\partial \gamma_{rpp}}{\partial s_l} + e_l\,\gamma_{rll}\,\gamma_{lpp} - e_r\,\gamma_{lrr}\,\gamma_{rpp} = 0,
$$

which is the form of (35.7) for the present case.

The characterization in invariant form of a V_n admitting an orthogonal ennuple of normal congruences is obtained by expressing the condition that equations (36.2), (36.3), (36.4), (30.2) and

$$
g_{ij}\,\lambda_{h|}{}^{i}\,\lambda_{h|}{}^{j} = e_h, \qquad g_{ij}\,\lambda_{h|}{}^{i}\,\lambda_{k|}{}^{j} = 0 \qquad (h \neq k)
$$

possess a solution in the n^2 quantities $\lambda_{h|}{}^{i}$ and the $n(n-1)$ quantities γ_{hkk}.

By means of the above theorem we are able to prove the following theorem:

If a tensor a_{ij} is such that the roots of (33.1) *are simple, a necessary and sufficient condition that the principal congruences determined by a_{ij} be normal is that the components of these congruences, as given by* (33.3), *satisfy the equations*

$$
(36.6)\qquad a_{ij,k}\,\lambda_{h|}{}^{i}\,\lambda_{l|}{}^{j}\,\lambda_{m|}{}^{k} = 0 \quad (h, l, m = 1, \cdots, n;\ h, l, m \neq).
$$

In fact, if we differentiate the first of (33.10) covariantly with respect to x^k, we have in consequence of (30.2), (30.3) and (33.10)

$$
a_{ij,k}\,\lambda_{h|}{}^{i}\,\lambda_{l|}{}^{j} + \sum_p e_p\,(\varrho_h - \varrho_l)\,\gamma_{lhp}\,\lambda_{p|k} = 0.
$$

Multiplying by $\lambda_{m|}{}^{k}$ and summing for k, we obtain

$$
(36.7)\qquad a_{ij,k}\,\lambda_{h|}{}^{i}\,\lambda_{l|}{}^{j}\,\lambda_{m|}{}^{k} = (\varrho_h - \varrho_l)\,\gamma_{hlm} \qquad (h \neq l),
$$

from which we obtain the theorem.*

---

* For a discussion of the case where the roots of (33.1) are not simple see *Eisenhart*, 1923, 6, pp. 263–280.

Proceeding in like manner with the second of (33.10), we obtain

$$a_{ij,k}\,\lambda_{h|}{}^{i}\,\lambda_{h|}{}^{j}\,\lambda_{l|}{}^{k} = e_h\,\lambda_{l|}{}^{k}\frac{\partial \varrho_h}{\partial x^k} \equiv e_h\frac{\partial \varrho_h}{\partial s_l}. \tag{36.8}$$

We observe that (36.7) and (36.8) hold whether the roots of (33.1) be simple or not.

37. N-tuply orthogonal systems of hypersurfaces in a space conformal to a flat space. When the congruences of a normal orthogonal ennuple are taken as parametric and we put

$$g_{ii} = e_i H_i^2, \quad g_{ij} = 0, \quad g^{ii} = \frac{e_i}{H_i^2}, \quad g^{ij} = 0 \qquad (i \neq j), \tag{37.1}$$

the functions H_i being defined by these equations, we have

$$\lambda_{i|}{}^{i} = \frac{1}{H_i}, \quad \lambda_{i|}{}^{j} = 0, \quad \lambda_{i|i} = e_i H_i, \quad \lambda_{i|j} = 0 \qquad (i \neq j). \tag{37.2}$$

From (30.1) and (15.7) we have

$$\gamma_{hii} = e_i\frac{1}{H_h H_i}\frac{\partial H_i}{\partial x^h} \qquad (h \neq i). \tag{37.3}$$

When expressions of this form are substituted in equations of the form (36.2), (36.3) and (36.4), we obtain

$$\begin{aligned}
R_{hijk} &= 0 \qquad (h, i, j, k \neq),\\
R_{hiik} &= e_i H_i\left(\frac{\partial^2 H_i}{\partial x^h \partial x^k} - \frac{\partial H_i}{\partial x^h}\frac{\partial \log H_h}{\partial x^k} - \frac{\partial H_i}{\partial x^k}\frac{\partial \log H_k}{\partial x^h}\right)\\
&\qquad (h, i, k \neq),\\
R_{hiih} &= H_h H_i\left[e_i\frac{\partial}{\partial x^h}\left(\frac{1}{H_h}\frac{\partial H_i}{\partial x^h}\right) + e_h\frac{\partial}{\partial x^i}\left(\frac{1}{H_i}\frac{\partial H_h}{\partial x_i}\right)\right.\\
&\qquad \left. + \sum_l{}' \frac{e_l e_i e_h}{H_l^2}\frac{\partial H_h}{\partial x^l}\frac{\partial H_i}{\partial x^l}\right],
\end{aligned} \tag{37.4}$$

where l is summed over the values $1, \cdots, n$ except h and i. These equations follow directly also from (15.8) by means of (37.1).

We introduce with Darboux* the functions β_{ij} defined by

$$\beta_{ij} = \frac{1}{H_i}\frac{\partial H_j}{\partial x^i} \qquad (i \neq j). \tag{37.5}$$

If the V_n is an S_n equations (37.4) become in this notation

$$\frac{\partial \beta_{hi}}{\partial x^k} - \beta_{hk}\beta_{ki} = 0, \tag{37.6}$$

$$e_i \frac{\partial \beta_{hi}}{\partial x^h} + e_h \frac{\partial \beta_{ih}}{\partial x^i} + \sum_l e_l e_i e_h \beta_{lh}\beta_{li} = 0 \qquad (h, i, k \neq).$$

Let y^i be the generalized cartesian coördinates of the S_n in terms of which the fundamental form is

$$\varphi = c_{ij}\, dy^i\, dy^j, \tag{37.7}$$

where c_{ii} are plus or minus one and $c_{ij} = 0$ $(i \neq j)$. If $Y_{i|}{}^j$ are the components in the y's of the vector $\lambda_{i|}{}^j$ in the x's, we have from the equations

$$Y_{j|}{}^i = \lambda_{j|}{}^k \frac{\partial y^i}{\partial x^k}$$

and (37.2)

$$\frac{\partial y^i}{\partial x^j} = H_j Y_{j|}{}^i. \tag{37.8}$$

For the present case equations (7.14) become

$$\frac{\partial^2 y^i}{\partial x^j \partial x^k} = \frac{\partial y^i}{\partial x^l}\begin{Bmatrix} l \\ jk \end{Bmatrix}_g.$$

Substituting from (37.8) and making use of (15.7), we obtain

$$\frac{\partial Y_{j|}{}^i}{\partial x^k} = \beta_{jk} Y_{k|}{}^i, \quad \frac{\partial Y_{j|}{}^i}{\partial x^j} = -\sum_l e_j e_l \beta_{lj} Y_{l|}{}^i \qquad (k \neq j). \tag{37.9}$$

From (37.8), (37.1) and equations of the form (7.10) we have

$$c_{ij} Y_{k|}{}^i Y_{l|}{}^j = e_{kl}, \tag{37.10}$$

* 1898, 1, p. 161.

where

(37.11) $$e_{kk} = e_k, \quad e_{kl} = 0 \qquad (k \neq l).$$

If the functions β_{jk} satisfy the conditions (37.6), equations (37.9) are completely integrable. Moreover it can be shown that any n sets of solutions satisfy the conditions See App. 12

$$c_{ij}\, Y_{k|}^{\ i}\, Y_{l|}^{\ j} = \text{const.}$$

Hence if we take any orthogonal ennuple of unit vectors at a point, there corresponds a solution of (37.9) satisfying (37.10) and (37.11), and having the given values at the point. If then there exists a set of functions H_i for which the right-hand members of (37.4) vanish, and consequently (37.5) and (37.6) are satisfied, there exist solutions of (37.9) defining an orthogonal ennuple in S_n determined by an arbitrary orthogonal ennuple at a point. Then by quadratures from (37.8) we can find the equations $y^i = \varphi^i(x^1, x^2, \cdots, x^n)$ defining an n-tuply orthogonal family of hypersurfaces $x^i =$ const. for which the fundamental tensor is given by (37.1).

The proof of the existence and generality of solutions of equations (37.6) has been given by Bianchi*. He has shown also that the solution of equations (37.5) for a given set of functions β_{ij} involves n arbitrary functions, each of a single x. Hence we have:

In a flat space of n dimensions any orthogonal ennuple of non-null directions at a point are tangent to the curves of intersection of the hypersurfaces of an n-tuply orthogonal system.

As a corollary we have:

If a V_n is conformal to a flat space, there exists an n-tuply orthogonal system of hypersurfaces whose curves of intersection have a given orientation at a point.†

We shall obtain a characteristic property of any V_n $(n > 3)$ conformal to an S_n. We have from (28.17) that for any orthogonal ennuple in such a V_n

---

* 1924, 3, pp. 625–629.

† Because of the generality of the functions β_{ij} and H_i satisfying (37.5) and (37.6) it is evident that the n-tuply orthogonal system is not uniquely determined by the given orientation.

$$R_{hijk}\,\lambda_{p|}{}^h\,\lambda_{q|}{}^i\,\lambda_{r|}{}^j\,\lambda_{s|}{}^k = 0 \qquad (p, q, r, s \neq), \tag{37.12}$$

that is [Cf. (30.6)],

$$\gamma_{pqrs} = 0 \qquad (p, q, r, s \neq). \tag{37.13}$$

We seek conversely the condition that (37.13) hold for every orthogonal ennuple. To this end we put

$$\begin{aligned} \bar{\lambda}_{\alpha|}{}^i &= e_p\,a\,\lambda_{p|}{}^i + e_q\,b\,\lambda_{q|}{}^i, \\ \bar{\lambda}_{\beta|}{}^i &= -b\,\lambda_{p|}{}^i + a\,\lambda_{q|}{}^i, \\ \bar{\lambda}_{\gamma|}{}^i &= e_r\,c\,\lambda_{r|}{}^i + e_s\,d\,\lambda_{s|}{}^i, \\ \bar{\lambda}_{\delta|}{}^i &= -d\,\lambda_{r|}{}^i + c\,\lambda_{s|}{}^i. \end{aligned} \tag{37.14}$$

Expressing the condition that $\bar{\gamma}_{\alpha\beta\gamma\delta} = 0$ for every a, b, c and d, we get

$$\begin{aligned} e_p\,\gamma_{sppr} - e_q\,\gamma_{sqqr} &= 0, \\ e_p\,e_r\,\gamma_{rppr} - e_r\,e_q\,\gamma_{rqqr} - e_p\,e_s\,\gamma_{spps} + e_q\,e_s\,\gamma_{sqqs} &= 0 \quad (p, q, r, s \neq). \end{aligned} \tag{37.15}$$

From the first of (37.15) we have

$$e_p\,\gamma_{sppr} = \frac{1}{n-2} \sum_q^{1,\ldots,n} e_q\,\gamma_{sqqr}. \tag{37.16}$$

In consequence of (29.5) we have from (30.6)

$$\sum_q e_q\,\gamma_{sqqr} = R_{hijk}\,\lambda_{s|}{}^h\,\lambda_{r|}{}^k\,g^{ij} = R_{hk}\,\lambda_{s|}{}^h\,\lambda_{r|}{}^k, \tag{37.17}$$

so that (37.16) becomes

$$e_p\,R_{hijk}\,\lambda_{s|}{}^h\,\lambda_{p|}{}^i\,\lambda_{p|}{}^j\,\lambda_{r|}{}^k = \frac{1}{n-2} R_{hk}\,\lambda_{s|}{}^h\,\lambda_{r|}{}^k. \tag{37.18}$$

If we write the second of (37.15) in the form

$$\begin{aligned} & e_{p_1}\,e_{p_2}\,\gamma_{p_1p_2p_2p_1} \\ &\quad = e_{p_1}\,e_{p_3}\,\gamma_{p_1p_3p_3p_1} + e_{p_2}\,e_{p_4}\,\gamma_{p_2p_4p_4p_2} - e_{p_3}\,e_{p_4}\,\gamma_{p_3p_4p_4p_3}, \end{aligned} \tag{37.19}$$

we can obtain $n-3$ other expressions for the term on the left by replacing p_3 and p_4 on the right by the respective pairs p_4, p_5; p_5, p_6; $\cdots$; p_{n-1}, p_n; p_n, p_3, where $p_1, p_2, \cdots, p_n$ is some permutation of the integers $1, \cdots, n$. Adding together these $n-2$ equations and adding $2e_{p_1} e_{p_2} \gamma_{p_1 p_2 p_2 p_1}$ to both sides of the resulting equation, we have in consequence of (37.17)

$$
\begin{aligned}
(37.20) \qquad n e_{p_1} e_{p_2} \gamma_{p_1 p_2 p_2 p_1} &= R_{hk}\,(e_{p_1} \lambda_{p_1|}{}^h \lambda_{p_1|}{}^k + e_{p_2} \lambda_{p_2|}{}^h \lambda_{p_2|}{}^k) \\
&\quad -(e_{p_3} e_{p_4} \gamma_{p_3 p_4 p_4 p_3} + \cdots + e_{p_n} e_{p_3} \gamma_{p_n p_3 p_3 p_n}).
\end{aligned}
$$

If we add to this the $n-1$ equations obtained by permuting the p's cyclicly in the sequence $p_1, p_2, \cdots, p_n$, the resulting equation is reducible by means of (29.5) to

$$
nP(e_{p_1} e_{p_2} \gamma_{p_1 p_2 p_2 p_1}) = 2R-(n-2)\,P(e_{p_1} e_{p_2} \gamma_{p_1 p_2 p_2 p_1}),
$$

where $P(\ )$ indicates the sum of the n terms obtained by the process indicated above. Hence

$$
(37.21) \qquad (n-1)\,P(e_{p_1} e_{p_2} \gamma_{p_1 p_2 p_2 p_1}) = R.
$$

The last expression in (37.20) is equal to

$$
\begin{aligned}
P(e_{p_1} e_{p_2} \gamma_{p_1 p_2 p_2 p_1}) - e_{p_1} e_{p_2} \gamma_{p_1 p_2 p_2 p_1} - e_{p_2 p_3} \gamma_{p_2 p_3 p_3 p_2} \\
- e_{p_n} e_{p_1} \gamma_{p_n p_1 p_1 p_n} + e_{p_n} e_{p_3} \gamma_{p_n p_3 p_3 p_n}.
\end{aligned}
$$

In consequence of an equation of the form (37.19) the last three terms of this expression are equal to $-e_{p_1} e_{p_2} \gamma_{p_1 p_2 p_2 p_1}$. Hence (37.20) can be written

$$
\begin{aligned}
(37.22) \qquad (n-2)\, e_{p_1} e_{p_2} R_{hijk}\, \lambda_{p_1|}{}^h \lambda_{p_2|}{}^i \lambda_{p_2|}{}^j \lambda_{p_1|}{}^k \\
= R_{hk}\,(e_{p_1} \lambda_{p_1|}{}^h \lambda_{p_1|}{}^k + e_{p_2} \lambda_{p_2|}{}^h \lambda_{p_2|}{}^k) - \frac{R}{n-1}.
\end{aligned}
$$

Consider now any point P in V_n and choose the coördinate system so that at P $g_{ii} = e_i$, $g_{ij} = 0$ $(i \neq j)$. The tangents to the parametric curves at P are mutually orthogonal, and the components of the unit vectors in these directions are $\lambda_{h|}{}^i = \delta_h^i$ $(h, i = 1, \cdots, n)$. From (37.12), (37.18) and (37.22) we have at P

$$R_{hijk} = 0, \qquad R_{hiik} = \frac{1}{n-2} e_i R_{hk} \qquad (h, i, j, k \neq),$$

$$R_{hiih} = \frac{1}{n-2}(e_h R_{ii} + e_i R_{hh}) - \frac{e_i e_h R}{(n-1)(n-2)}.$$

From (28.17) it follows that at P all the components of the conformal tensor are zero. Since P is any point, we have:

A necessary and sufficient condition that (37.12) *be satisfied for every orthogonal ennuple in a* $V_n (n > 3)$ *is that the* V_n *be conformal to an* S_n.*

Exercises.

1. If φ is any function of the x's, the coefficients of $\varrho^{n-1}, \varrho^{n-2}, \cdots, \varrho$ and ϱ^0 in the determinant equation $\frac{1}{g} | \varphi_{,ij} - \varrho g_{ij} | = 0$ are invariants of degrees $1, \cdots, n$ respectively in the second derivatives of φ; the first of these is $\Delta_2 \varphi$.

Ricci and *Levi-Civita*, 1901, 1, p. 164.

2. Show that equations (33.3) can be written in any of the forms

$$(a_i^{\ j} - \varrho_h \delta_i^j) \lambda_{h|}^{\ i} = 0, \qquad (a_i^{\ j} - \varrho_h \delta_i^j) \lambda_{h|j} = 0, \qquad (a^{ij} - \varrho_h g^{ij}) \lambda_{h|i} = 0,$$

where $a_i^{\ j}$ and a^{ij} are associate to a_{ij} by means of g_{ij}.

3. If in accordance with (29.7) the components of a symmetric tensor a_{ij} are expressed in the form

$$a_{ij} = \sum_{r,s}^{1,\cdots,n} c_{rs} e_r e_s \lambda_{r|i} \lambda_{s|j},$$

a necessary and sufficient condition that the orthogonal ennuple $\lambda_{r|}^{\ i}$ consist of the principal directions determined by a_{ij} is that $c_{rs} = 0 \ (r \neq s)$.

4. If there exists for a V_n a symmetric tensor a_{ij} other than g_{ij}, whose first covariant derivative is zero and the corresponding equation (33.1) has simple elementary divisors, then the roots of this equation are constant.

Eisenhart, 1923, 5, p. 299.

5. If $\lambda_{h|}^{\ i}$ and $\lambda_{k|}^{\ i}$ are the components of congruences determined by different roots in Ex. 4, then $\gamma_{hkl} = 0$ for $l = 1, \cdots, n$. Show also that if $\lambda_{1|}^{\ i}, \cdots, \lambda_{m|}^{\ i}$ are components of mutually orthogonal congruences corresponding to a multiple root of order m, then the equations

$$\lambda_{k|}^{\ i} \frac{\partial f}{\partial x^i} = 0 \qquad (k = m+1, \cdots, n)$$

are completely integrable. *Eisenhart*, 1923, 5, p. 300.

6. If $\lambda_{h|}^{\ i}$ for $h, i = 1, \cdots, n$ are the components of n mutually orthogonal normal congruences and

$$\lambda^i = a \lambda_{1|}^{\ i} + b \lambda_{2|}^{\ i}$$

---

* *Schouten*, 1924, 1, p. 170.

are the components of a normal congruence, so also are

$$\lambda^i = -a\lambda_{1|}{}^i + b\lambda_{2|}{}^i.$$

Schouten, 1924, 1, p. 213.

7. If $\lambda_{h|}{}^i$ are the components of an orthogonal ennuple, a necessary and sufficient condition that the equations

$$\lambda_{k|}{}^i \frac{\partial f}{\partial x^i} = 0 \qquad (k = p+1, \cdots, n)$$

form a complete system is that

$$\gamma_{ijk} - \gamma_{kji} = 0 \qquad \begin{pmatrix} j = 1, \cdots, p; \\ i, k = p+1, \cdots, n \end{pmatrix}.$$

In particular, if the congruences $\lambda_{j|}{}^i$ for $j = 1, \cdots, p$ are normal, these conditions are satisfied.

38. Congruences canonical with respect to a given congruence. In § 13 we showed that there are $\infty^{(n-1)(n-2)/2}$ sets of $n-1$ mutually orthogonal congruences orthogonal to a given non-null congruence. In this section we define a particular set of $n-1$ such congruences which was discovered by Ricci,* and called by him the *congruences canonical with respect to the given congruence.*

Let $\lambda_{n|i}$ be the components of the given congruence and put

$$X_{ij} = \frac{1}{2}(\lambda_{n|i,j} + \lambda_{n|j,i}). \tag{38.1}$$

We consider the system of $n+1$ equations in the $n+1$ quantities λ^i $(i = 1, \cdots, n)$ and ϱ

$$\begin{aligned} \lambda_{n|i}\lambda^i &= 0, \\ (X_{ij} - \omega g_{ij})\lambda^i + \varrho\lambda_{n|j} &= 0, \end{aligned} \tag{38.2}$$

of which the determinant equation is

$$\Delta(\omega) = \begin{vmatrix} X_{11} - \omega g_{11} & \cdots & X_{1n} - \omega g_{1n} & \lambda_{n|1} \\ \cdot & \cdot & \cdot & \cdot \\ X_{1n} - \omega g_{1n} & \cdots & X_{nn} - \omega g_{nn} & \lambda_{n|n} \\ \lambda_{n|1} & \cdots & \lambda_{n|n} & 0 \end{vmatrix} = 0. \tag{38.3}$$

If the rank of this determinant for a root ω is $n-r+1$, ω is an r-tuple root in accordance with the general algebraic conditions for a multiple root.

* 1895, 1, p. 301; also *Ricci* and *Levi-Civita*, 1901, 1, p. 154.

We shall show conversely that the rank of Δ is $n - r + 1$ for an r-tuple root of (38.3), when the fundamental form of V_n is definite. To this end we choose a coördinate system so that at a point P $g_{1i} = 0$ and $\lambda_{n|i} = 0$ for $i = 2, \cdots, n$. At P we have

$$\Delta = \begin{vmatrix} 0 & 1 & 0 \cdots 0 \\ 1 & X_{11} - \omega g_{11} & X_{12} \cdots X_{1n} \\ 0 & X_{12} & \\ \cdot & \cdot \quad \cdot \quad \cdot & \cdot \quad \cdot \quad \cdot \\ 0 & X_{1n} \cdots\cdots & X_{nn} - \omega g_{nn} \end{vmatrix} \tag{38.4}$$

$$= - \begin{vmatrix} X_{22} - \omega g_{22} & \cdots & X_{2n} - \omega g_{2n} \\ \cdot & \cdot & \cdot \\ \cdot & \cdot & \cdot \\ X_{2n} - \omega g_{2n} & \cdots & X_{nn} - \omega g_{nn} \end{vmatrix}.$$

Since by hypothesis the fundamental form of V_n is definite, so also is the form $g_{\alpha\beta}\, dx^\alpha\, dx^\beta$ for $\alpha, \beta = 2, \cdots, n$. From the second form of Δ in (38.4) it follows (§ 33) that the roots ω are real and that for an r-tuple root the rank of this form is $n - r - 1$, and for the first form of Δ in (38.4) the rank is $n - r + 1$, as was to be proved. If the fundamental form is indefinite and $X_{ij}\, dx^i\, dx^j$ is definite, the same argument applies.

In consequence of this result, it follows that for a simple root equations (38.2) define a unique congruence orthogonal to $\lambda_{n|}{}^i$, and for an r-tuple root ∞^r congruences the components of any one of which are expressible linearly in terms of the components of r mutually orthogonal congruences orthogonal to $\lambda_{n|}{}^i$ (cf. § 33). Let ω_h and ω_k be two different roots of (38.3) and denote by $\lambda_{h|}{}^i$ and $\lambda_{k|}{}^i$ the components of congruences corresponding to these roots. In this case from the second of (38.2) we have

$$(X_{ij} - \omega_h\, g_{ij})\, \lambda_{h|}{}^i + \varrho_h\, \lambda_{n|j} = 0. \tag{38.5}$$

Multiplying by $\lambda_{k|}{}^j$ and summing for j, we have

$$(X_{ij} - \omega_h\, g_{ij})\, \lambda_{h|}{}^i\, \lambda_{k|}{}^j = 0.$$

Interchanging h and k and subtracting the resulting equation from the former, we obtain

$$(38.6) \qquad g_{ij}\,\lambda_{h|}{}^{i}\,\lambda_{k|}{}^{j} = 0, \qquad X_{ij}\,\lambda_{h|}{}^{i}\,\lambda_{k|}{}^{j} = 0 \qquad (h \neq k).$$

Consequently, the congruences corresponding to two different roots of (38.3) are orthogonal to one another. Hence:

When either the fundamental form of V_n *or the form* $X_{ij}\,dx^i\,dx^j$ *is definite, the roots of* (38.3) *are real and equations* (38.2) *define* $n-1$ *mutually orthogonal real congruences orthogonal to the given congruence* $\lambda_{n|}{}^{i}$; *the congruences corresponding to a multiple root are not uniquely determined**.

We have also the following theorem:

When neither the fundamental form of V_n *nor the form* $X_{ij}\,dx^i\,dx^j$ *is definite, a necessary and sufficient condition that equations* (38.2) *define* $n-1$ *mutually orthogonal real congruences orthogonal to a given congruence is that the roots of* (38.3) *be real and the rank of* Δ *be* $n-r+1$ *for an r-tuple root.*

The congruences so defined are said to be *canonical* with respect to the given congruence. When we take them and $\lambda_{n|}{}^{i}$ for an orthogonal ennuple and apply (30.2) to the definition (38.1) of X_{ij}, equations (38.5) become

$$(38.7) \qquad \frac{1}{2}\sum_{m} e_m\,(\gamma_{nhm}+\gamma_{nmh})\,\lambda_{m|j} - \omega_h\,\lambda_{h|j} + \varrho_h\,\lambda_{n|j} = 0.$$

Multiplying by $\lambda_{k|}{}^{j}$ for $k \neq h$, $k \neq n$ and summing for j, we get

$$(38.8) \qquad \gamma_{nhk} + \gamma_{nkh} = 0 \qquad (h, k = 1, \cdots, n-1;\ h \neq k).$$

From (38.7) follow also

$$(38.9) \qquad \omega_h = e_h\,\gamma_{nhh}, \qquad \varrho_h = \frac{1}{2}\,e_n\,\gamma_{hnn}.$$

Conversely, if (38.8) are satisfied, the $n-1$ congruences of components $\lambda_{h|}{}^{i}$ for $h = 1, \cdots, n-1$ are canonical with respect to $\lambda_{n|}{}^{i}$. Hence:

* *Ricci*, 1895, 1, p. 302; *Ricci* and *Levi-Civita*, 1901, 1, p. 155.

A necessary and sufficient condition that the congruences $\lambda_{h|}^{\;i}$ *for* $h = 1, \cdots, n-1$ *of an orthogonal ennuple be canonical with respect to the congruence* $\lambda_{n|}^{\;i}$ *is that* (38.8) *be satisfied.*

From (38.8) and (35.4) follows the theorem:

A necessary and sufficient condition that $n-1$ *non-null mutually orthogonal congruences* $\lambda_{h|}^{\;i}$ *for* $h = 1, \cdots, n-1$ *orthogonal to a normal congruence be canonical with respect to the latter is that*

$$\gamma_{nhk} = 0 \qquad (h, k = 1, \cdots, n-1;\ h \neq k). \tag{38.10}$$

As a corollary we have:

When a space V_n *admits an orthogonal ennuple of normal congruences, any* $n-1$ *of these congruences is canonical with respect to the other one.*

39. Spaces for which the equations of geodesics admit a first integral. If each integral of the equations (17.8) of the geodesics of a space satisfies the condition

$$a_{r_1 \cdots r_m} \frac{dx^{r_1}}{ds} \cdots \frac{dx^{r_m}}{ds} = \text{const.}, \tag{39.1}$$

the equations (17.8) are said to admit a first integral of the mth order. From the form of (39.1) it is seen that there is no loss of generality in assuming that the tensor $a_{r_1 \cdots r_m}$ is symmetric in all the subscripts. If we differentiate (39.1) covariantly with respect to x^k, multiply by $\frac{dx^k}{ds}$, sum for k and make use of the equations of the geodesics in the form (17.11), we obtain

$$a_{r_1 \cdots r_m, k} \frac{dx^{r_1}}{ds} \cdots \frac{dx^{r_m}}{ds} \frac{dx^k}{ds} = 0.$$

Since the equation must be satisfied identically (otherwise we should have the solutions of (17.8) satisfying a differential equation of the first order), we must have

$$P(a_{r_1 \cdots r_m, k}) = 0, \tag{39.2}$$

where P indicates the sum of the $m+1$ terms obtained by permuting the subscripts cyclically.

In particular, if (39.1) is of the first order, that is,

$$a_i \frac{dx^i}{ds} = \text{const.}, \tag{39.3}$$

the condition (39.2) is

$$a_{i,j} + a_{j,i} = 0. \tag{39.4}$$

The question of integrals of the first order is considered in § 71.

In this section we are interested primarily in the case when (39.1) is quadratic, that is,

$$a_{ij} \frac{dx^i}{ds} \frac{dx^j}{ds} = \text{const.}, \tag{39.5}$$

for which the condition (39.2) is

$$a_{ij,k} + a_{jk,i} + a_{ki,j} = 0. \tag{39.6}$$

We consider the case when a_{ij} are such that the elementary divisors of (33.1) are simple, and make use of the orthogonal ennuple defined by (33.3). We observe furthermore that equations (39.6) are equivalent to the equations

$$(a_{ij,k} + a_{jk,i} + a_{ki,j}) \lambda_{p|}{}^i \lambda_{q|}{}^j \lambda_{r|}{}^k = 0 \quad (p, q, r = 1, \cdots, n), \tag{39.7}$$

since the determinant of the λ's is not zero. By means of (36.7) and (36.8), according as $p, q, r \neq$, $r = p \neq q$ and $p = q = r$, equations (39.7) become

$$(\varrho_p - \varrho_q) \gamma_{pqr} + (\varrho_q - \varrho_r) \gamma_{qrp} + (\varrho_r - \varrho_p) \gamma_{rpq} = 0 \quad (p, q, r \neq), \tag{39.8}$$

$$e_p \frac{\partial \varrho_p}{\partial s_q} + 2 (\varrho_q - \varrho_p) \gamma_{qpp} = 0 \qquad (p \neq q), \tag{39.9}$$

$$\frac{\partial \varrho_p}{\partial s_p} = 0. \tag{39.10}$$

Conversely, when equations (39.8), (39.9) and (39.10) are satisfied, then a_{ij} defined by (33.12) satisfy the conditions (39.6). The problem of finding all V_n's admitting a quadratic integral consists in finding a tensor g_{ij} and an orthogonal ennuple $\lambda_{h|}{}^i$ for which the coefficients of rotation γ_{pqr} and $\lambda_{h|}{}^i$ satisfy the conditions obtained by the elimination of the ϱ's from (39.8), (39.9) and (39.10). The general solution has not been obtained, but we shall consider two particular solutions of the problem.

If all the ϱ's are equal, equations (39.8) are satisfied identically, and from (39.9) and (39.10) it follows that the common value of the ϱ's is constant. Then from (33.12) and (29.3) we have $a_{ij} = \varrho g_{ij}$. This is the result obtained in § 17, namely, that (17.9) is a quadratic first integral of the equations of the geodesics.

If we assume that all of the ϱ's are different and the principal congruences determined by a_{ij} are normal, it follows from (36.1) that (39.8) are satisfied identically. When we take the normal congruences for the parametric curves, and make use of (37.1), (37.2) and (37.3), we have from (39.10) that ϱ_i is independent
See App. 13 of x^i, and from (39.9) that $H_i^2/(\varrho_i - \varrho_j)$ is independent of x^j.

A solution of this problem has been given by Stäckel* as follows: Let φ_{ij} for $j = 1, \cdots, n$ be arbitrary functions of x^i alone such that the determinant Φ of these n^2 functions φ_{ij} is not zero. If φ^{ij} is the cofactor of φ_{ij} in Φ divided by Φ, then

$$H_i^2 = \frac{1}{\varphi^{i1}}, \qquad \varrho_i = \frac{\varphi^{ik}}{\varphi^{i1}} \tag{39.11}$$

for a given value of k different from 1 satisfy the conditions above stated. From (33.12) and (37.2) we have

$$a_{ii} = e_i\,\varrho_i\,H_i^2 = e_i\frac{\varphi^{ik}}{(\varphi^{i1})^2}, \qquad a_{ij} = 0 \qquad (i \neq j). \tag{39.12}$$

Since k can take the values $2, \cdots, n$, there are $n-1$ quadratic first integrals other than the fundamental form.

We recall that the conditions of the problem are that the ϱ's be different, that ϱ_i be independent of x^i and that

$$\begin{aligned} H_i^2 &= f_{1i}\ (\varrho_i - \varrho_1) = \cdots = f_{i-1\,i}(\varrho_i - \varrho_{i-1}) \\ &= f_{i+1\,i}(\varrho_i - \varrho_{i+1}) = \cdots = f_{ni}\ (\varrho_i - \varrho_n), \end{aligned} \tag{39.13}$$

where f_{ki} is a function independent of x^k for $i, k = 1, \cdots, n$; $i \neq k$. From (39.13) for a given i and from

* 1893, 1, p. 486.

$$
\begin{aligned}
(39.14) \qquad H_j^2 &= f_{1j}(\varrho_j - \varrho_1) = \cdots = f_{j-1j}(\varrho_j - \varrho_{j-1}) = \cdots \\
&\cdots = f_{nj} \quad (\varrho_j - \varrho_n)
\end{aligned}
$$

for a given j, we get pairs of equations of the form

$$\frac{f_{ji}}{f_{ki}} = \frac{\varrho_i - \varrho_k}{\varrho_i - \varrho_j}, \qquad \frac{f_{ij}}{f_{kj}} = \frac{\varrho_j - \varrho_k}{\varrho_j - \varrho_i} \qquad (i, j, k \neq),$$

from which follows

$$\frac{f_{ji}}{f_{ki}} + \frac{f_{ij}}{f_{kj}} = 1. \tag{39.15}$$

Again eliminating $(\varrho_i - \varrho_j)$ from (39.13) and (39.14), we obtain $H_i^2 f_{ij} + H_j^2 f_{ji} = 0$. Replacing i, j by j, k and k, i respectively and eliminating H_i^2, H_j^2 and H_k^2, we get

$$\frac{f_{ij}}{f_{ik}} \frac{f_{jk}}{f_{ji}} \frac{f_{ki}}{f_{kj}} = -1. \tag{39.16}$$

The problem reduces to the solution of these two sets of functional equations. Di Pirro* has shown that (39.11) and (39.12) give the general solution of the problem for $n = 3$.

40. Spaces with corresponding geodesics. From equations (17.7) it follows that the equations of the geodesics in a space V_n in terms of any parameter t are

$$
\begin{aligned}
&\frac{dx^j}{dt}\frac{d^2x^i}{dt^2} - \frac{dx^i}{dt}\frac{d^2x^j}{dt^2} \\
&\qquad + \left(\begin{Bmatrix} i \\ l\ m \end{Bmatrix}\frac{dx^j}{dt} - \begin{Bmatrix} j \\ l\ m \end{Bmatrix}\frac{dx^i}{dt}\right)\frac{dx^l}{dt}\frac{dx^m}{dt} = 0.
\end{aligned} \tag{40.1}
$$

If $\overline{V}_n$ is a second space with the fundamental form

$$\overline{\varphi} = \overline{g}_{ij}\, dx^i\, dx^j, \tag{40.2}$$

the equations of its geodesics are analogous to (40.1), and are obtained by replacing $\begin{Bmatrix} i \\ l\ m \end{Bmatrix}$ in (40.1) by the Christoffel symbols

---

* 1896, 1, pp. 318–322; he states without proof that the same is true for any n and considers also the case when the roots are not simple. The reader is referred to this paper and to *Levi-Civita*, 1896, 2, p. 292.

$\overline{\left\{ {i \atop l\,m} \right\}}$ formed with respect to (40.2). In order that every set of solutions of (40.1) define a geodesic in $\overline{V}_n$, the equations

$$(40.3)\quad \left[\left(\overline{\left\{ {i \atop l\,m} \right\}} - \left\{ {i \atop l\,m} \right\}\right)\frac{dx^j}{dt} - \left(\overline{\left\{ {j \atop l\,m} \right\}} - \left\{ {j \atop l\,m} \right\}\right)\frac{dx^i}{dt}\right]\frac{dx^l}{dt}\frac{dx^m}{dt} = 0$$

must be satisfied identically.

If we subtract equations (8.1) from the corresponding equations for $\overline{V}_n$, the resulting equations may be written

$$(40.4)\quad \overline{\left\{ {\lambda \atop \mu\,\sigma} \right\}}' - \left\{ {\lambda \atop \mu\,\sigma} \right\}' = \left(\overline{\left\{ {l \atop i\,j} \right\}} - \left\{ {l \atop i\,j} \right\}\right)\frac{\partial x^i}{\partial x'^\mu}\frac{\partial x^j}{\partial x'^\sigma}\frac{\partial x'^\lambda}{\partial x^l}.$$

Hence if we put

$$(40.5)\quad \overline{\left\{ {l \atop i\,j} \right\}} = \left\{ {l \atop i\,j} \right\} + a^l_{\ ij},$$

the quantities $a^l_{\ ij}$ are the components of a tensor, symmetric in i and j.

When the expressions (40.5) are substituted in (40.3), the latter can be written

$$(\delta^j_k\, a^i_{\ lm} - \delta^i_k\, a^j_{\ lm})\frac{dx^k}{dt}\frac{dx^l}{dt}\frac{dx^m}{dt} = 0.$$

Since these equations must be satisfied identically (cf. § 39), we must have

$$\delta^j_k\, a^i_{\ lm} + \delta^j_l\, a^i_{\ mk} + \delta^j_m\, a^i_{\ kl} = \delta^i_k\, a^j_{\ lm} + \delta^i_l\, a^j_{\ mk} + \delta^i_m\, a^j_{\ kl}.$$

Contracting for j and m, we get

$$a^i_{\ kl} = \delta^i_k\, \psi_l + \delta^i_l\, \psi_k,$$

where ψ_l is the vector $a^j_{\ lj}/(n+1)$. Hence in order that equations (40.3) be satisfied identically, it is necessary and sufficient that

$$(40.6)\quad \overline{\left\{ {l \atop i\,j} \right\}} = \left\{ {l \atop i\,j} \right\} + \delta^l_i\, \psi_j + \delta^l_j\, \psi_i,$$

where ψ_i are the components of a vector.* If now we contract for l and j we have in consequence of (7.9)

$$(40.7) \qquad \frac{\partial \log \overline{g}}{\partial x^i} = \frac{\partial \log g}{\partial x^i} + 2(n+1)\psi_i,$$

where $\overline{g} = |\overline{g}_{ij}|$. Hence ψ_i is the gradient of a function ψ, that is, $\psi_{,i}$, since $\overline{g}/g$ is an invariant.

Expressing the condition that the covariant derivative of $\overline{g}_{ik}$ with respect to x^j and the form (40.2) is zero, and replacing the symbols $\left\{ {l \atop ij} \right\}$ by their expressions (40.6), we get the following equations equivalent to (40.6):

$$(40.8) \qquad \overline{g}_{ik,j} = 2\overline{g}_{ik}\psi_{,j} + \overline{g}_{jk}\psi_{,i} + \overline{g}_{ij}\psi_{,k},$$

where $\overline{g}_{ik,j}$ is the covariant derivative of $\overline{g}_{ik}$ with respect to x^j and the fundamental tensor g_{ij}. The conditions of integrability (11.15) of these equations are reducible to

$$(40.9) \qquad \overline{g}_{mk} R^m{}_{ijl} + \overline{g}_{im} R^m{}_{kjl} = \overline{g}_{ij}\psi_{kl} - \overline{g}_{il}\psi_{kj} + \overline{g}_{kj}\psi_{il} - \overline{g}_{kl}\psi_{ij},$$

where we have put

$$(40.10) \qquad \psi_{ij} = \psi_{,ij} - \psi_{,i}\psi_{,j}.$$

If we denote by $\overline{R}^m{}_{ijl}$ the Riemann tensor for $\overline{g}_{ij}$, we have from (40.6) and (8.3)

$$(40.11) \qquad \overline{R}^m{}_{ijl} = R^m{}_{ijl} + \delta^m_l \psi_{ij} - \delta^m_j \psi_{il}.$$

From these equations it follows that (40.9) is equivalent to the identity $\overline{R}_{kijl} + \overline{R}_{ikjl} = 0$.

When V_n is of constant curvature K_0, we have from (27.1)

$$(40.12) \qquad R^m{}_{ijl} = K_0(\delta^m_j g_{il} - \delta^m_l g_{ij}).$$

In this case (40.9) and (40.11) reduce respectively to

$$(40.13) \qquad \begin{aligned} \overline{g}_{jk} A_{il} - \overline{g}_{kl} A_{ij} + \overline{g}_{ij} A_{kl} - \overline{g}_{il} A_{jk} &= 0, \\ \overline{R}_{hijl} &= \overline{g}_{hj} A_{il} - \overline{g}_{hl} A_{ij}, \end{aligned}$$

* Cf. *Weyl*, 1921, 4, p. 100; also *Eisenhart*, 1922, 6, p. 234 and *Veblen*, 1922, 7, p. 349.

where

(40.14) $$A_{ij} = K_0 g_{ij} - \psi_{ij}.$$

Multiplying the first of (40.13) by $\bar{g}^{jk}$ and summing for j and k, we find that

(40.15) $$A_{ij} = \varrho \bar{g}_{ij},$$

where ϱ is an invariant. Hence the second of (40.13) becomes $\bar{R}_{hijl} = \varrho(\bar{g}_{hj}\bar{g}_{il} - \bar{g}_{hl}\bar{g}_{ij})$ and from § 26 it follows that ϱ is a constant and $\bar{V}_n$ also is a space of constant curvature. Hence we have the theorem of Beltrami:*

The only spaces whose geodesics correspond to the geodesics of a space of constant curvature are spaces of constant curvature.

From (40.8), (40.10), (40.14) and (40.15) we have for $\varrho \neq 0$

(40.16) $$\begin{aligned}\psi_{,ikj} = 2\,(\psi_{,i}\psi_{,jk} + \psi_{,j}\psi_{,ki} + \psi_{,k}\psi_{,ij}) - 4\,\psi_{,i}\psi_{,j}\psi_{,k} \\ - K_0\,(2\,g_{ik}\psi_{,j} + g_{jk}\psi_{,i} + g_{ij}\psi_{,k}).\end{aligned}$$

In consequence of (40.12) the conditions of integrability (11.14) of (40.16) are of the form

(40.17) $$\psi_{,ijk} - \psi_{,ikj} = K_0\,(\psi_{,j}g_{ik} - \psi_{,k}g_{ij}),$$

which are satisfied identically by (40.16).

For $\varrho = 0$ we have from (40.15), (40.14) and (40.10)

(40.18) $$\psi_{,ij} = \psi_{,i}\psi_{,j} + K_0 g_{ij},$$

which are readily shown to satisfy the conditions (40.17). Hence according as we have a solution ψ of (40.16) or (40.18) we can find a space of constant curvature different from or equal to zero with geodesics corresponding to those of V_n. In the former case $\bar{g}_{ij}$ is given directly by (40.15) and in the latter by the solution of (40.8).

When ϱ in (40.15) is K_0, $\bar{V}_n$ has the same curvature as V_n. From the considerations of § 27 we may think of (40.15) and (40.14) for a given solution of (40.16) as defining a correspondence of V_n with itself such that geodesics correspond.

---

* 1868, 1, p. 232; also *Struik*, 1922, 8, p. 140 and *Schouten*, 1924, 1, p. 204.

Contracting (40.11) for m and l, we have

$$\bar{R}_{ij} = R_{ij} + (n-1)\psi_{ij}. \tag{40.19}$$

If the expressions for ψ_{ij} from (40.19) are substituted in (40.11), we find that

$$\bar{W}^l{}_{ijk} = W^l{}_{ijk},$$

where

$$W^l{}_{ijk} = R^l{}_{ijk} - \frac{1}{n-1}(\delta^l_k R_{ij} - \delta^l_j R_{ik}). \tag{40.20}$$

This tensor was discovered by Weyl* and called by him the *projective curvature tensor*.

In order that the components of $W^l{}_{ijk}$ be zero, in which case Weyl calls V_n a *projective plane* space, it is necessary and sufficient that

$$R_{lijk} = \frac{1}{n-1}(g_{kl} R_{ij} - g_{jl} R_{ik}). \tag{40.21}$$

Since we must have $R_{iijk} = 0$, we find that for $n > 2$

$$R_{ij} = \varrho g_{ij}$$

and consequently V_n is of constant Riemannian curvature.†

41. Certain spaces with corresponding geodesics. We return to the consideration of equations (40.8). If we put $\psi = -\frac{1}{2}\log\mu$, the equations become

$$2\mu \bar{g}_{ik,j} + 2\bar{g}_{ik}\mu_{,j} + \bar{g}_{jk}\mu_{,i} + \bar{g}_{ji}\mu_{,k} = 0, \tag{41.1}$$

and from (40.7) we have

$$\mu = C\left(\frac{g}{\bar{g}}\right)^{\frac{1}{n+1}}, \tag{41.2}$$

where C is an arbitrary constant.

* 1921, 4, p. 101.

† Cf. *Weyl*, 1921, 4, p. 110.

We assume that the elementary divisors of

$$(41.3) \qquad |\bar{g}_{ij} - \varrho g_{ij}| = 0$$

are simple and denote by $\lambda_{h|}{}^{i}$ the components of the orthogonal ennuple defined by equations of the form (33.3). Equations (41.1) are equivalent to the system obtained by multiplying (41.1) by $\lambda_{p|}{}^{i}\lambda_{q|}{}^{k}\lambda_{r|}{}^{j}$ and summing for i, j and k, for $p, q, r = 1, \cdots, n$ (cf. § 39). According as we take $p, q, r \neq$, $p = q \neq r$, $p \neq q = r$ and $p = q = r$, these equations are reducible by means of equations analogous to (36.7) and (36.8) to the respective equations

$$(41.4) \qquad \begin{aligned} (\varrho_p - \varrho_q)\gamma_{pqr} &= 0 && (p, q, r \neq), \\ \frac{\partial}{\partial s_q}(\mu\varrho_p) &= 0 && (p \neq q), \\ 2\mu(\varrho_p - \varrho_q)\gamma_{pqq} + e_q \frac{\partial \mu}{\partial s_p}\varrho_q &= 0 && (p \neq q), \\ \frac{\partial}{\partial s_p}(\mu^2\varrho_p) &= 0. \end{aligned}$$

We consider the case when the roots of (41.3) are simple.* From the first of (41.4) it follows that $\gamma_{pqr} = 0$ for $p, q, r \neq$, and consequently the principal congruences are normal [cf. (36.1)]. If we choose these curves as parametric, equations (41.4) reduce, in consequence of (37.1) (37.2) and (37.3), to

$$(41.5) \qquad \begin{aligned} \frac{\partial}{\partial x^j}(\mu\varrho_i) &= 0, && (i \neq j), \\ 2(\varrho_i - \varrho_j)\frac{\partial \log H_i}{\partial x^j} + \frac{\partial \varrho_i}{\partial x^j} &= 0, && (i \neq j), \\ \frac{\partial}{\partial x^i}(\mu^2\varrho_i) &= 0. \end{aligned}$$

From the first of these equations we have

$$(41.6) \qquad \mu\varrho_i = \frac{1}{\varphi_i},$$

* This case and the case of multiple roots when the fundamental forms of V_n and $\bar{V}_n$ are definite have been treated by *Levi-Civita*, 1896, 2, pp. 255–300. We refer the reader to this paper for the case of multiple roots.

where φ_i is a function of x^i alone, and from the third and (41.6) that μ/φ_i is independent of x^i. Hence

$$\mu = c\varphi_1 \cdots \varphi_n, \tag{41.7}$$

where c is an arbitrary constant. From (41.6) and (41.7) it follows that the second of (41.5) becomes

$$\frac{\partial \log H_i^2}{\partial x^j} = \frac{\partial}{\partial x^j} \log(\varphi_j - \varphi_i). \tag{41.8}$$

Hence if $\prod_j'(\varphi_j - \varphi_i)$ denotes the product of the factors $(\varphi_j - \varphi_i)$ for $j = 1, \cdots, n$ $(j \neq i)$, we have that $H_i^2/\prod_j'(\varphi_j - \varphi_i)$ is at most a function of x^i alone. Consequently the coördinates x^i can be chosen so that, in consequence of (37.1),

$$g_{ii} = e_i H_i^2 = e_i \left|\prod_j'(\varphi_j - \varphi_i)\right|, \qquad g_{ij} = 0. \tag{41.9}$$

These expressions for H_i^2 are not changed if we replace φ_i by $\varphi_i + a$, where a is an arbitrary constant, for $i = 1, \cdots, n$. Then from (33.12), (41.6), (41.7) and (37.2) we have

$$\bar{g}_{ii} = \frac{e_i}{c(\varphi_1 + a)\cdots(\varphi_n + a)} \frac{1}{\varphi_i + a} \left|\prod_j'(\varphi_j - \varphi_i)\right|, \tag{41.10}$$
$$\bar{g}_{ij} = 0.$$

If we put

$$a_{ij} = \mu^2 \bar{g}_{ij},$$

from (41.1) it follows that a_{ij} satisfies the condition (39.6). Consequently

$$\sum_i^{1,\cdots,n} e_i(\varphi_1 + a)\cdots(\varphi_{i-1} + a)(\varphi_{i+1} + a)\cdots \tag{41.11}$$
$$\cdots(\varphi_n + a)\left|\prod_j'(\varphi_j - \varphi_i)\right|\left(\frac{dx^i}{ds}\right)^2 = \text{const.}$$

is a first integral of the equations of the geodesics of V_n with the fundamental tensor g_{ij}. Since (41.11) must be a quadratic first integral whatever be a and the left-hand member is a polynomial of degree $n-1$ in a a, it follows that the equations of the geodesics admit n distinct quadratic first integrals.*

In the case just considered corresponding parametric hypersurfaces of V_n and $\overline{V}_n$ are n-tuply orthogonal. We shall obtain other solutions satisfying this condition. From (15.7) and (40.6) in which ψ_i is replaced by the gradient of $-\frac{1}{2}\log\mu$, we have the following set of conditions:

$$\frac{1}{\overline{g}_{jj}}\frac{\partial \overline{g}_{ii}}{\partial x^j} = \frac{1}{g_{jj}}\frac{\partial g_{ii}}{\partial x^j} \qquad (i \neq j),$$

$$(41.12) \qquad \frac{\partial}{\partial x^j}\log\overline{g}_{ii} = \frac{\partial}{\partial x^j}\log\frac{g_{ii}}{\mu} \qquad (i \neq j),$$

$$\frac{\partial}{\partial x^i}\log\overline{g}_{ii} = \frac{\partial}{\partial x^i}\log\frac{g_{ii}}{\mu^2}.$$

We consider first the case when every g_{ii} is a function of all the coördinates. Expressing the condition of integrability of the last two of (41.12), we find that μ must be of the form (41.7), and then from these equations we have

$$(41.13) \qquad \overline{g}_{ii} = \frac{g_{ii}}{\varphi_i\,\mu},$$

to within negligible constant factors. Then from the first of (41.12) we have

$$\frac{\partial}{\partial x^j}\log g_{ii} = \frac{\partial}{\partial x^j}\log(\varphi_j - \varphi_i).$$

Comparing this equation with (41.8), we obtain equations (41.9) and (41.10).

Suppose now that $g_{\alpha\alpha}$ for $\alpha = 1, \cdots, m$ are independent of x^σ for $\sigma = m+1, \cdots, n$, then from the first of (41.12) it follows that $\overline{g}_{\alpha\alpha}$ are independent of x^σ. Proceeding as before, we find

$$(41.14) \quad \mu = c\,\varphi_1\varphi_2\cdots\varphi_m, \quad g_{\alpha\alpha} = e_\alpha\left|\textstyle\prod'_\beta(\varphi_\beta - \varphi_\alpha)\right|, \quad \overline{g}_{\alpha\alpha} = \frac{g_{\alpha\alpha}}{\varphi_\alpha\,\mu}$$

$$(\alpha, \beta = 1, \cdots, m).$$

* Cf. *Levi-Civita*, 1896, 2, p. 287.

For the other g's we have from the first of (41.12) and (41.14)

$$\begin{aligned} \frac{\partial \bar{g}_{\sigma\sigma}}{\partial x^\alpha} &= \frac{1}{\varphi_\alpha \mu} \frac{\partial g_{\sigma\sigma}}{\partial x^\alpha}, \\ \frac{1}{\bar{g}_{\tau\tau}} \frac{\partial \bar{g}_{\sigma\sigma}}{\partial x^\tau} &= \frac{1}{g_{\tau\tau}} \frac{\partial g_{\sigma\sigma}}{\partial x^\tau} \qquad (\sigma, \tau = m+1, \cdots, n;\ \tau \neq \sigma), \end{aligned} \tag{41.15}$$

and from the second and third of (41.12) we have $\bar{g}_{\sigma\sigma} = c_\sigma \dfrac{g_{\sigma\sigma}}{\mu}$. From the second of (41.15) it follows that all the constants c_σ must be equal, say $1/c$. Then from the first of (41.15) we have

$$\frac{\partial \log g_{\sigma\sigma}}{\partial x^\alpha} = \frac{\partial}{\partial x^\alpha} \log(\varphi_\alpha - c).$$

Hence

$$g_{\sigma\sigma} = \prod_\alpha^{1,\cdots,m} (\varphi_\alpha - c) f_\sigma \qquad (\sigma = m+1, \cdots, n), \tag{41.16}$$

where f_σ are arbitrary functions of $x^{m+1}, \cdots, x^n$.

From these results the general form (40.2) is obtained similarly to (41.10) by replacing φ_i by $\varphi_i + a$ in the expression for μ.

Exercises.

1. Solve equations (40.8) for the case where V_n is of constant Riemannian curvature $K_0 \neq 0$ and $\bar{V}_n$ is a flat space.

2. Determine solutions of (41.12) other than those given in § 41.

3. Show that if λ^i are the components of a geodesic congruence, then

$$\lambda^i (\lambda_{i,j} + \lambda_{j,i}) = 0,$$

and consequently the determinant $|\lambda_{i,j} + \lambda_{j,i}|$ is zero.

4. If $\lambda_{n|i}$ are the components of a geodesic congruence, the congruences canonical with respect to it are given by [Cf. (38.2)]

$$(X_{ij} - \omega g_{ij}) \lambda^i = 0.$$

In particular, the congruence $\lambda_{n|}{}^i$ satisfies this equation for $\omega = 0$.

Ricci, 1895, 1, p. 304.

5. If $\lambda_{h|i}$ are the components of an orthogonal ennuple in a V_n, a necessary and sufficient condition that the congruence of components $\mu_i = a^h \lambda_{h|i}$ be geodesic is that the invariants a^h satisfy the equations

$$a^k \lambda_{k|}{}^i \frac{\partial a^h}{\partial x^i} = e_h \gamma_{hij} a^i a^j.$$

Ricci, 1924, 6.

6. A necessary and sufficient condition that the congruences $\lambda_{\alpha|}{}^i$ for $\alpha = 1, \cdots, n - k$ of an orthogonal ennuple be normal to ∞^{n-k} sub-spaces V_k is that

$$\gamma_{\sigma\alpha\tau} = \gamma_{\tau\alpha\sigma} \qquad (\alpha = 1, \cdots, n - k;\ \sigma, \tau = n - k + 1, \cdots, n).$$

Levy, 1925, 8, p. 41.

7. If every set of $n - k$ congruences of an orthogonal ennuple are normal to ∞^{n-k} sub-spaces V_k, then all the congruences of the ennuple are normal.

Levy, 1925, 8, p. 42.

8. If ϱ_h is a multiple root of order m of equation (33.1) and all the elementary divisors of this equation are simple, in order that m mutually orthogonal congruences corresponding to ϱ_h be normal,[*] it is necessary that any m independent congruences $\lambda_{r|}{}^i$ for $r = 1, \cdots, m$ corresponding to this root and any $n - m$ independent congruences corresponding to the other roots satisfy the equations

$$a_{ij,k}\,\lambda_{h|}{}^i\,(\lambda_{r|}{}^j\,\lambda_{r|}{}^k - \lambda_{s|}{}^j\,\lambda_{s|}{}^k) = 0 \qquad \begin{pmatrix} r, s = 1, \cdots, m; \\ h = m+1, \cdots, n \end{pmatrix}.$$

Eisenhart, 1923, 6, p. 265.

9. If the roots of equation (33.1) are simple or double and the elementary divisors are simple, a necessary and sufficient condition that there exist a normal orthogonal ennuple whose components satisfy (33.3) is that any orthogonal ennuple satisfying (33.3) shall satisfy (36.1) and (36.6) in which h and k, h and l respectively do not correspond to the same root, that the equations of Ex. 8 be satisfied and that (36.2) be satisfied, when l and p refer to the same double root, and q and r to any other root or roots. *Eisenhart*, 1923, 6, p. 267.

10. If the congruences $\lambda_{\alpha|}{}^i$ for $\alpha = 1, \cdots, n - 1$ of an orthogonal ennuple are normal, they are canonical with respect to the congruence $\lambda_{n|}{}^i$.

Ricci, 1895, 1, p. 308.

11. If for a V_4 the equation $|R_{ij} + \varrho g_{ij}| = 0$ admits a simple root ϱ_1 and a triple root ϱ_2, the elementary divisors being simple, and the principal directions corresponding to ϱ_1 and ϱ_2 satisfy the respective conditions

$$g_{ij}\,\lambda_{1|}{}^i\,\lambda_{1|}{}^j = 1, \qquad g_{ij}\,\lambda_{h|}{}^i\,\lambda_{h|}{}^j = -1 \qquad (h = 2, 3, 4),$$

then

$$R_{ij} - \frac{1}{2}\,g_{ij}\,R = (\varrho_2 - \varrho_1)\,\lambda_{1|i}\,\lambda_{1|j} + \frac{1}{2}\,(\varrho_1 + \varrho_2)\,g_{ij}.$$

Such a V_4 may be interpreted as the space-time continuum of a perfect fluid in the general theory of relativity, the congruence $\lambda_{1|}{}^i$ consisting of the lines of flow. *Eisenhart*, 1924, 4, p. 209.

12. When the fundamental form is defined by (39.11), the determination of the equations of the geodesics in finite form is reducible to quadratures (cf. Ex. 8, p. 60).

Stäckel, 1893, 2, p. 1284.

13. Show that the quantities

$$\prod_{jk}^i = \left\{ \begin{matrix} i \\ j\ k \end{matrix} \right\} - \frac{1}{n+1}\,\delta_j^i \left\{ \begin{matrix} l \\ l\ k \end{matrix} \right\} - \frac{1}{n+1}\,\delta_k^i \left\{ \begin{matrix} l \\ l\ j \end{matrix} \right\}$$

---

[*] and that this applies to every root.

have the same values at corresponding points of two spaces in geodesic correspondence, and that for a new set of coördinates x'^i the corresponding functions $\prod'^{\alpha}_{\beta\gamma}$ are given by

$$\frac{\partial^2 x^i}{\partial x'^\alpha \partial x'^\beta} = \prod'^{\sigma}_{\alpha\beta} \frac{\partial x^i}{\partial x'^\sigma} - \prod^i_{jk} \frac{\partial x^j}{\partial x'^\alpha} \frac{\partial x^k}{\partial x'^\beta}$$

$$+ \frac{1}{n+1} \left(\frac{\partial \log \Delta}{\partial x'^\alpha} \frac{\partial x^i}{\partial x'^\beta} + \frac{\partial \log \Delta}{\partial x'^\beta} \frac{\partial x^i}{\partial x'^\alpha} \right),$$

where Δ is the Jacobian $\left| \frac{\partial x^i}{\partial x'^\alpha} \right|$. *T. Y. Thomas*, 1925, 9, p. 200.

14. By expressing integrability conditions of the second set of equations in Ex. 13, derive the tensor W^i_{ijk} defined by (40.20).

J. M. Thomas, 1925, 10, p. 207.

15. For the parameter t, defined along any geodesic by

$$t = \int e^{-\frac{2}{n+1} \int \left\{ {i \atop l i} \right\} dx^l} ds,$$

the differential equations of the geodesics are

$$\frac{d^2 x^i}{dt^2} + \prod^i_{jk} \frac{dx^j}{dt} \frac{dx^k}{dt} = 0,$$

where the functions $\prod^i_{jk}$ are defined in Ex. 13.

T. Y. Thomas, 1925, 9, p. 200.

16. Show that the parameter t in Ex. 15 is the same for spaces in geodesic correspondence.

17. Show that at corresponding points of two spaces in geodesic correspondence a coördinate system y^i can be established such that the equations of the geodesics through the given points in the two spaces are given by $y^i = \eta^i t$, where η^i are constants and t is the parameter defined in Ex. 15; show also that the equations

$$P^i_{jk} y^j y^k = 0$$

are satisfied identically, where P^i_{jk} are the functions for the y's analogous to $\prod^i_{jk}$ in the x's defined in Ex. 13, (Cf. § 18).

Veblen and *Thomas*, 1925, 11, p. 205.

18. Show that the quantities $\prod^i_{jk}$ in Ex. 13 behave like the components of a tensor under linear fractional transformations of the coördinates, and under them alone. *Veblen* and *Thomas*, 1925, 11, p. 206.

19. A necessary and sufficient condition that there exist for a V_n a symmetric tensor $\bar{g}_{ij}$, where $|\bar{g}_{ij}| \neq 0$, whose first covariant derivatives are zero, is that the equations of the geodesics of V_n admit the first integral $\bar{g}_{ij} \frac{dx^i}{ds} \frac{dx^j}{ds} =$ const. and that the $\bar{V}_n$ with $\bar{g}_{ij}$ as fundamental tensor admit geodesic representation on V_n.

Levy, 1926, 1.

20. For a space of constant curvature $\neq 0$ the only tensor $\bar{g}_{ij}$, where $|\bar{g}_{ij}| \neq 0$, whose first covariant derivatives are zero is given by $\bar{g}_{ij} = \varrho g_{ij}$, where ϱ is a constant.

Levy, 1926, 1.

21. A necessary and sufficient condition that a Riemannian space admit a symmetric tensor a_{ij} other than g_{ij}, whose first covariant derivative is zero and such that the elementary divisors of the corresponding equation (33.1) are simple, is that its fundamental form be reducible to the sum of forms $\varphi_\alpha = g_{\alpha|ij}\, dx^i dx^j$, where $g_{\alpha|ij}$ are functions at most of the x's of that form; then

$$a_{rs}\, dx^r dx^s = \sum_\alpha \varrho_\alpha \varphi_\alpha,$$

where the ϱ's are constants. (Cf. Exs. 4 and 5, p. 124.) *Eisenhart*, 1923, 5, p. 303.

22. The congruence corresponding to each simple root of equation (33.1) of Ex. 21 is normal, and the tangents to the curves of the congruence form a field of parallel vectors. *Eisenhart*, 1923, 5, p. 303.

CHAPTER IV

The geometry of sub-spaces

42. The normals to a space V_n immersed in a space V_m. Let V_n be a space with the fundamental quadratic form

$$\varphi = g_{ij}\, dx^i\, dx^j \qquad (i, j = 1, \cdots, n) \tag{42.1}$$

immersed in a space V_m with the quadratic form

$$\varphi = a_{\alpha\beta}\, dy^\alpha\, dy^\beta \qquad (\alpha, \beta = 1, \cdots, m),* \tag{42.2}$$

V_n being defined by equations of the form (cf. § 16)

$$y^\alpha = f^\alpha(x^1, \cdots, x^n), \tag{42.3}$$

where the rank of the Jacobian matrix $\left\| \frac{\partial f^\alpha}{\partial x^i} \right\|$ is n.

For displacements in V_n we have

$$a_{\alpha\beta}\, dy^\alpha\, dy^\beta = g_{ij}\, dx^i\, dx^j, \tag{42.4}$$

and consequently

$$a_{\alpha\beta} \frac{\partial y^\alpha}{\partial x^i} \frac{\partial y^\beta}{\partial x^j} = g_{ij}. \tag{42.5}$$

Since the y's are invariants for transformations of coördinates in V_n, their first derivatives with respect to the x's are the same as their first covariant derivatives with respect to (42.1). Hence we may write (42.5) in the form

$$a_{\alpha\beta}\, y^\alpha_{\ ,i}\, y^\beta_{\ ,j} = g_{ij}. \tag{42.6}$$

If λ^α are the components of a vector-field in V_m normal to V_n at points of the latter, we must have (§ 16)

$$a_{\alpha\beta}\, y^\alpha_{\ ,i}\, \lambda^\beta = 0. \tag{42.7}$$

---

* In this section Greek indices take the values $1, \cdots, m$ and Latin $1, \cdots, n$, unless stated otherwise.

Since the matrix of these equations in λ^β is the product of the matrix $\|y^\alpha{}_{,i}\|$ and the determinant

$$a = |a_{\alpha\beta}|, \tag{42.8}$$

which we assume to be different from zero, it follows that this matrix is of rank n,* and consequently equations (42.7) admit $m-n$ linearly independent sets of solutions; that is, there are $m-n$ independent vectors normal to V_n at a point.

We consider first the case when $m = n+1$ and prove the theorem:

A necessary and sufficient condition that the normals to a V_n immersed in a V_{n+1} form a null vector system is that the determinant g for V_n be zero.

In accordance with the theorem of § 31 it follows from (42.6) that the determinant g is the sum of the products of corresponding n-row determinants of the two matrices $\|a_{\alpha\beta}\, y^\alpha{}_{,i}\|$ and $\|y^\alpha{}_{,j}\|$. If (42.7) is written in the form

$$y^\alpha{}_{,i}\, \lambda_\alpha = 0,$$

it follows from this equation and (42.7) that corresponding determinants of these matrices are proportional to λ^β and λ_β respectively, and consequently $g = \varrho\sigma\lambda^\beta\lambda_\beta$, where ϱ and σ are factors of proportionality. From this expression for g the theorem follows at once (§ 12; cf. § 14).

We consider now the case $m > n+1$ and indicate by $\lambda_{\sigma|}{}^\alpha$ for $\sigma = n+1, \cdots, m$ the contravariant components of $m-n$ independent vectors normal to V_n. If we put

$$\xi_{\tau|}{}^\alpha = t_\tau^\sigma \lambda_{\sigma|}{}^\alpha \qquad (\sigma, \tau = n+1, \cdots, m), \tag{42.9}$$

where t_τ^σ are functions of the x's, the vectors with components $\xi_{\tau|}{}^\alpha$ are normal to V_n. In order that they be orthogonal to one another, the functions t_τ^σ must satisfy the conditions

$$a_{\alpha\beta}\, \xi_{\tau|}{}^\alpha\, \xi_{\varrho|}{}^\beta = a_{\alpha\beta}\, \lambda_{\mu|}{}^\alpha\, \lambda_{\nu|}{}^\beta\, t_\tau^\mu\, t_\varrho^\nu = 0$$
$$(\mu, \nu, \tau, \varrho = n+1, \cdots, m;\ \tau \neq \varrho),$$

---

* *Bôcher*, 1907, 1, p. 79.

which we write

(42.10) $$c_{\mu\nu} t^{\mu}_{\tau} t^{\nu}_{\sigma} = 0.$$

The problem of finding $m-n$ sets of functions t^{μ}_{τ} satisfying this condition is equivalent to the algebraic problem of finding a self-polar polyhedron (§ 13) with respect to

(42.11) $$c_{\mu\nu} t^{\mu} t^{\nu} = 0.$$

When the determinant $|c_{\mu\nu}|$ is different from zero, there can be found $m-n$ sets of t's satisfying (42.10), none of which satisfies (42.11). Consequently $m-n$ sets of mutually orthogonal vectors normal to V_n exist, none of which is a null vector.

If $|c_{\mu\nu}| = 0$ and the rank of the determinant is $m-n-p$, there are p linearly independent vertices of the hyperquadric (42.11),* and consequently p linearly independent null vectors are given by (42.9) and $m-n-p$ other vectors, which are not null vectors, orthogonal to the former. Thus there are $m-n$ independent vectors $\xi_{\sigma|}^{\alpha}$ normal to V_n, of which p are null vectors. For any one of these null vectors, say $\xi_{1|}^{\alpha}$, we have See App. 14

$$a_{\alpha\beta}\, \xi_{1|}^{\alpha}\, \xi_{\sigma|}^{\beta} = 0 \qquad a_{\alpha\beta}\, \xi_{1|}^{\alpha}\, y^{\beta}_{,i} = 0 \qquad \begin{pmatrix} \sigma = n+1, \cdots, m; \\ i = 1, \cdots, n \end{pmatrix}.$$

Since $|a_{\alpha\beta}| \neq 0$ by hypothesis, we cannot have $a_{\alpha\beta}\, \xi_{1|}^{\alpha} = 0$ for $\beta = 1, \cdots, m$. Hence there must exist relations of fhe form

$$a^{\sigma}\, \xi_{\sigma|}^{\alpha} + b^{i}\, y^{\alpha}_{,i} = 0,$$

where all the b's cannot be zero, otherwise the $m-n$ vectors $\xi_{\sigma|}^{\alpha}$ would not be linearly independent. Multiplying by $a_{\alpha\beta}\, y^{\beta}_{,j}$ and summing for α, we have $b^{i} g_{ij} = 0$. Since all the b's cannot vanish, we must have $g = 0$. Therefore the case $|c_{\mu\nu}| = 0$ is possible only when $g = 0$, and hence:†

When the determinant g of the fundamental form of a V_n immersed in a space V_m is different from zero, $m-n$ real mutually orthogonal vectors normal to V_n can be found none of which is a null vector.

* Cf. *Bôcher*, 1907, 1, p. 130.

† *Ricci*, 1922, 9, 10.

Suppose now that $\lambda_{\sigma|}^{\alpha}$ are the components of $m-n$ such mutually orthogonal vectors normal to V_n. The magnitudes of these components can be chosen so that

$$a_{\alpha\beta}\,\lambda_{\sigma|}^{\alpha}\,\lambda_{\sigma|}^{\beta} = e_{\sigma} \quad (\sigma = n+1, \cdots, m),$$

where the quantities e_{σ} are plus or minus one. Then $c_{\mu\nu} = 0$ in (42.10) for $\mu \neq \nu$ and $c_{\mu\mu} = e_{\mu}$, so that (42.10) reduces to $\sum_{\mu} e_{\mu}\, t_{\tau}^{\mu}\, t_{\varrho}^{\mu} = 0$ for $\mu, \varrho, \tau = n+1, \cdots, m$ $(\varrho \neq \tau)$. The problem of finding such functions t is that of finding an orthogonal ennuple in a space S_{m-n} (§ 26). Each such ennuple determines by means of (42.9) a new set of mutually orthogonal non-null vectors normal to V_n. Hence we have:

When $m-n$ mutually orthogonal unit vectors in V_m normal to a V_n immersed in V_m are known, linear combinations of their components, whose coefficients are the components of any orthogonal ennuple in a certain flat space of $m-n$ dimensions, are the components of another set of mutually orthogonal normal vectors.

From the results of § 13 it follows that any one of these linear combinations can be chosen arbitrarily, provided that the functions t^{σ} are such that $\sum_{\sigma} e_{\sigma}(t^{\sigma})^2 \neq 0$.

43. The Gauss and Codazzi equations for a hypersurface. Consider a space V_{n+1} of coördinates y^{α} and a hypersurface V_n of coördinates x^i defined by the equations

$$y^{\alpha} = f^{\alpha}(x^1, \cdots, x^n).^{*} \tag{43.1}$$

We take (42.1) and (42.2) for the fundamental forms of V_n and V_{n+1} respectively, and consequently have the relations

$$a_{\alpha\beta}\, y^{\alpha}{}_{,i}\, y^{\beta}{}_{,j} = g_{ij} \tag{43.2}$$

between the components of the two fundamental tensors.

From the first theorem of § 42 it follows that the normal vector to V_n is not a null vector, since it is assumed that $g \neq 0$.

* In this and subsequent sections Greek indices take the values $1, \cdots, n+1$ and Latin $1, \cdots, n$.

If ξ^α are the components of the unit normal vector, we have from (42.7)

$$a_{\alpha\beta}\, y^\alpha{}_{,i}\, \xi^\beta = 0, \qquad a_{\alpha\beta}\, \xi^\alpha\, \xi^\beta = e. \tag{43.3}$$

If equation (43.2) be differentiated covariantly with respect to x^k and the g's, we have

$$\frac{\partial\, a_{\alpha\beta}}{\partial\, y^\gamma} y^\alpha{}_{,i}\, y^\beta{}_{,j}\, y^\gamma{}_{,k} + a_{\alpha\beta}\,(y^\alpha{}_{,ik}\, y^\beta{}_{,j} + y^\beta{}_{,jk}\, y^\alpha{}_{,i}) = 0.$$

If we subtract this equation from the sum of the two equations obtained from it by interchanging i and k and j and k respectively, we obtain, in consequence of (11.12),

$$a_{\alpha\beta}\, y^\alpha{}_{,k}\, y^\beta{}_{,ij} + [\alpha\beta, \gamma]_a\, y^\alpha{}_{,i}\, y^\beta{}_{,j}\, y^\gamma{}_{,k} = 0,$$

where the Christoffel symbols of the first kind are formed with respect to $a_{\alpha\beta}$ and evaluated at points of V_n. When this equation is written in the form

$$a_{\alpha\beta}\, y^\beta{}_{,k} \left(y^\alpha{}_{,ij} + \left\{ {\alpha \atop \mu\nu} \right\}_a y^\mu{}_{,i}\, y^\nu{}_{,j} \right) = 0,$$

it follows from the first of (43.3), since the Jacobian $\| y^\alpha{}_{,i} \|$ is of rank n by hypothesis, that

$$y^\alpha{}_{,ij} = -\left\{ {\alpha \atop \mu\nu} \right\}_a y^\mu{}_{,i}\, y^\nu{}_{,j} + e\, \Omega_{ij}\, \xi^\alpha, \tag{43.4}$$

where the functions Ω_{ij} are thus defined. If these equations be multiplied by $a_{\alpha\beta}\, \xi^\beta$ and summed for α, we obtain

$$\Omega_{ij} = a_{\alpha\beta}\, \xi^\beta\, y^\alpha{}_{,ij} + [\mu\nu, \beta]_a\, y^\mu{}_{,i}\, y^\nu{}_{,j}\, \xi^\beta. \tag{43.5}$$

Since $a_{\alpha\beta}$, ξ^β and $[\mu\nu, \beta]_a$ are invariants for transformations of coördinates x^i in V_n, it follows from (43.5) that Ω_{ij} are the components of a symmetric covariant tensor in the x's.

If the first of (43.3) be differentiated covariantly with respect to x^j and the g's, we have

$$(43.6)\quad a_{\alpha\beta}\, y^\alpha{}_{,ij}\, \xi^\beta + a_{\alpha\beta}\, y^\alpha{}_{,i}\, \xi^\beta{}_{,j} = -y^\alpha{}_{,i}\, y^\nu{}_{,j}\, \xi^\beta \frac{\partial a_{\alpha\beta}}{\partial y^\nu}$$
$$= -y^\alpha{}_{,i}\, y^\nu{}_{,j}\, \xi^\beta ([\alpha\nu, \beta]_a + [\beta\nu, \alpha]_a).$$

in consequence of (7.4). By means of this result equations (43.5) are equivalent to

$$(43.7)\qquad \Omega_{ij} = -a_{\alpha\beta}\, y^\alpha{}_{,i}\, \xi^\beta{}_{,j} - [\beta\nu, \mu]_a\, y^\mu{}_{,i}\, y^\nu{}_{,j}\, \xi^\beta.$$

See App. 15 These equations can be written in the form

$$(43.8)\qquad a_{\alpha\beta}\, y^\alpha{}_{,i} \left(\xi^\beta{}_{,j} + \left\{ {\beta \atop \mu\nu} \right\}_a y^\mu{}_{,j}\, \xi^\nu\right) = -\Omega_{ij}.$$

If the second of equations (43.3) be differentiated with respect to x^j, the resulting equation is reducible by considerations similar to those used in (43.6) to

$$(43.9)\qquad a_{\alpha\beta}\, \xi^\alpha \left(\xi^\beta{}_{,j} + \left\{ {\beta \atop \mu\nu} \right\}_a y^\mu{}_{,j}\, \xi^\nu\right) = 0.$$

From this equation and the first of (43.3) it follows that

$$\xi^\beta{}_{,j} + \left\{ {\beta \atop \mu\nu} \right\}_a y^\mu{}_{,j}\, \xi^\nu = A^k{}_j\, y^\beta{}_{,k},$$

where the A's are determined by substitution in (43.8); in consequence of (43.2) we have

$$g_{ik}\, A^k{}_j = -\Omega_{ij}, \qquad A^k{}_j = -\Omega_{ij}\, g^{ki}.$$

Hence we have

$$(43.10)\qquad \xi^\beta{}_{,j} = -\Omega_{lj}\, g^{lm}\, y^\beta{}_{,m} - \left\{ {\beta \atop \mu\nu} \right\}_a y^\mu{}_{,j}\, \xi^\nu.$$

In order to obtain the conditions of integrability of (43.4), we make use of the Ricci identity (§ 11)

$$(43.11)\qquad y^\alpha{}_{,ijk} - y^\alpha{}_{,ikj} = y^\alpha{}_{,m}\, g^{mh}\, R_{hijk},$$

where R_{hijk} are the Riemann symbols of the first kind formed with respect to the g's. Substituting from (43.4) and making use of (43.4) and (43.10) in the reduction, we obtain

$$y^{\alpha}{}_{,m}\, g^{mh}\, [R_{hijk} - e(\Omega_{hj}\,\Omega_{ik} - \Omega_{hk}\,\Omega_{ij})] - e\,\xi^{\alpha}\,(\Omega_{ij,k} - \Omega_{ik,j}) - \overline{R}^{\alpha}{}_{\mu\nu\lambda}\, y^{\mu}{}_{,i}\, y^{\nu}{}_{,j}\, y^{\lambda}{}_{,k} = 0,$$

where the components $\overline{R}^{\alpha}{}_{\mu\nu\lambda}$ are formed with respect to $a_{\alpha\beta}$ and evaluated at points of V_n. If this equation be multiplied by $a_{\alpha\beta}\, y^{\beta}{}_{,l}$ and summed for α, and again by $a_{\alpha\beta}\,\xi^{\beta}$, we obtain the two sets of equations (after changing the indices)

$$(43.12)\quad R_{ijkl} = e(\Omega_{ik}\,\Omega_{jl} - \Omega_{il}\,\Omega_{jk}) + \overline{R}_{\alpha\beta\gamma\delta}\, y^{\alpha}{}_{,i}\, y^{\beta}{}_{,j}\, y^{\gamma}{}_{,k}\, y^{\delta}{}_{,l},$$

$$(43.13)\quad \Omega_{ij,k} - \Omega_{ik,j} = \overline{R}_{\alpha\beta\gamma\delta}\, y^{\alpha}{}_{,i}\, y^{\gamma}{}_{,j}\, y^{\delta}{}_{,k}\, \xi^{\beta}.$$

In consequence of these equations the conditions of integrability of (43.10) are satisfied.

When V_{n+1} is a euclidean 3-space and the y's are cartesian coördinates, equations (43.4) become

$$(43.14)\quad y^{\alpha}{}_{,ij} = \Omega_{ij}\,\xi^{\alpha}.$$

These are the Gauss equations* for the surface, where in accordance with the customary notation

$$(43.15)\quad x^1 = u,\quad x^2 = v,\quad \Omega_{11} = D,\quad \Omega_{12} = D',\quad \Omega_{22} = D''.$$

In this case equations (43.12) reduce to the single equation

$$(43.16)\quad R_{1212} = DD'' - D'^2,$$

the equation of Gauss, and (43.13) to the equations of Codazzi

$$(43.17)\quad \Omega_{ij,k} - \Omega_{ik,j} = 0.\dagger$$

* 1909, 1, p. 154.
† 1909, 1, p. 155.

Accordingly (43.12) and (43.13) are called the *equations of Gauss and Codazzi* for the hypersurface V_n; they were established first by Voss.* Also the quadratic form

$$\psi = \Omega_{ij}\, dx^i\, dx^j \tag{43.18}$$

is called the *second fundamental form* of V_n.

When V_{n+1} is a space of constant curvature K_0, we have from (27.1)

$$\overline{R}_{\alpha\beta\gamma\delta} = K_0\,(a_{\alpha\gamma}\, a_{\beta\delta} - a_{\alpha\delta}\, a_{\beta\gamma}). \tag{43.19}$$

Because of (43.2) and (43.3) equations (43.12) and (43.13) reduce to

$$R_{ijkl} = e\,(\Omega_{ik}\,\Omega_{jl} - \Omega_{il}\,\Omega_{jk}) + K_0\,(g_{ik}\, g_{jl} - g_{il}\, g_{jk}) \tag{43.20}$$

and

$$\Omega_{ij,k} - \Omega_{ik,j} = 0. \tag{43.21}$$

44. Curvature of a curve in a hypersurface. Consider a non-minimal† curve C lying in a V_n and defined by the x's as functions of the arc. When these expressions are substituted in (43.1), we have the y's of the enveloping space V_{n+1} as functions of s. Consequently

$$\frac{dy^\alpha}{ds} = y^\alpha{}_{,i}\,\frac{dx^i}{ds}.$$

See App. 16 Since the left-hand member is an invariant in V_n, we have by covariant differentiation with respect to x^j

$$\left(\frac{dy^\alpha}{ds}\right)_{,j} = y^\alpha{}_{,ij}\,\frac{dx^i}{ds} + y^\alpha{}_{,i}\left(\frac{dx^i}{ds}\right)_{,j}.$$

Substituting for $y^\alpha{}_{,ij}$ the expression from (43.4), multiplying by $\frac{dx^j}{ds}$ and summing for j, we have

---

* 1880, 1, p. 146; cf. also *Bianchi*, 1902, 1, p. 361.

† For the method of proceedure when C is minimal see the first foot-note of § 24.

$$(44.1)\qquad \begin{aligned} &\frac{d^2 y^\alpha}{ds^2} + \begin{Bmatrix} \alpha \\ \mu\nu \end{Bmatrix}_a \frac{dy^\mu}{ds}\frac{dy^\nu}{ds} \\ &\qquad = e\,\Omega_{ij}\,\xi^\alpha \frac{dx^i}{ds}\frac{dx^j}{ds} + y^\alpha_{,i}\left(\frac{d^2 x^i}{ds^2} + \begin{Bmatrix} i \\ jk \end{Bmatrix}_g \frac{dx^j}{ds}\frac{dx^k}{ds}\right). \end{aligned}$$

From § 20 it follows that the left-hand member of this equation is the component η^α of the principal normal of C in V_{n+1}, and the expression in parenthesis on the right is the component μ^i of the principal normal in V_n. The first curvatures of C in V_n and in V_{n+1} respectively are given by [cf. (20.3)]

$$(44.2)\qquad \frac{1}{\varrho_g} = \sqrt{|g_{ij}\,\mu^i\,\mu^j|}, \qquad \frac{1}{\varrho} = \sqrt{|a_{\alpha\beta}\,\eta^\alpha\,\eta^\beta|}.$$

The former of these is called the *relative curvature* of C with respect to V_n.

If we put

$$(44.3)\qquad \frac{1}{R} = \Omega_{ij}\frac{dx^i}{ds}\frac{dx^j}{ds},$$

it follows from (44.1) that $1/R$ is the component normal to V_n of the first curvature of C in V_{n+1}. Its value at a point P is the same for all curves of V_n through P with the same direction. Accordingly it is called the *normal curvature* of V_n at P for a given direction. From (44.1) we have:

*The normal curvature of a hypersurface for a direction is the first curvature in the enveloping space of the geodesic of the hypersurface in this direction.**

If we denote by $\bar{\eta}^\alpha$ the components in the y's of the vector μ^i, that is,

$$(44.4)\qquad \bar{\eta}^\alpha = \mu^i y^\alpha_{,i},$$

equations (44.1) can be written

$$(44.5)\qquad \eta^\alpha = e\frac{\xi^\alpha}{R} + \bar{\eta}^\alpha.$$

The vector $\bar{\eta}^\alpha$ is called the *relative curvature vector.*

---

* These results and those which follow are immediate generalizations of well-known ideas in the theory of surfaces in euclidean 3-space. Cf. 1909, 1, pp. 131–133.

If the vectors η^α and $\overline{\eta}^\alpha$ are not null vectors, in consequence of (44.2), equations (44.5) can be written

$$\frac{\eta^\alpha}{\varrho} = e\frac{\xi^\alpha}{R} + \frac{\overline{\eta}^\alpha}{\varrho_g}, \tag{44.6}$$

where now η^α and $\overline{\eta}^\alpha$ are the components of the unit vectors in their respective directions.

Since the vector of components $\overline{\eta}^\alpha$ lies in V_n, we have

$$a_{\alpha\beta}\,\xi^\alpha\,\overline{\eta}^\beta = 0,$$

and from (44.6) it follows that the principal normal in V_{n+1} is one of the directions in the pencil of directions formed by the orthogonal vectors ξ^α and $\overline{\eta}^\alpha$. If we put

$$a_{\alpha\beta}\,\xi^\alpha\,\eta^\beta = \cos\sigma, \qquad a_{\alpha\beta}\,\eta^\alpha\,\overline{\eta}^\beta = \cos\overline{\sigma},$$

we have from (44.6)

$$\frac{1}{R} = \frac{\cos\sigma}{\varrho}, \qquad \frac{1}{\varrho_g} = \frac{\bar{e}\cos\overline{\sigma}}{\varrho}, \tag{44.7}$$

where $a_{\alpha\beta}\,\overline{\eta}^\alpha\,\overline{\eta}^\beta = \bar{e}$.

If the fundamental form for V_{n+1} is positive definite, we have $\bar{e} = 1$, $\cos\overline{\sigma} = \sin\sigma$, and consequently

$$\frac{1}{R} = \frac{\cos\sigma}{\varrho}, \qquad \frac{1}{\varrho_g} = \frac{\sin\sigma}{\varrho}. \tag{44.8}$$

The first of these equations is the generalization of Meusnier's theorem to curved spaces of any order and the second shows that the curvature of C relative to V_n is a generalization of the geodesic curvature of C.*

45. Principal normal curvatures of a hypersurface and lines of curvature. The principal directions in V_n determined by Ω_{ij} are given by

$$(R_h\,\Omega_{ij} - g_{ij})\,\lambda_{h|}{}^{i} = 0, \tag{45.1}$$

* 1909, 1, p. 118.

where R_h are the roots of the determinant equation

$$|R\,\Omega_{ij} - g_{ij}| = 0. \tag{45.2}$$

From § 33 it follows that R_h are the maxima and minima values of the radii of normal curvature defined by

$$\frac{1}{R} = \frac{\Omega_{ij}\,\lambda^i\lambda^j}{g_{ij}\,\lambda^i\lambda^j}, \tag{45.3}$$

and $\lambda_{h|}^{\;i}$ defined by (45.1) are the corresponding directions. The roots of (45.2) are called the *principal radii of normal curvature* of V_n. The curves of the congruences determined by $\lambda_{h|}^{\;i}$ are called *lines of curvature* of V_n. If the roots of (45.2) are simple, there are n uniquely determined families of lines of curvature, and their directions at any point are mutually orthogonal (§ 33). If a root is of order r and the elementary divisors are simple, the corresponding principal directions are linearly expressible in terms of r directions, orthogonal to one another and to the directions corresponding to the other roots. If the elementary divisors are not simple, which can happen only for certain cases when the fundamental quadratic form of V_n is indefinite, it is not possible to find n families of lines of curvature whose directions at a point are mutually orthogonal. The lines of curvature corresponding to a real root are always real. When the fundamental form is definite, all the roots are real. This is not necessarily the case when the form is indefinite.

Suppose that the elementary divisors of (45.2) are simple, in which case none of the vectors defined by (45.1) is a null vector (§ 33). Hence there exist n mutually orthogonal unit vectors $\lambda_{h|}^{\;i}$ satisfying (45.1) such that

$$g_{ij}\,\lambda_{h|}^{\;i}\lambda_{h|}^{\;j} = e_h, \qquad g_{ij}\,\lambda_{h|}^{\;i}\lambda_{k|}^{\;j} = 0 \qquad (h \neq k). \tag{45.4}$$

Any unit vector-field in V_n, say λ^i, is defined by

$$\lambda^i = e_1 \cos\alpha_1\,\lambda_{1|}^{\;i} + \cdots + e_n \cos\alpha_n\,\lambda_{n|}^{\;i}, \tag{45.5}$$

where (§ 13)

$$\cos\alpha_r = g_{ij}\,\lambda^i\,\lambda_{r|}{}^j, \qquad g_{ij}\,\lambda^i\,\lambda^j = \bar{e}.$$

Now (45.3) becomes

$$\frac{1}{R} = \bar{e}\,\Omega_{ij}\,\lambda^i\,\lambda^j, \tag{45.6}$$

and from (45.1) we have

$$\frac{1}{R_h} = e_h\,\Omega_{ij}\,\lambda_{h|}{}^i\,\lambda_{h|}{}^j. \tag{45.7}$$

Substituting in (45.6) from (45.5) and making use of (45.7), we obtain

$$\frac{\bar{e}}{R} = \frac{e_1\cos^2\alpha_1}{R_1} + \cdots + \frac{e_n\cos^2\alpha_n}{R_n}, \tag{45.8}$$

which is the generalization of Euler's formula.*

We shall prove the following theorem:

The congruences canonical with respect to a normal congruence are the lines of curvature of the hypersurfaces normal to the congruence.

Let ξ^α be the components of the congruence of normals to a V_n in a V_{n+1}, and $\xi_{h|}{}^\alpha$ for $h = 1, \cdots, n$ the components of the congruences canonical with respect to the congruence ξ^α. From (38.2) we have

$$\left[\frac{1}{2}(\xi_{\alpha,\beta} + \xi_{\beta,\alpha}) - \omega_h\,a_{\alpha\beta}\right]\xi_{h|}{}^\beta + \varrho_h\,\xi_\alpha = 0, \tag{45.9}$$

where the covariant differentiation is with respect to the fundamental form of V_{n+1}.

Since

$$\xi_{\alpha,\beta}\,y^\beta{}_{,j} = y^\beta{}_{,j}\left(\frac{\partial\xi_\alpha}{\partial y^\beta} - \xi_\gamma\left\{{\gamma \atop \alpha\beta}\right\}_a\right) = \frac{\partial\xi_\alpha}{\partial x^j} - [\alpha\beta, \nu]_a\,y^\beta{}_{,j}\,\xi^\nu$$

and from (43.10) we have

$$\frac{\partial\xi_\alpha}{\partial x^j} = \frac{\partial}{\partial x^j}(a_{\alpha\beta}\,\xi^\beta) = -\Omega_{lj}\,g^{lm}\,y^\beta{}_{,m}\,a_{\alpha\beta} + [\alpha\mu, \nu]_a\,y^\mu{}_{,j}\,\xi^\nu,$$

it follows that

$$\xi_{\alpha,\beta}\,y^\beta{}_{,j} = -\Omega_{lj}\,g^{lm}\,y^\beta{}_{,m}\,a_{\alpha\beta}. \tag{45.10}$$

---

* Cf. *Voss*, 1880, 1, p. 151; *Bianchi*, 1902, 1, p. 370; also 1909, 1, p. 124.

From $\xi_\beta y^\beta_{,j} = 0$ we have by covariant differentiation with respect to the g's and by means of (43.4)

$$\xi_{\beta,\alpha} y^\beta_{,j} y^\alpha_{,i} + \Omega_{ij} = 0. \tag{45.11}$$

If (45.9) be multiplied by $y^\alpha_{,i}$ and summed for α, and $\xi_{h|}^\beta$ be replaced by $\lambda_{h|}^j y^\beta_{,j}$, we obtain

$$\left[\frac{1}{2}(\xi_{\alpha,\beta} + \xi_{\beta,\alpha}) - \omega_h a_{\alpha\beta}\right] y^\alpha_{,i} y^\beta_{,j} \lambda_{h|}^j = 0. \tag{45.12}$$

Because of (45.10), (45.11) and (42.5) this reduces to

$$(\Omega_{ij} + \omega_h g_{ij}) \lambda_{h|}^i = 0,$$

which proves the theorem.

As a consequence of this result and the last theorem of § 38 we have the following generalization of the theorem of Dupin:*

When a space V_n admits an n-tuply orthogonal system of hypersurfaces, any hypersurface is cut by the hypersurfaces of the other families in the lines of curvature of the former.

46. Properties of the second fundamental form. Conjugate directions. Asymptotic directions. If $P(x^i)$ and $P'(x^i + dx^i)$ are nearby points of a hypersurface V_n, and C is the geodesic in V_n determined by these points, it follows from (44.5) that $|R|$ as given by (44.3) is the radius of first curvature of C at P. From (20.6) it follows that p given by

$$2p = \Omega_{ij}\, dx^i\, dx^j \tag{46.1}$$

is the distance from P' to the geodesic of V_{n+1} tangent to C at P, to within terms of higher order.† This is the well-known property of the second fundamental form of a surface immersed in euclidean 3-space.‡ Hence we have:

* 1909, 1, p. 449.

† Since the principal normal to C is normal to V_n and consequently is not a null vector, the exceptional case treated in § 20 does not arise in this instance.

‡ 1909, 1, p. 114.

If $\overline{V}_n$ is the locus of geodesics of V_{n+1} tangent to a V_n at a point $P(x^i)$, the distance from a point $P'(x^i + dx^i)$ of V_n to $\overline{V}_n$ is one-half the value of the second fundamental form for the given dx^i, to within terms of higher order.

Generalizing a concept* of the theory of surfaces, we say that two directions at a point P determined by dx^i and δx^i are *conjugate*, if

$$\Omega_{ij}\, dx^i\, \delta x^j = 0. \tag{46.2}$$

From § 45 and (33.10) we have:

The directions of two lines of curvature at a point of a hypersurface are conjugate.

Also we have the more general theorem:

A vector at a point of a hypersurface whose components are linear combinations of the components of p vectors tangent to lines of curvature is conjugate to the vector whose components are linear combinations of the remaining $n - p$ vectors tangent to lines of curvature.

A direction which is self-conjugate is called *asymptotic*. Hence:

The directions at a point of a hypersurface defined by

$$\Omega_{ij}\, dx^i\, dx^j = 0 \tag{46.3}$$

are asymptotic.

From (44.5) and (20.6) we have:

A geodesic of a hypersurface in an asymptotic direction at a point P has contact of the second or higher order with the geodesic of the enveloping space in this direction at P.

By definition an *asymptotic line* is one whose direction at every point is asymptotic. From (44.5) we have:

When an asymptotic line is a geodesic of a hypersurface, it is a geodesic of the enveloping space, and conversely.

If $\lambda_{h|}{}^j$ and $\xi_{h|}{}^\alpha$ are the components in the x's and y's respectively of a vector-field in V_n, we have

$$\xi_{h|}{}^\alpha = \lambda_{h|}{}^j\, y^\alpha{}_{,j}. \tag{46.4}$$

If equations (43.10) be multiplied by $\lambda_{h|}{}^j$ and summed for j, we have in consequence of (46.4)

---

* 1909, 1, p. 127.

$$(46.5) \qquad \xi_{h|}^{\alpha}\, \xi^{\beta}{}_{;\alpha} = -\,\Omega_{lj}\, g^{li}\, y^{\beta}{}_{,i}\, \lambda_{h|}{}^{j},$$

where $\xi^{\beta}{}_{;\alpha}$ is the covariant derivative with respect to the fundamental tensor of V_{n+1}. From the form of (46.5) it is seen that the right-hand member is the associate direction in V_{n+1} for the displacement of the normal vector in the direction $\xi_{h|}^{\alpha}$, unless the normal is parallel along the curve (cf. Ex. 5, p. 158). In order that this associate direction coincide with the direction $\xi_{h|}^{\alpha}$, the right-hand member of (46.5) must equal $\varrho\, \xi_{h|}^{\beta}$. The resulting equation is reducible by means of (46.4) to

$$(\Omega_{lj}\, g^{li}\, \lambda_{h|}{}^{j} + \varrho\, \lambda_{h|}{}^{i})\, y^{\beta}{}_{,i} = 0.$$

Multiplying by $a_{\alpha\beta}\, y^{\alpha}{}_{,k}$ and summing for β, we have, in consequence of (42.5),

$$(\Omega_{kj} + \varrho\, g_{kj})\, \lambda_{h|}{}^{j} = 0.$$

Comparing this equation with (45.1) we have:

A necessary and sufficient condition that the associate direction (when it exists) of the normal vector to a hypersurface for a curve in the hypersurface be tangent to the curve is that the curve be a line of curvature.

In order that the associate direction be orthogonal to the curve, we must have

$$a_{\alpha\beta}\, \xi_{h|}^{\alpha}\, \Omega_{lj}\, g^{li}\, y^{\beta}{}_{,i}\, \lambda_{h|}{}^{j} = 0,$$

which is reducible by (46.4) and (42.5) to

$$\Omega_{jk}\, \lambda_{h|}{}^{j}\, \lambda_{h|}{}^{k} = 0.$$

Hence we have:

*A necessary and sufficient condition that the associate direction (when it exists) of the normal to a hypersurface for a curve in the hypersurface be orthogonal to the curve is that the curve be an asymptotic line.**

Exercises.

1. When the elementary divisors of equation (45.2) are simple for a hypersurface V_n of a space of constant Riemannian curvature K_0, the scalar curvature R of V_n is given by

---

* These two theorems are generalizations of well-known theorems in the theory of surfaces in euclidean 3-space. Cf. 1909, 1, pp. 143, 144.

$$R = e\left[\sum_i \frac{1}{R_i^2} - \left(\sum_i \frac{1}{R_i}\right)^2\right] + K_0 n(1-n),$$

where R_i are the radii of principal normal curvature.

2. Let V_n be a given hypersurface of a V_{n+1} and refer the latter to a coördinate system x^α in which the hypersurfaces $x^{n+1} =$ const. are geodesically parallel to V_n (§ 19), x^{n+1} being the arc of the geodesics normal to these hypersurfaces measured from V_n; then

$$\varphi = e(dx^{n+1})^2 + c_{ij}\,dx^i\,dx^j \qquad (i,j = 1,\cdots,n),$$

and $g_{ij} = (c_{ij})_{x^{n+1}=0}$. Show that in this coördinate system the components of the normal to V_n are $\xi^i = 0$ $(i = 1,\cdots,n)$, $\xi^{n+1} = 1$, and by means of (43.4) that

$$\Omega_{ij} = -\frac{1}{2}\left(\frac{\partial c_{ij}}{\partial x^{n+1}}\right)_{x^{n+1}=0}.$$

Bianchi, 1902, 1, p. 359.

3. When a V_n admits an n-tuply orthogonal system of hypersurfaces $x^i =$ const., the components in the x's of the tensor Ω_{ij} for the hypersurface $x^n =$ const. are

$$\Omega_{ii} = -e_i \frac{H_i}{H_n}\frac{\partial H_i}{\partial x_n}, \quad \Omega_{ij} = 0 \qquad (i,j = 1,\cdots,n-1;\ i \neq j),$$

as follows from (37.1), (37.2) and (43.4); and the radii of principal normal curvature are

$$\frac{1}{R_{ni}} = -\frac{1}{H_i H_n}\frac{\partial H_i}{\partial x^n} = -e_i \gamma_{nii}.$$

Bianchi, 1902, 1, p. 378.

4. When a V_n admits an n-tuply orthogonal system of hypersurfaces $x^i =$ const., the first curvature of the curves of parameter x^h is given by [cf. (30.18) and Ex. 3]

$$\frac{1}{\varrho_h^2} = \left|\sum_r^{i,\cdots,n} \frac{e_r}{(R_{rh})^2}\right| \qquad (r \neq h),$$

where R_{rh} is the radius of principal normal curvature of $x^r =$ const. for the curve of parameter x^h. *Bianchi*, 1902, 1, p. 379.

5. In order that the normals to a hypersurface along a curve of it be parallel with respect to the curve in the enveloping space, it is necessary and sufficient that

$$\Omega_{ij}\frac{dx^j}{dt} = 0,$$

where t is a parameter along the curve; show also that such a curve is an asymptotic line.

6. For a V_3 the functions β^{rs}, defined by (cf. § 31)

$$\beta^{rs} = \frac{1}{4g}\varepsilon^{rhi}\varepsilon^{sjk} R_{hijk},$$

are the components of a symmetric contravariant tensor. Show that on taking indices as equivalent which are congruent modulo three

$$g\,\beta^{rs} = R_{r+1\,r+2\,s+1\,s+2}.$$

Ricci, 1895, 1, p. 292.

7. In a V_3 the Riemannian curvature at a point for an orientation orthogonal to the vector λ_i is given by

$$K = \frac{\beta^{ij}\lambda_i\lambda_j}{g^{ij}\lambda_i\lambda_j},$$

where β^{ij} is defined in Ex. 6. Hence the principal directions determined by β^{ij} are those for which K has maximum and minimum values; these are given by the roots of $|\beta^{ij} - \varrho\, g^{ij}| = 0$. *Bianchi*, 1902, 1, p. 354.

8. For a hypersurface of a space V_4 of constant curvature K_0 the lines of curvature are the directions for which the Riemannian curvature are maximum and minimum, and these are given by

$$K_i = K_0 + \frac{e}{R_j R_k} \qquad (i, j, k = 1, 2, 3;\; i, j, k \neq).$$

Bianchi, 1902, 1, p. 371.

47. Equations of Gauss and Codazzi for a V_n immersed in a V_m.

Given a V_n of coördinates x^i in a V_m of coördinates y^α; let the fundamental tensors of V_n and V_m be taken in the forms (42.1) and (42.2) respectively*. As shown in § 42 there exist $\infty^{(m-n)(m-n-1)/2}$ systems of real unit vectors in V_m mutually orthogonal to one another and normal to V_n. We choose a particular system of such normal vectors and denote their components by $\xi_{\sigma|}^{\alpha}$ for $\sigma = n+1, \cdots, m$; then we have

$$(47.1) \qquad a_{\alpha\beta}\,\xi_{\sigma|}^{\alpha}\,\xi_{\sigma|}^{\beta} = e_\sigma, \qquad a_{\alpha\beta}\,\xi_{\sigma|}^{\alpha}\,\xi_{\tau|}^{\beta} = 0 \qquad (\sigma, \tau = n+1, \cdots, m;\; \sigma \neq \tau),$$

where e_σ is plus or minus unity. These components satisfy equations (42.7), that is,

$$(47.2) \qquad a_{\alpha\beta}\,y^{\alpha}{}_{,i}\,\xi_{\sigma|}^{\beta} = 0.$$

If (42.6) be differentiated covariantly with respect to the quadratic form (42.1), we have

$$(47.3) \qquad \frac{\partial a_{\alpha\beta}}{\partial y^\gamma}\,y^{\alpha}{}_{,i}\,y^{\beta}{}_{,j}\,y^{\gamma}{}_{,k} + a_{\alpha\beta}\,(y^{\alpha}{}_{,ik}\,y^{\beta}{}_{,j} + y^{\beta}{}_{,jk}\,y^{\alpha}{}_{,i}) = 0.$$

---

* In this and subsequent sections Greek indices take the values $1, \cdots, m$, unless stated otherwise, and Latin $1, \cdots, n$.

If we subtract this equation from the sum of the two equations obtained from it by interchanging i and k, and j and k respectively, we obtain

$$a_{\alpha\beta}\, y^\alpha{}_{,k}\, y^\beta{}_{,ij} + [\alpha\,\beta, \gamma]_a\, y^\alpha{}_{,i}\, y^\beta{}_{,j}\, y^\gamma{}_{,k} = 0,$$

where the Christoffel symbols are formed with respect to the form (42.2) for V_m and evaluated at points of V_n. This equation may be written

$$a_{\alpha\beta}\, y^\beta{}_{,k} \left(y^\alpha{}_{,ij} + \left\{ {\alpha \atop \mu\nu} \right\}_a y^\mu{}_{,i}\, y^\nu{}_{,j} \right) = 0.$$

Since any solution of (42.7) is expressible linearly in terms of the $m-n$ vectors $\xi_{\sigma|}{}^\alpha$, there must exist functions $\Omega_{\sigma|ij}$ such that

$$(47.4) \qquad y^\alpha{}_{,ij} = -\left\{ {\alpha \atop \mu\nu} \right\}_a y^\mu{}_{,i}\, y^\nu{}_{,j} + \sum_\sigma e_\sigma\, \Omega_{\sigma|ij}\, \xi_{\sigma|}{}^\alpha$$
$$(\sigma = n+1, \cdots, m).$$

From these equations we have in consequence of (47.1)

$$(47.5) \qquad a_{\alpha\beta}\, y^\alpha{}_{,ij}\, \xi_{\sigma|}{}^\beta = -[\mu\,\nu, \beta]_a\, y^\mu{}_{,i}\, y^\nu{}_{,j}\, \xi_{\sigma|}{}^\beta + \Omega_{\sigma|ij}.$$

The functions $\xi_{\sigma|}{}^\beta$ and $[\mu\nu, \beta]_a$ are invariants for transformations of coördinates x^i in V_n, $y^\alpha{}_{,ij}$ are the components of a symmetric covariant tensor of the second order in the x's and $y^\mu{}_{,i}$ are components of a vector. Hence it follows from (47.5), that for each value of σ the quantities $\Omega_{\sigma|ij}$ are the components of a symmetric tensor in V_n.

See App. 17 Differentiating (47.2) covariantly with respect to x^j, and making use of (47.5), we have

$$(47.6) \qquad a_{\alpha\beta}\, y^\alpha{}_{,i}\, \xi_{\sigma|}{}^\beta{}_{,j} = -\Omega_{\sigma|ij} - [\mu\,\beta, \nu]_a\, y^\nu{}_{,i}\, y^\mu{}_{,j}\, \xi_{\sigma|}{}^\beta.$$

If we define functions $\mu_{\tau\sigma|j}$ by the equations

$$(47.7) \qquad a_{\alpha\beta}\, \xi_{\tau|}{}^\alpha\, \xi_{\sigma|}{}^\beta{}_{,j} + [\mu\,\nu, \beta]_a\, y^\mu{}_{,j}\, \xi_{\sigma|}{}^\nu\, \xi_{\tau|}{}^\beta = \mu_{\tau\sigma|j},$$

then for each value of τ and σ the quantities $\mu_{\tau\sigma|j}$ are components of a vector, since the term on the left of (47.7) is the component of a vector. Moreover, if the second of equations (47.1) be differentiated with respect to x^j, we have from the resulting equation and (47.7) that

(47.8) $$\mu_{\tau\sigma|j} + \mu_{\sigma\tau|j} = 0, \qquad \mu_{\sigma\sigma|j} = 0.$$

For a given value of j the quantities $\xi_{\sigma|}{}^{\beta}{}_{,j}$ are the contravariant components of a vector in V_m. Accordingly we write

$$\xi_{\sigma|}{}^{\beta}{}_{,j} = A^k y^{\beta}{}_{,k} + \sum_{\tau} B_{\tau}\, \xi_{\tau|}{}^{\beta} \qquad (\tau = n+1, \cdots, m),$$

where the A's and B's are to be determined by substituting this expression in (47.6) and (47.7). This gives

$$A^k g_{ik} = -\Omega_{\sigma|ij} - [\mu\beta, \nu]_a\, y^{\nu}{}_{,i}\, y^{\mu}{}_{,j}\, \xi_{\sigma|}{}^{\beta},$$
$$B_{\tau} = e_{\tau}\, \mu_{\tau\sigma|j} - e_{\tau}\, [\mu\nu, \beta]_a\, y^{\mu}{}_{,j}\, \xi_{\sigma|}{}^{\nu}\, \xi_{\tau|}{}^{\beta}.$$

From the first of these we get, on multiplying by g^{il} and summing for i,

$$A^l = -\Omega_{\sigma|ij}\, g^{il} - [\mu\beta, \nu]_a\, y^{\nu}{}_{,i}\, y^{\mu}{}_{,j}\, \xi_{\sigma|}{}^{\beta}\, g^{il}.$$

If $\lambda_{h|}{}^{i}$ are the components of any mutually orthogonal unit vectors in V_n, we have from (29.5)

$$\sum_h e_h\, \lambda_{h|}{}^{i}\, \lambda_{h|}{}^{l} = g^{il} \qquad (h = 1, \cdots, n).$$

If $\xi_{h|}{}^{\alpha}$ are the components of these vectors in the y's, we have $\xi_{h|}{}^{\alpha} = \lambda_{h|}{}^{i}\, y^{\alpha}{}_{,i}$ and consequently

$$A^l y^{\beta}{}_{,l} = -\Omega_{\sigma|ij}\, g^{il}\, y^{\beta}{}_{,l} - [\mu\lambda, \nu]_a\, y^{\mu}{}_{,j}\, \xi_{\sigma|}{}^{\lambda} \sum_h e_h\, \xi_{h|}{}^{\nu}\, \xi_{h|}{}^{\beta}.$$

Substituting these expressions in the above equation for $\xi_{\sigma|}{}^{\beta}{}_{,j}$ and making use of an equation of the form (29.5) for V_m, we have (on changing indices)

(47.9) $$\xi_{\sigma|}{}^{\beta}{}_{,j} = -\Omega_{\sigma|lj}\, g^{lk}\, y^{\beta}{}_{,k} - \left\{ {\beta \atop \mu\nu} \right\}_a y^{\mu}{}_{,j}\, \xi_{\sigma|}{}^{\nu} + \sum_{\tau} e_{\tau}\, \mu_{\tau\sigma|j}\, \xi_{\tau|}{}^{\beta}$$
$$(\sigma, \tau = n+1, \cdots, m).$$

In order to obtain the conditions of integrability of (47.4), we make use of the Ricci identity (cf. § 11)

$$y^\alpha{}_{,ijk} - y^\alpha{}_{,ikj} = y^\alpha{}_{,l}\, g^{lh} R_{hijk}, \tag{47.10}$$

where the Riemann symbols R_{hijk} are formed with respect to (42.1). Substituting from (47.4) and making use of (47.4) and (47.9) in the reduction, we obtain

$$\begin{aligned} & y^\alpha{}_{,l}\, g^{lh} \big[R_{hijk} - \sum_\sigma e_\sigma (\Omega_{\sigma|hj}\Omega_{\sigma|ik} - \Omega_{\sigma|hk}\,\Omega_{\sigma|ij})\big] \\ & \quad - \sum_\sigma e_\sigma\, \xi_{\sigma|}{}^\alpha \Big[\Omega_{\sigma|ij,k} - \Omega_{\sigma|ik,j} - \sum_\tau e_\tau (\mu_{\tau\sigma|k}\,\Omega_{\tau|ij} - \mu_{\tau\sigma|j}\,\Omega_{\tau|ik})\Big] \\ & \quad - \overline{R}^\alpha{}_{\mu\nu\lambda}\, y^\mu{}_{,i}\, y^\nu{}_{,j}\, y^\lambda{}_{,k} = 0, \end{aligned}$$

where $\overline{R}^\alpha{}_{\mu\nu\lambda}$ is the Riemann tensor with respect to the fundamental form (42.2) of V_m evaluated at points of V_n. If this equation be multiplied by $a_{\alpha\beta}\, y^\beta{}_{,l}$ and summed for α and again by $a_{\alpha\beta}\, \xi_{\sigma|}{}^\beta$, we obtain the two sets of equations

$$\begin{aligned} R_{ijkl} = & \sum_\sigma e_\sigma (\Omega_{\sigma|ik}\,\Omega_{\sigma|jl} - \Omega_{\sigma|il}\,\Omega_{\sigma|jk}) \\ & + \overline{R}_{\alpha\beta\gamma\delta}\, y^\alpha{}_{,i}\, y^\beta{}_{,j}\, y^\gamma{}_{,k}\, y^\delta{}_{,l} \end{aligned} \tag{47.11}$$

and

$$\begin{aligned} \Omega_{\sigma|ij,k} - \Omega_{\sigma|ik,j} = & \sum_\tau e_\tau (\mu_{\tau\sigma|k}\,\Omega_{\tau|ij} - \mu_{\tau\sigma|j}\,\Omega_{\tau|ik}) \\ & + \overline{R}_{\alpha\beta\gamma\delta}\, y^\alpha{}_{,i}\, y^\gamma{}_{,j}\, y^\delta{}_{,k}\, \xi_{\sigma|}{}^\beta \quad (\sigma, \tau = n+1, \cdots, m). \end{aligned} \tag{47.12}$$

Since $\xi_{\sigma|}{}^\beta{}_{,jk} = \xi_{\sigma|}{}^\beta{}_{,kj}$, the conditions of integrability of (47.9) are reducible by means of (47.12) to

$$\begin{aligned} & \sum_\tau e_\tau (\mu_{\tau\sigma|j,k} - \mu_{\tau\sigma|k,j})\, \xi_{\tau|}{}^\beta + \sum_{\varrho,\tau} e_\varrho\, e_\tau\, (\mu_{\tau\sigma|j}\, \mu_{\varrho\tau|k} - \mu_{\tau\sigma|k}\, \mu_{\varrho\tau|j})\, \xi_{\varrho|}{}^\beta \\ & + \sum_\tau e_\tau\, g^{lh} (\Omega_{\sigma|lk}\,\Omega_{\tau|hj} - \Omega_{\sigma|lj}\,\Omega_{\tau|hk})\, \xi_{\tau|}{}^\beta + \overline{R}^\beta{}_{\lambda\mu\nu}\, y^\mu{}_{,j}\, y^\nu{}_{,k}\, \xi_{\sigma|}{}^\lambda \\ & - g^{lh}\, y^\beta{}_{,l}\, \overline{R}_{\alpha\lambda\mu\nu}\, y^\alpha{}_{,l}\, y^\mu{}_{,j}\, y^\nu{}_{,k}\, \xi_{\sigma|}{}^\lambda = 0. \end{aligned} \tag{47.13}$$

Multiplying this equation by $\xi_{\tau|\beta}$ and summing for β, we obtain

$$\begin{aligned}(47.14)\quad &\mu_{\tau\sigma|j,k} - \mu_{\tau\sigma|k,j} + \sum_{\varrho} e_{\varrho} (\mu_{\varrho\tau|j}\, \mu_{\varrho\sigma|k} - \mu_{\varrho\tau|k}\, \mu_{\varrho\sigma|j}) \\ &+ g^{lh} (\Omega_{\tau|lj} \Omega_{\sigma|hk} - \Omega_{\tau|lk}\, \Omega_{\sigma|hj}) + \overline{R}^{\beta}{}_{\lambda\mu\nu}\, y^{\mu}{}_{,j}\, y^{\nu}{}_{,k}\, \xi_{\sigma|}{}^{\lambda}\, \xi_{\tau|\beta} = 0 \\ &(\varrho, \sigma, \tau = n+1, \cdots, m).\end{aligned}$$

When $m = n+1$, the quantities $\mu_{11|j}$ are zero, as follows from (47.8). Then (47.11) and (47.12) reduce to (43.12) and (43.13), and (47.14) are satisfied identically. Hence we call (47.11) and (47.12) the *equations of Gauss and Codazzi* of a V_n in a V_m.*

If in accordance with § 42 we take another set of real mutually orthogonal vectors normal to V_n defined by

$$(47.15)\qquad \overline{\xi}_{\nu|}{}^{\alpha} = t^{\sigma}_{\nu}\, \xi_{\sigma|}{}^{\alpha},$$

the functions t^{σ}_{ν} satisfy the conditions

$$(47.16)\qquad \sum_{\sigma} e_{\sigma}\, t^{\sigma}_{\nu}\, t^{\sigma}_{\varrho} = 0, \qquad \sum_{\sigma} e_{\sigma} (t^{\sigma}_{\nu})^2 = \overline{e}_{\nu}$$
$$(\sigma, \nu, \varrho = n+1, \cdots, m;\ \nu \neq \varrho).$$

In consequence of the results of § 29 we have

$$(47.17)\qquad \sum_{\nu} \overline{e}_{\nu}\, t^{\sigma}_{\nu}\, t^{\tau}_{\nu} = 0, \qquad \sum_{\nu} \overline{e}_{\nu}\, (t^{\sigma}_{\nu})^2 = e_{\sigma}.$$

From equations similar to (47.5) and (47.7) by means of (47.15) we have respectively

$$(47.18)\qquad \overline{\Omega}_{\nu|ij} = t^{\sigma}_{\nu}\, \Omega_{\sigma|ij},$$

$$(47.19)\qquad \overline{\mu}_{\nu\varrho|j} = t^{\sigma}_{\nu}\, t^{\tau}_{\varrho}\, \mu_{\sigma\tau|j} + \sum_{\lambda} e_{\lambda}\, t^{\lambda}_{\nu}\, t^{\lambda}_{\varrho,j}$$
$$(\lambda, \nu, \varrho, \sigma, \tau = n+1, \cdots, m).$$

---

* These results for positive definite forms are due to *Voss*, 1880, 1, p. 139 and to *Ricci*, 1902, 2, p. 357.

When these expressions are substituted in equations similar to (47.11), (47.12) and (47.14), these equations are found to be consistent with the latter in consequence of (47.17).

48. Normal and relative curvatures of a curve in a V_n immersed in a V_m. In § 24 we considered the vectors of a field in V_n at points of a curve in a V_n immersed in a V_m, the components of the vector being λ^i in the x's of V_n and ξ^β in the y's of V_m, and we obtained the following expressions for the components η^β of the associate direction for V_m along the curve:

$$\eta^\beta = \frac{d\lambda^j}{ds}\frac{\partial y^\beta}{\partial x^j} + \lambda^j \frac{dx^i}{ds}\left(\frac{\partial^2 y^\beta}{\partial x^i \partial x^j} + \left\{ {\beta \atop \alpha\gamma} \right\}_a \frac{\partial y^\alpha}{\partial x^i}\frac{\partial y^\gamma}{\partial x^j}\right).$$

In consequence of (47.4) this can be written

$$\eta^\alpha = \mu^j y^\alpha{}_{,j} + \sum_\sigma e_\sigma \Omega_{\sigma|ij}\frac{dx^i}{ds}\lambda^j \xi_{\sigma|}{}^\alpha \quad (\sigma = n+1, \cdots, m), \tag{48.1}$$

where μ^j are the components of the associate direction in V_n for the vector λ^i and are given by (24.2).

The associate curvature of the vector λ^i in V_n is given by (24.4) which now we denote by $1/r_g$, and analogously the associate curvature in V_m is defined by

$$\frac{1}{r_a} = \sqrt{|a_{\alpha\beta}\eta^\alpha\eta^\beta|}. \tag{48.2}$$

From these definitions and (47.1) we have, in consequence of (47.1-2),

$$\frac{e_a}{r_a^2} = \frac{e_g}{r_g^2} + \sum_\sigma e_\sigma \Omega_{\sigma|ij}\Omega_{\sigma|kl}\frac{dx^i}{ds}\frac{dx^k}{ds}\lambda^j\lambda^l, \tag{48.3}$$

where e_a and e_g are plus or minus one when the respective associate directions are not null vectors. From (48.1) it is seen that the component in V_n of the associate vector for V_m is in the associate direction for V_n and its magnitude is $1/r_g$.

When λ^i are the components of the unit vector tangent to the curve,* equations (48.1) can be written

* For the method of procedure when the curve is minimal see the first footnote of § 24.

$$\eta^\alpha = \mu^j y^\alpha{}_{,j} + \zeta^\alpha = \bar{\eta}^\alpha + \zeta^\alpha, \tag{48.4}$$

where η^α and $\bar{\eta}^\alpha$ are the components in the y's of the principal normals of the curve in V_m and V_n respectively, and by definition

$$\zeta^\alpha = \sum_\sigma e_\sigma \Omega_{\sigma|ij} \frac{dx^i}{ds} \frac{dx^j}{ds} \xi_{\sigma|}{}^\alpha, \tag{48.5}$$

which evidently is a vector normal to V_n. Its magnitude $1/R$ is given by

$$\frac{1}{R^2} = \left| \sum_\sigma e_\sigma \Omega_{\sigma|ij} \Omega_{\sigma|kl} \frac{dx^i}{ds} \frac{dx^j}{ds} \frac{dx^k}{ds} \frac{dx^l}{ds} \right|; \tag{48.6}$$

it is the component normal to V_n of the first curvature of the curve in V_m. Its value at a point P is the same for all curves of V_n through P in the same direction. We call it the *normal curvature* of V_n at P for the given direction and the vector ζ^α defined by (48.5) the *normal curvature vector.* When the curve is the geodesic through P, we have $\bar{\eta}^\alpha = 0$, and consequently:

The normal curvature of a V_n, immersed in a V_m, at a point and for a direction is the first curvature in V_m of the geodesic of V_n through the point in the given direction.

The first curvatures of the curve in V_m and V_n are given by equations of the form (44.2); $1/\varrho_g$ so defined is called the *relative curvature* of the curve with respect to V_n, and the vector $\bar{\eta}^\alpha$ defined by (48.4) the *relative curvature vector.* In this case equation (48.3) reduces to

$$\frac{e_a}{\varrho^2} = \frac{e_g}{\varrho_g^2} + \frac{e}{R^2}, \tag{48.7}$$

where e_a, e_g and e are plus or minus one, when the respective vectors η^α, $\bar{\eta}^\alpha$ and ζ^α are not null vectors. When all of these vectors are not null vectors, equations (48.4) can be written in the form

$$\frac{\eta^\alpha}{\varrho} = \frac{\zeta^\alpha}{R} + \frac{\bar{\eta}^\alpha}{\varrho_g}, \tag{48.8}$$

where η^α, ζ^α and $\bar{\eta}^\alpha$ are components of unit vectors.*

* Cf. the results of this section with those of § 44.

49. The second fundamental form of a V_n in a V_m. Conjugate and asymptotic directions. Consider the biquadratic differential form

$$(49.1)\quad \psi = \sum_\sigma e_\sigma \Omega_{\sigma|ij} \Omega_{\sigma|kl}\, dx^i\, dx^j\, dx^k\, dx^l \quad (\sigma = n+1, \cdots, m).$$

When $m = n+1$, the expression $e\psi$ is the square of the second fundamental form of V_n (§ 43). Accordingly when $m > n+1$ we call (49.1) the *second fundamental form* of V_n. From (48.6) and the geometrical interpretation of R it follows that the form ψ is independent of the choice of the $m-n$ mutually orthogonal vectors in V_m normal to V_n, in terms of which the functions $\Omega_{\sigma|ij}$ are defined by (47.5).

Let C be a geodesic of V_n through a point P, and consider first the case when the principal normal of C in V_m is not a null vector, the components of the principal normal being defined by (48.4). From the theorem of § 48 and equations (20.6), (48.6) and (49.1) it follows that the distance from a nearby point of C to the geodesic of V_m tangent to C at P is one-half the square root of the absolute value of ψ for the direction of C, to within terms of higher order. When the principal normal of C is a null vector, we have $1/R = 0$ so that the distance is of the third or higher order as follows from (20.6). Hence:

If $\overline{V}_n$ is the locus of geodesics of V_m tangent to a sub-space V_n at a point $P(x^i)$, the distance from a point $P'(x^i + dx^i)$ of V_n to $\overline{V}_n$ is equal to one half the square root of the absolute value of ψ for the given values of dx^i, to within terms of higher order.

Generalizing the concepts of conjugate and asymptotic directions of a hypersurface (§ 46), we say that two directions at a point determined by dx^i and δx^i are *conjugate*, when

$$(49.2)\qquad \sum_\sigma e_\sigma \Omega_{\sigma|ij} \Omega_{\sigma|kl}\, dx^i\, \delta x^j\, dx^k\, \delta x^l = 0;$$

and *asymptotic*, or self-conjugate, directions are defined by

$$(49.3)\qquad \sum_\sigma e_\sigma \Omega_{\sigma|ij} \Omega_{\sigma|kl}\, dx^i\, dx^j\, dx^k\, dx^l = 0.^*$$

* Cf. *Voss*, 1880, 1, p. 151.

From (48.6) we have:

The normal curvature of a V_n in an asymptotic direction is zero. From this result, the theorem of § 48 and (20.6) we have:

A geodesic of V_n in an asymptotic direction at a point P has contact of the second, or higher, order with the geodesic of V_m in the direction at P.

An *asymptotic line* is by definition a curve whose direction at every point of the curve is asymptotic. From (48.8) we have:

When an asymptotic line is a geodesic in V_n, it is a geodesic in V_m or its principal normal in V_m is a null vector; and conversely, when a geodesic in V_n is a geodesic in V_m, it is an asymptotic line in V_n.

From equation (48.3) and § 24 we have:

When a vector in V_n is displaced parallel to itself in V_n along a curve whose direction is conjugate to that of the given vector, it moves parallel to itself in V_m, or its associate direction in V_m is a null vector.

In order that a vector displaced parallel to itself in V_n shall move parallel to itself in V_m, it is necessary that the direction of displacement be conjugate to the vector in V_n.

From (48.4) and (48.5) it follows that the components of the principal normal in V_m of any curve of V_n through a point P are expressible linearly in terms of n mutually orthogonal vectors $\xi_{h|}^{\alpha}$ for $h = 1, \cdots, n$ in V_n at P and the $n(n+1)/2$ vectors $\sum_{\sigma} e_{\sigma} \Omega_{\sigma|ij} \xi_{\sigma|}^{\alpha}$ for $\sigma = n+1, \cdots, m$ normal to V_n at P. We denote by τ the number of linearly independent vectors in these combined systems. Evidently $\tau \leqq m$, and also $\tau \leqq n(n+3)/2$. If it is less than $n(n+3)/2$, there must exist linear and homogeneous relations between the functions $\Omega_{\sigma|ij}$ at P. We denote by G_{τ} the variety of order τ consisting of all the geodesics of V_m through P in directions determined by the τ independent vectors. From the last theorem of § 20 it follows that G_{τ} has contact of the second order with every curve of V_n through P. Hence we call G_{τ} the *osculating geodesic variety* of V_n at P.*

50. Lines of curvature and mean curvature. The principal directions determined by each of the $m-n$ tensors $\Omega_{\sigma|ij}$ corres-

* Cf. *Bompiani*, 1921, 5, p. 1122.

ponding to a given set of $m-n$ mutually orthogonal unit vectors normal to a V_n in a V_m define an orthogonal ennuple of congruences analogous to the lines of curvature of a hypersurface (§ 45). We call them the *lines of curvature* of V_n for the corresponding normal $\xi_{\sigma|}^{\alpha}$. In order to obtain a geometric characterization of these lines, we multiply equations (47.9) by $\lambda_{h|}{}^{j}$ and sum for j. Making use of (46.4), we obtain

$$\xi_{h|}{}^{\alpha}\, \xi_{\sigma|}{}^{\beta}{}_{;\alpha} = -\Omega_{\sigma|lj}\, g^{lk}\, y^{\beta}{}_{,k}\, \lambda_{h|}{}^{j} + \sum_{\tau} e_{\tau}\, \mu_{\tau\sigma|j}\, \xi_{\tau|}{}^{\beta}\, \lambda_{h|}{}^{j}. \tag{50.1}$$

See App. 18 Proceeding with this equation in a manner similar to that followed in the case of (46.5), we get the theorem:

A necessary and sufficient condition that the associate direction of a normal vector to a V_n for a curve in the V_n be tangent to the curve is that the curve be a line of curvature for the given normal.

Any unit vector ξ^{α} normal to a V_n is expressible linearly in terms of $m-n$ mutually orthogonal unit vectors normal to V_n, as in (47.15), and the corresponding tensor Ω_{ij} is given by

$$\Omega_{ij} = (a_{\alpha\beta}\, y^{\alpha}{}_{,ij} + [\mu\nu, \beta]_a\, y^{\mu}{}_{,i}\, y^{\nu}{}_{,j})\, \xi^{\beta}, \tag{50.2}$$

as follows from (47.5), (47.15) and (47.18). When the normal vector is a null vector, its components ξ^{α} involve an arbitrary factor and consequently the corresponding Ω_{ij} is determined by (50.2) only to within a factor.

From equation (45.2) it follows that the sum of the principal normal curvatures of a hypersurface is

$$\Omega = g^{ij}\, \Omega_{ij}. \tag{50.3}$$

This is the generalization of the mean curvature of a surface* and is called the *mean curvature* of the hypersurface. In a similar manner we call $\Omega_{\sigma|}$, defined by

$$\Omega_{\sigma|} = g^{ij}\, \Omega_{\sigma|ij}, \tag{50.4}$$

the *mean curvature* of V_n for the normal direction $\xi_{\sigma|}{}^{\alpha}$.

* 1909, 1, p. 123.

Consider the vector normal to V_n whose components ξ^α are given by

$$(50.5)\qquad \xi^\alpha = \sum_\sigma e_\sigma \Omega_{\sigma|ij}\, g^{ij}\, \xi_{\sigma|}{}^\alpha .$$

Its magnitude M is given by

$$(50.6)\qquad M^2 = \left| \sum_\sigma e_\sigma \Omega_{\sigma|ij}\, \Omega_{\sigma|kl}\, g^{ij}\, g^{kl} \right| .$$

From (47.15) and (47.18) it follows that the vector ξ^α is independent of the choice of the $m-n$ mutually orthogonal vectors $\xi_{\sigma|}{}^\alpha$ normal to V_n.

Since the rank of the matrix $\| \xi_{\sigma|}{}^\alpha \|$ is $m-n$, the components of the above vector vanish, when, and only when,

$$(50.7)\qquad \Omega_{\sigma|ij}\, g^{ij} = 0 \qquad (\sigma = n+1, \cdots, m).$$

The invariant M is zero in this case, and also when the vector is a null vector.*

Suppose now that $M \neq 0$ and write (50.5) in the form

$$(50.8)\qquad M \xi^\alpha = \sum_\sigma e_\sigma \Omega_{\sigma|ij}\, g^{ij}\, \xi_{\sigma|}{}^\alpha ,$$

ξ^α being the components of the unit vector. Then from (50.2) and (47.5) we have for the components of the tensor Ω_{ij} corresponding to the vector ξ^α

$$(50.9)\qquad M \Omega_{ij} = \sum_\sigma e_\sigma \Omega_{\sigma|ij}\, \Omega_{\sigma|kl}\, g^{kl}.$$

From this equation and (50.6) it follows that the mean curvature of V_n for the direction ξ^α, that is, $\Omega_{ij}\, g^{ij}$, is equal to M, to within sign at most. Moreover, if the vector ξ^α is a null vector, we find that the mean curvature for this normal is zero. If we call M the *mean curvature* of V_n and the vector defined by (50.5) the *mean curvature normal*, we have:

* Cf. § 52.

The mean curvature of a V_n immersed in a V_m for the mean curvature normal is the mean curvature of V_n, to within sign at most.

Also we have in view of the above results:

A necessary and sufficient condition that the mean curvature of a V_n be zero is that the mean curvature with respect to every normal to V_n be zero, or that the mean curvature normal be a null vector.*

Let η^α be the components of any vector normal to V_n; then $\eta^\alpha = t^\tau \xi_{\tau|}{}^\alpha$. From (50.2) and (47.18) it follows that the mean curvature for this direction is $t^\sigma \Omega_{\sigma|ij} g^{ij}$. From (50.5) we have

$$\begin{aligned} a_{\alpha\beta} \xi^\alpha \eta^\beta &= a_{\alpha\beta} t^\tau \xi_{\tau|}{}^\beta \sum_\sigma e_\sigma \Omega_{\sigma|ij} g^{ij} \xi_{\sigma|}{}^\alpha \\ &= t^\sigma \Omega_{\sigma|ij} g^{ij}. \end{aligned}$$

Consequently we have:

The mean curvature of a V_n for any normal orthogonal to the mean curvature normal is zero†.

51. The fundamental equations of a V_n in a V_m in terms of invariants and an orthogonal ennuple. In a V_n immersed in a V_m of coördinates y^α we choose an orthogonal ennuple of unit vectors of components $\lambda_{h|}{}^i$, that is,

$$(51.1)\qquad g_{ij} \lambda_{h|}{}^i \lambda_{h|}{}^j = e_h, \quad g_{ij} \lambda_{h|}{}^i \lambda_{k|}{}^j = 0 \quad (h, k = 1, \cdots, n;\ h \neq k).$$

Since $y^\alpha{}_{,i}$ are the components of a covariant vector in V_n for each value of α, a set of invariants $\xi_{h|}{}^\alpha$ are defined by (cf. § 29)

$$(51.2)\qquad y^\alpha{}_{,i} = \sum_h e_h \xi_{h|}{}^\alpha \lambda_{h|i} \qquad (h, i = 1, \cdots, n;\ \alpha = 1, \cdots, m).$$

These equations are equivalent to

$$(51.3)\qquad \xi_{h|}{}^\alpha = y^\alpha{}_{,i} \lambda_{h|}{}^i.$$

From the latter we have

$$(51.4)\qquad a_{\alpha\beta} \xi_{h|}{}^\alpha \xi_{h|}{}^\beta = e_h, \qquad a_{\alpha\beta} \xi_{h|}{}^\alpha \xi_{k|}{}^\beta = 0 \qquad (h \neq k).$$

---

* The second alternative does not arise, when the fundamental form of V_m is definite.

† Cf. *Bompiani*, 1921, 5, p. 1134.

From (51.3) it follows that $\xi_{h|}^{\alpha}$ are the components in the y's of the vector whose components in the x's are $\lambda_{h|}^{i}$, and from (51.4) we find that the e's are the same for a congruence whether given in the y's or x's.

In similar manner, if we put

$$(51.5) \qquad \Omega_{\sigma|ij} = \sum_{h,k} e_h e_k \omega_{\sigma|hk} \lambda_{h|i} \lambda_{k|j} \qquad (\sigma = n+1, \cdots, m),$$

the quantities $\omega_{\sigma|hk}$ are invariants in V_n, which are given by

$$(51.6) \qquad \omega_{\sigma|hk} = \Omega_{\sigma|ij} \lambda_{h|}^{i} \lambda_{k|}^{j},$$

and thus

$$(51.7) \qquad \omega_{\sigma|hk} = \omega_{\sigma|kh}.$$

We recall from § 30 the formulas

$$(51.8) \qquad \lambda_{l|i,j} = \sum_{h,k} e_h e_k \gamma_{lhk} \lambda_{h|i} \lambda_{k|j}, \qquad \gamma_{lhk} = \lambda_{l|i,j} \lambda_{h|}^{i} \lambda_{k|}^{j}.$$

From (47.4), (51.3) and (51.6) we have

$$(51.9) \qquad \lambda_{h|}^{i} \lambda_{k|}^{j} y^{\alpha}_{,ij} = \sum_{\sigma} e_\sigma \omega_{\sigma|hk} \xi_{\sigma|}^{\alpha} - \left\{ {\alpha \atop \mu\nu} \right\}_a \xi_{h|}^{\mu} \xi_{k|}^{\nu}$$

$$(i, j, h, k = 1, \cdots, n; \quad \alpha = 1, \cdots, m; \quad \sigma = n+1, \cdots, m).$$

Differentiating (51.3) covariantly with respect to x^j and the fundamental form of V_n, we have

$$\xi_{h|}^{\alpha}{}_{,j} = y^{\alpha}_{,ij} \lambda_{h|}^{i} + y^{\alpha}_{,i} \lambda_{h|}^{i}{}_{,j}.$$

Multiplying by $\lambda_{k|}^{j}$ and summing for j, we have, in consequence of (51.8) and (51.9),

$$(51.10) \qquad \frac{\partial \xi_{h|}^{\alpha}}{\partial s_k} = \sum_{\sigma} e_\sigma \omega_{\sigma|hk} \xi_{\sigma|}^{\alpha} - \left\{ {\alpha \atop \mu\nu} \right\}_a \xi_{h|}^{\mu} \xi_{k|}^{\nu} - \sum_{l} e_l \gamma_{lhk} \xi_{l|}^{\alpha},$$

where

$$\frac{\partial \xi_{h|}^{\alpha}}{\partial s_k} \equiv \lambda_{k|}^{j} \xi_{h|}^{\alpha}{}_{,j}.$$

Since $\xi_{h|\alpha} = a_{\alpha\beta}\, \xi_{h|}{}^{\beta}$, it follows that we have also

$$(51.11)\quad \frac{\partial \xi_{h|\alpha}}{\partial s_k} = \sum_{\sigma} e_\sigma\, \omega_{\sigma|hk}\, \xi_{\sigma|\alpha} + [\alpha\nu, \mu]_a\, \xi_{h|}{}^{\mu}\, \xi_{k|}{}^{\nu} - \sum_{l} e_l\, \gamma_{lhk}\, \xi_{l|\alpha}.$$

Equations (47.11) may be written in the form

$$(51.12)\quad \begin{aligned} R_{ijkl}\, \lambda_{p|}{}^{i}\, \lambda_{q|}{}^{j}\, \lambda_{r|}{}^{k}\, \lambda_{s|}{}^{l} &= \sum_{\sigma} e_\sigma \left(\omega_{\sigma|pr}\, \omega_{\sigma|qs} - \omega_{\sigma|ps}\, \omega_{\sigma|qr}\right) \\ &\quad + \overline{R}_{\alpha\beta\gamma\delta}\, \xi_{p|}{}^{\alpha}\, \xi_{q|}{}^{\beta}\, \xi_{r|}{}^{\gamma}\, \xi_{s|}{}^{\delta}. \end{aligned}$$

When the expressions for $\Omega_{\sigma|ij,k}$ and $\Omega_{\sigma|ik,j}$ as obtained from (51.5) are substituted in (47.12) and this equation is multiplied by $\lambda_{p|}{}^{i}\, \lambda_{q|}{}^{j}\, \lambda_{r|}{}^{k}$ and summed for i, j, k, we have

$$(51.13)\quad \begin{aligned} &\frac{\partial \omega_{\sigma|pq}}{\partial s_r} - \frac{\partial \omega_{\sigma|pr}}{\partial s_q} + \sum_{h} e_h \left[\omega_{\sigma|hq}\, \gamma_{hpr} - \omega_{\sigma|hr}\, \gamma_{hpq} + \omega_{\sigma|hp}(\gamma_{hqr} - \gamma_{hrq})\right] \\ &= \sum_{\tau} e_\tau \left(\lambda_{r|}{}^{k}\, \mu_{\tau\sigma|k}\, \omega_{\tau|pq} - \lambda_{q|}{}^{j}\, \mu_{\tau\sigma|j}\, \omega_{\tau|pr}\right) + \overline{R}_{\lambda\beta\mu\nu}\, \xi_{p|}{}^{\lambda}\, \xi_{q|}{}^{\mu}\, \xi_{r|}{}^{\nu}\, \xi_{\sigma|}{}^{\beta}.^{*} \end{aligned}$$

The ξ's defined by (51.3) and those of § 47 are the components of an orthogonal ennuple in V_m. Analogous to (51.8) for V_n we have for V_m

$$(51.14)\quad \xi_{\varrho|\alpha;\beta} = \sum_{\mu,\nu} e_\mu\, e_\nu\, \overline{\gamma}_{\varrho\mu\nu}\, \xi_{\mu|\alpha}\, \xi_{\nu|\beta},$$

and

$$(51.15)\quad \xi_{\varrho|}{}^{\alpha}{}_{;\beta} = \sum_{\mu,\nu} e_\mu\, e_\nu\, \overline{\gamma}_{\varrho\mu\nu}\, \xi_{\mu|}{}^{\alpha}\, \xi_{\nu|\beta}.$$

If we substitute in (51.15) for the y's their expressions in terms of the x's, multiply by $y^{\beta}{}_{,i}$ and sum for β, we have for points of V_n

$$\frac{\partial \xi_{\varrho|}{}^{\alpha}}{\partial x^i} + \xi_{\varrho|}{}^{\gamma}\, y^{\beta}{}_{,i} \left\{ {\alpha \atop \beta\gamma} \right\}_a = \sum_{\mu,\nu} e_\mu\, e_\nu\, \overline{\gamma}_{\varrho\mu\nu}\, \xi_{\mu|}{}^{\alpha}\, \xi_{\nu|\beta}\, y^{\beta}{}_{,i}.$$

Multiplying by $\lambda_{k|}{}^{i}$ and summing for i, we have, in consequence of (51.3) and (51.4),

* *Ricci,* 1902, 2, p. 359.

$$\frac{\partial \xi_{\varrho|}{}^{\alpha}}{\partial s_k} = \sum_{\mu} e_\mu \overline{\gamma}_{\varrho\mu k} \xi_{\mu|}{}^{\alpha} - \begin{Bmatrix} \alpha \\ \beta\gamma \end{Bmatrix}_a \xi_{\varrho|}{}^{\gamma} \xi_{k|}{}^{\beta} \tag{51.16}$$
$$(\alpha, \beta, \gamma, \mu, \varrho = 1, \cdots, m;\ k = 1, \cdots, n).$$

If we compare these equations for $\varrho = 1, \cdots, n$ with (51.10), we have

$$\overline{\gamma}_{lhk} = \gamma_{lhk}, \quad \overline{\gamma}_{\sigma hk} = -\omega_{\sigma|hk} \quad \begin{pmatrix} h, k, l = 1, \cdots, n; \\ \sigma = n+1, \cdots, m \end{pmatrix}. \tag{51.17}$$

Equations (51.10) can be written

$$\xi_{k|}{}^{\beta} \xi_{h|}{}^{\alpha}{}_{;\beta} = \sum_{\sigma} e_\sigma \omega_{\sigma|hk} \xi_{\sigma|}{}^{\alpha} - \sum_{l} e_l \gamma_{hlk} \xi_{l|}{}^{\alpha} \tag{51.18}$$
$$(\alpha, \beta = 1, \cdots, m;\ \sigma = n+1, \cdots, m;\ h, l, k = 1, \cdots, n).$$

When $k = h$, we have on comparing (51.18) with (48.4) that $\omega_{\sigma|hh}$ are the invariants of the normal curvature vector of the curve whose tangential vector has the components $\xi_{h|}{}^{\alpha}$ in the y's and that γ_{lhh} are the invariants of the relative curvature vector.

When $k \neq h$, it is seen from (51.18) that $\omega_{\sigma|hk}$ are the invariants of the normal component of the associate curvature vector for V_m of the vector $\xi_{h|}{}^{\alpha}$ for the direction $\xi_{k|}{}^{\alpha}$ and that γ_{lhk} are the invariants of the component relative to V_n. Since $\omega_{\sigma|hk} = \omega_{\sigma|kh}$, it follows that the components normal to V_n of the associate curvatures for V_m of $\xi_{h|}{}^{\alpha}$ in the direction $\xi_{k|}{}^{\alpha}$ and of $\xi_{k|}{}^{\alpha}$ in the direction $\xi_{h|}{}^{\alpha}$ are equal in magnitude and direction.

In order to give another interpretation to these invariants, we consider the case when $m = n+1$, indicating by ξ^α the components of the vector normal to V_n. In this case, because of (51.17), equations (51.16) for $\varrho = n+1$ become [cf. (30.4)]

$$\xi_{k|}{}^{\beta} \xi^{\alpha}{}_{;\beta} = -\sum_{h} e_h \omega_{hk} \xi_{h|}{}^{\alpha}, \tag{51.19}$$

where

$$\omega_{hk} = \Omega_{ij} \lambda_{h|}{}^{i} \lambda_{k|}{}^{j}. \tag{51.20}$$

If the curve whose tangent vector is $\xi_{k|}{}^{\beta}$ is a geodesic in V_n, ξ^α are the components of its principal normal. Comparing (51.19) with (32.16) for $p = 2$, we have again that ω_{kk} is the first curvature, and ω_{hk} for $h \neq k$ are the invariants of the second curvature vector. This second curvature is evidently the generalization of the geodesic torsion of the curve in V_n of direction $\xi_{k|}{}^{\alpha}$*. Hence in the case $m > n+1$, if at any point P in V_n we take the flat 3-space determined by two directions $\xi_{h|}{}^{\alpha}$ and $\xi_{k|}{}^{\alpha}$ in V_n and by a normal $\xi_{\sigma|}{}^{\alpha}$ to V_n, the invariants $\omega_{\sigma|hh}$ and $\omega_{\sigma|kk}$ are the normal curvatures of the curves of direction $\xi_{h|}{}^{\alpha}$ and $\xi_{k|}{}^{\alpha}$ respectively, and $\omega_{\sigma|hk}$ their geodesic torsions to within sign at most.

Since the left-hand member of (51.12) and the second term of the right-hand member do not involve normal directions, it follows that the value of the first term on the right is independent of the choice of the $m-n$ mutually orthogonal vectors normal to V_n.

For $r = p$, $s = q$ equations (51.12) become

$$(51.21)\qquad \begin{aligned} R_{ijkl}\,\lambda_{p|}{}^{i}\,\lambda_{q|}{}^{j}\,\lambda_{p|}{}^{k}\,\lambda_{q|}{}^{l} &= \sum_{\sigma} e_{\sigma}\,(\omega_{\sigma|pp}\,\omega_{\sigma|qq} - \omega^{2}_{\sigma|pq}) \\ &\quad + \overline{R}_{\alpha\beta\gamma\delta}\,\xi_{p|}{}^{\alpha}\,\xi_{q|}{}^{\beta}\,\xi_{p|}{}^{\gamma}\,\xi_{q|}{}^{\delta}. \end{aligned}$$

If $m = n+1$ and the V_{n+1} is an S_{n+1}, we have

$$R_{ijkl}\,\lambda_{p|}{}^{i}\,\lambda_{q|}{}^{j}\,\lambda_{p|}{}^{k}\,\lambda_{q|}{}^{l} = e\,(\omega_{pp}\,\omega_{qq} - \omega^{2}_{pq}).$$

Hence each of the terms $e_{\sigma}(\omega_{\sigma|pp}\,\omega_{\sigma|qq} - \omega^{2}_{\sigma|pq})$ in (51.21) multiplied by $e_p e_q$ may be interpreted as the Riemannian curvature at a point P for the orientation O_{pq} determined by $\lambda_{p|}{}^{i}$ and $\lambda_{q|}{}^{i}$ in the flat 3-space defined by these two vectors and the direction $\xi_{\sigma|}{}^{\alpha}$ at P. Accordingly Ricci† calls the first term in the right-hand member of (51.21) the *relative curvature* of O_{pq} and equation (51.21) may be interpreted as follows:

The Riemannian curvature for an orientation in V_n is the sum of the relative curvature and the Riemannian curvature of the orientation in V_m.

By means of (30.6) equations (51.21) are expressible in the form.

$$(51.22)\qquad \gamma_{pqpq} = \sum_{\sigma} e_{\sigma}\,(\omega_{\sigma|pp}\,\omega_{\sigma|qq} - \omega^{2}_{\sigma|pq}) + \overline{\gamma}_{pqpq},$$

and the preceding theorem gives the interpretation of these invariants

---

* 1909, 1, p. 138.

† 1902, 2, p. 361.

If we multiply equations (51.21) by e_p and sum for p from $1, \cdots, n$, we have in consequence of (29.5)

$$-R_{jl}\lambda_{q|}{}^j\lambda_{q|}{}^l = \sum_{\sigma, p} e_p e_\sigma(\omega_{\sigma|pp}\omega_{\sigma|qq} - \omega^2_{\sigma|pq}) + \overline{R}_{\alpha\beta\gamma\delta}\sum_p e_p\xi_{p|}{}^\alpha\xi_{q|}{}^\beta\xi_{p|}{}^\gamma\xi_{q|}{}^\delta.$$

Multiplying by e_q and summing for q, we have

$$\begin{aligned} -R &= \sum_{\sigma, q} e_p e_q e_\sigma(\omega_{\sigma|pp}\omega_{\sigma|qq} - \omega^2_{\sigma|pq}) \\ &\quad + \overline{R}_{\alpha\beta\gamma\delta}\sum_{p,q} e_p e_q \xi_{p|}{}^\alpha \xi_{q|}{}^\beta \xi_{p|}{}^\gamma \xi_{q|}{}^\delta, \end{aligned} \tag{51.23}$$

or

$$-R = \sum_{\sigma, p, q} e_p e_q e_\sigma(\omega_{\sigma|pp}\omega_{\sigma|qq} - \omega^2_{\sigma|pq}) + \sum_{p,q} e_p e_q \overline{\gamma}_{pqpq}. \tag{51.24}$$

Exercises.

1. A necessary and sufficient condition that the principal normals of a curve in a V_n and for an enveloping V_m coincide is that

$$\Omega_{\sigma|ij}\frac{dx^i}{ds}\frac{dx^j}{ds} = 0 \qquad (\sigma = n+1, \cdots, m).$$

2. If the functions (42.3) defining a V_n in a V_m satisfy the equations

$$\frac{\partial^2 y^\alpha}{\partial x^{i^2}} - \begin{Bmatrix} j \\ i\,i \end{Bmatrix}_g \frac{\partial y^\alpha}{\partial x^j} + \begin{Bmatrix} \alpha \\ \beta\,\gamma \end{Bmatrix}_a \frac{\partial y^\beta}{\partial x^i}\frac{\partial y^\gamma}{\partial x^i} = 0 \qquad \begin{pmatrix} \alpha, \beta, \gamma = 1, \cdots, m; \\ i, j = 1, \cdots, n \end{pmatrix},$$

the parametric curves in V_n are asymptotic lines.

3. If the functions (42.3) defining a V_n in a V_m satisfy the equations

$$\frac{\partial^2 y^\alpha}{\partial x^i \partial x^j} - \begin{Bmatrix} k \\ i\,j \end{Bmatrix}_g \frac{\partial y^\alpha}{\partial x^k} + \begin{Bmatrix} \alpha \\ \beta\,\gamma \end{Bmatrix}_a \frac{\partial y^\beta}{\partial x^i}\frac{\partial y^\gamma}{\partial x^j} = 0 \qquad \begin{pmatrix} \alpha, \beta, \gamma = 1, \cdots, m; \\ i, j, k = 1, \cdots, n;\ i \neq j \end{pmatrix},$$

the directions of any two parametric lines at a point in V_n are conjugate.

4. If the functions $y^\alpha = f^\alpha(x^1, x^2)$ defining a V_2 in a V_m satisfy the equations

$$\frac{\partial^2 y^\alpha}{\partial x^1 \partial x^2} + \begin{Bmatrix} \alpha \\ \beta\,\gamma \end{Bmatrix}_a \frac{\partial y^\beta}{\partial x^1}\frac{\partial y^\gamma}{\partial x^2} = 0,$$

the parametric curves in V_2 form a conjugate system of lines, such that the tangents to the curves of either family where they meet a curve of the other family are parallel in V_2 with respect to the latter curve (cf. Ex. 3 and 4, p. 79); these are a generalization of surfaces of translation in euclidean 3-space.

Bompiani, 1919, 2, p. 841.

5. If a V_2 in a V_m admits a conjugate system of lines, the osculating geodesic variety (§ 49) of V_2 is at most of order four.

6. If a V_2 in a V_m admits two families of asymptotic lines, the osculating geodesic variety of V_2 is at most of order three.

7. Show directly by means of (47.18) and (51.5) that the first term of the right-hand member of (51.12) is independent of the choice of the $m-n$ mutually orthogonal congruences normal to V_n.

52. Minimal varieties. Consider any V_m and a V_n immersed in it, defined by the equations

$$y^\alpha = \varphi^\alpha(x^1, \cdots, x^n). \tag{52.1}$$

Let V_{n-1} be a given closed sub-space of V_n bounding a region R_n of the latter and consider the integral

$$I = \int_{R_n} L\, dx^1\, dx^2 \cdots dx^n \tag{52.2}$$

extended over R_n, where L is a function of the y's and their first derivatives $y^\alpha_{,i}$.*

Let $\omega^\alpha(x^1, \cdots, x^n)$ be a set of arbitrary functions such that

$$\omega^\alpha = 0 \quad \text{for } V_{n-1}. \tag{52.3}$$

Then

$$\bar{y}^\alpha = \varphi^\alpha + \varepsilon\,\omega^\alpha, \tag{52.4}$$

where ε is an infinitesimal, define another variety $\bar{V}_n$ containing the given V_{n-1} and nearby V_n. Substituting these expressions in the function L in (52.2) and expanding by Taylor's theorem, we have for the corresponding integral

$$\bar{I} = I + \varepsilon \int_{R_n} \left(\omega^\alpha \frac{\partial L}{\partial y^\alpha} + \omega^\alpha_{,i} \frac{\partial L}{\partial y^\alpha_{,i}}\right) dx^1\, dx^2 \cdots dx^n + \varrho,$$

where ϱ involves terms of the second and higher orders in ε. If we write

* It is understood that L and its first and second derivatives with respect to the arguments are continuous in the domain and on the boundary.

(52.5) $$\delta I = \varepsilon \int_{R_n} \left(\omega^\alpha \frac{\partial L}{\partial y^\alpha} + \omega^\alpha{}_{,i} \frac{\partial L}{\partial y^\alpha{}_{,i}} \right) dx^1 \, dx^2 \cdots dx^n,$$

and integrate the second term of (52.5) by parts, we have, in consequence of (52.3),

$$\delta I = \varepsilon \int_{R_n} \omega^\alpha \left[\frac{\partial L}{\partial y^\alpha} - \frac{\partial}{\partial x^i} \left(\frac{\partial L}{\partial y^\alpha{}_{,i}} \right) \right] dx^1 \cdots dx^n.$$

In order that the integral I be stationary, that is, that $\delta I = 0$, for every set of functions ω^α satisfying (52.3), it is necessary and sufficient that L be a function such that

(52.6) $$\frac{\partial}{\partial x^i} \left(\frac{\partial L}{\partial y^\alpha{}_{,i}} \right) - \frac{\partial L}{\partial y^\alpha} = 0.$$

These are the *generalized equations of Euler* (cf. § 17).

The element of area of a surface in euclidean 3-space is $\sqrt{EG - F^2}\, du\, dv$* in terms of the customary notation. Generalizing this expression to a V_n in a V_m defined by equations of the form (52.1), we have $\sqrt{g}\, dx^1 dx^2 \cdots dx^n$ as the *element of volume* of V_n. If we consider the region R_n of V_n bounded by a closed V_{n-1}, its volume is defined by the integral

(52.7) $$I = \int_{R_n} \sqrt{g}\, dx^1 \, dx^2 \cdots dx^n.$$

Generalizing the definition of minimal surfaces,† we say that V_n is a *minimal* variety in V_m, if for a given V_{n-1} the integral I is stationary.‡

In order to determine the characteristic property of minimal varieties in terms of the functions defined in § 47, we consider the equations

$$\frac{\partial}{\partial x^i} \left(\frac{\partial \sqrt{g}}{\partial y^\alpha{}_{,i}} \right) - \frac{\partial \sqrt{g}}{\partial y^\alpha} = 0.$$

From

$$g = |g_{ij}|, \qquad g_{ij} = a_{\alpha\beta}\, y^\alpha{}_{,i}\, y^\beta{}_{,j},$$

---

* 1909, 1, p. 75.

† 1909, 1, p. 251.

‡ This generalized problem was considered by *Lipschitz*, 1874, 1; in this paper he obtained equations (52.6).

we have

$$\frac{\partial \sqrt{g}}{\partial y^{\alpha}_{,i}} = \frac{1}{2\sqrt{g}} \frac{\partial g}{\partial g_{jk}} \frac{\partial g_{jk}}{\partial y^{\alpha}_{,i}} = \sqrt{g}\, g^{ij} a_{\alpha\beta} y^{\beta}_{,j}.$$

Making use of equations of the form (7.6) and (7.9) for the g's and (7.4) for the a's, we obtain

$$\frac{\partial}{\partial x^i}\left(\frac{\partial \sqrt{g}}{\partial y^{\alpha}_{,i}}\right) = \sqrt{g}\, g^{ij} \left[a_{\alpha\beta} y^{\beta}_{,ij} + ([\alpha\gamma, \beta]_a + [\beta\gamma, \alpha]_a) y^{\gamma}_{,i} y^{\beta}_{,j}\right].$$

Also we have

$$\frac{\partial \sqrt{g}}{\partial y^{\alpha}} = \frac{1}{2\sqrt{g}} \frac{\partial g}{\partial g_{jk}} \frac{\partial g_{jk}}{\partial y^{\alpha}} = \sqrt{g}\, g^{ij} [\alpha\gamma, \beta]_a y^{\gamma}_{,i} y^{\beta}_{,j}.$$

From these expressions and (47.4), we have

$$\begin{aligned}\frac{\partial}{\partial x^i}\left(\frac{\partial \sqrt{g}}{\partial y^{\alpha}_{,i}}\right) - \frac{\partial \sqrt{g}}{\partial y^{\alpha}} &= \sqrt{g}\, g^{ij} a_{\alpha\beta}\left(y^{\beta}_{,ij} + \left\{ {\beta \atop \mu\nu} \right\}_a y^{\mu}_{,i} y^{\nu}_{,j}\right) \\ &= \sqrt{g}\, a_{\alpha\beta} \sum_{\sigma} e_{\sigma} \Omega_{\sigma|ij} g^{ij} \xi_{\sigma|}^{\beta}.\end{aligned}$$

Since these expressions must vanish for all values of α, we have $\sum_{\sigma} e_{\sigma} \Omega_{\sigma|ij} g^{ij} \xi_{\sigma|}^{\beta} = 0$. Hence we have:

*A necessary and sufficient condition that a V_n be a minimal variety for a V_m is that its mean curvature normal vanish.**

From the results of § 50 we have also:

A necessary and sufficient condition that a V_n be a minimal variety for an enveloping V_m is that its mean curvature in every normal direction be zero.

This is an evident generalization of a characteristic property of minimal surfaces.†

Suppose that a V_{n+1} admits ∞^1 minimal hypersurfaces. If these be taken for $y^{n+1} =$ const. and their orthogonal trajectories for the curves of parameter y^{n+1}, we have

$$a_{n+1\,i} = 0, \quad a_{n+1\,n+1} = eH^2_{n+1}; \quad \xi^i = 0, \quad \xi^{n+1} = \frac{1}{H_{n+1}}$$
$$(i = 1, \cdots, n).$$

---

* *Lipschitz*, 1874, 1, p. 31.

† 1909, 1, p. 251.

From (43.5) it follows that

$$\Omega_{ij} = -\frac{1}{2H_{n+1}} \frac{\partial a_{ij}}{\partial y^{n+1}}.$$

In this case $g_{ij} = a_{ij}$ for any $y^{n+1} =$ const. and $g = |a_{ij}|$ for $i, j = 1, \cdots, n$. Consequently

$$M = g^{ij}\Omega_{ij} = -\frac{1}{2gH_{n+1}} \frac{\partial g}{\partial y^{n+1}}.$$

Hence we have:

*A necessary and sufficient condition that an infinity of hypersurfaces of a space be minimal is that their orthogonal trajectories determine a correspondence between them which preserves volume.**

53. Hypersurfaces with indeterminate lines of curvature. A hypersurface V_n with indeterminate lines of curvature is a generalization of a plane or sphere in euclidean 3-space.† As in the latter case, when the lines of curvature at a point P are indeterminate, P is called an *umbilical point.*

From (45.3) it follows that a necessary condition that every point of a V_n immersed in a V_{n+1} be an umbilical point is that $\Omega_{ij} = \varrho g_{ij}$, where ϱ is an invariant. If we multiply this equation by g^{ij} and sum for i and j, we obtain

$$\Omega \equiv \Omega_{ij} g^{ij} = n\varrho, \tag{53.1}$$

and consequently the condition is

$$\Omega_{ij} = \frac{\Omega}{n} g_{ij}. \tag{53.2}$$

In this case we have from (51.20)

$$\omega_{hh} = \frac{1}{n} e_h \Omega, \qquad \omega_{hk} = 0 \qquad (h \neq k). \tag{53.3}$$

The case when $\Omega = 0$, that is, when $1/R = 0$, will be treated in § 54.

* For a V_3 this result is due to *Bianchi*, 1903, 1, p. 578; for any V_n to *Bompiani*, 1921, 5, p. 1141.

† Cf. 1909, 1, p. 116.

When $m = n+1$,* equations (51.12) and (51.13) become

$$(53.4)\quad R_{ijkl}\lambda_{p|}^{\ i}\lambda_{q|}^{\ j}\lambda_{r|}^{\ k}\lambda_{s|}^{\ l} = e(\omega_{pr}\,\omega_{qs} - \omega_{ps}\omega_{qr}) + \overline{R}_{\alpha\beta\gamma\delta}\,\xi_{p|}^{\ \alpha}\,\xi_{q|}^{\ \beta}\,\xi_{r|}^{\ \gamma}\,\xi_{s|}^{\ \delta}$$

and

$$(53.5)\quad \begin{aligned} &\frac{\partial\,\omega_{pq}}{\partial\,s_r} - \frac{\partial\,\omega_{pr}}{\partial\,s_q} + \sum_h e_h\,[\omega_{hq}\,\gamma_{hpr} - \omega_{hr}\,\gamma_{hpq} + \omega_{hp}\,(\gamma_{hqr} - \gamma_{hrq})] \\ &\qquad = \overline{R}_{\lambda\beta\mu\nu}\,\xi_{p|}^{\ \lambda}\,\xi_{q|}^{\ \mu}\,\xi_{r|}^{\ \nu}\,\xi^{\beta}.\dagger \end{aligned}$$

For the values (53.3) we have from (53.5)

$$(53.6)\qquad \overline{R}_{\lambda\beta\mu\nu}\,\xi_{p|}^{\ \lambda}\,\xi_{q|}^{\ \mu}\,\xi_{r|}^{\ \nu}\,\xi^{\beta} = 0 \qquad (p,\,q,\,r \neq),$$

and for $p = r \neq q$

$$(53.7)\qquad \left(\frac{1}{n}\,e_p\,\frac{\partial\,\Omega}{\partial\,y^{\mu}} + \overline{R}_{\lambda\beta\mu\nu}\,\xi_{p|}^{\ \lambda}\,\xi_{p|}^{\ \nu}\,\xi^{\beta}\right)\xi_{q|}^{\ \mu} = 0.$$

When $\Omega =$ const., equations (53.7) become

$$(53.8)\qquad \overline{R}_{\lambda\beta\mu\nu}\,\xi_{p|}^{\ \lambda}\,\xi_{p|}^{\ \nu}\,\xi^{\beta}\,\xi_{q|}^{\ \mu} = 0 \qquad (p \neq q).$$

This equation is satisfied identically, if we take $p = q$; also if we replace $\xi_{p|}^{\ \lambda}$ by ξ^{λ}. Hence if (53.8) be multiplied by e_p and the resulting equation be summed for p, we have in consequence of (29.5)

$$(53.9)\qquad \overline{R}_{\lambda\beta\mu\nu}\,a^{\lambda\nu}\,\xi^{\beta}\,\xi_{q|}^{\ \mu} = \overline{R}_{\beta\mu}\,\xi^{\beta}\,\xi_{q|}^{\ \mu} = 0.$$

If we put $\overline{R}_{\beta\mu}\,\xi^{\beta}\,\xi^{\mu} = e\varrho$, where e is defined by (43.3), it follows from (53.9) and $a_{\beta\mu}\,\xi_{q|}^{\ \mu}\,\xi^{\beta} = 0$ that

$$(\overline{R}_{\beta\mu} - \varrho\,a_{\beta\mu})\,\xi^{\beta}\,\xi_{q|}^{\ \mu} = 0, \qquad (\overline{R}_{\beta\mu} - \varrho\,a_{\beta\mu})\,\xi^{\beta}\,\xi^{\mu} = 0,$$

and since the $n+1$ vector-fields $\xi_{q|}^{\ \alpha}$ and ξ^{α} are independent, we have

$$(\overline{R}_{\beta\mu} - \varrho\,a_{\beta\mu})\,\xi^{\beta} = 0.$$

---

* From (47.8) it follows that the functions μ are zero in this case.

† In this section and the next Greek indices take the values $1, \cdots, n+1$ and Latin $1, \cdots, n$.

Hence (§ 34) we have:

*If a V_{n+1} admits a hypersurface V_n with indeterminate lines of curvature and Ω is constant, the normal to V_n at a point is a Ricci principal direction for V_{n+1} at points of V_n.**

We seek the canonical form of the fundamental tensor of V_{n+1}, in order that there may exist ∞^1 hypersurfaces with indeterminate lines of curvature. To this end we choose the coördinate system so that $y^{n+1} =$ const. are the hypersurfaces and we choose their orthogonal trajectories for the curves of parameter y^{n+1}. Then we have

$$(53.10)\qquad a_{n+1\,i} = 0, \qquad a_{n+1\,n+1} = e_{n+1} H_{n+1}^2 \qquad (i = 1, \cdots, n).$$

The contravariant components of the normal vector are

$$(53.11)\qquad \xi^i = 0, \qquad \xi^{n+1} = \frac{1}{H_{n+1}}.$$

At points of any hypersurface $y^{n+1} =$ const., $y^i = x^i$ and $a_{ij} = g_{ij}$ for $i, j = 1, \cdots, n$. Consequently equations (43.10) become in this case

$$(53.12)\qquad \xi^\beta{}_{,j} = -\frac{\Omega}{n} g_{lj}\, g^{lm}\, \delta^\beta_m - \left\{ {\beta \atop j\, n+1} \right\}_a \frac{1}{H_{n+1}}.$$

For $\beta = n+1$ these equations are satisfied identically. For $\beta = 1, \cdots, n$ we obtain

$$\frac{\Omega}{n} \delta^i_j + \left\{ {i \atop j\, n+1} \right\}_a \frac{1}{H_{n+1}} = 0.$$

From these equations and (53.10) we have

$$[j\, n+1, k]_a = a_{\alpha k} \left\{ {\alpha \atop j\, n+1} \right\}_a = -\frac{\Omega}{n} a_{jk} H_{n+1},$$

and consequently

$$(53.13)\qquad \frac{\partial \log \sqrt{a_{jk}}}{\partial y^{n+1}} = -\frac{\Omega}{n} H_{n+1}.$$

* For $\Omega = 0$ this theorem is due to *Ricci*, 1904, 2, p. 1239; for $\Omega \neq 0$ to *Struik*, 1922, 8, p. 143.

Conversely, when (53.10) and (53.13) are satisfied, equations (43.10) for the values (53.11) lead to (53.2).

When $\Omega \neq 0$, it follows from (53.13) that the ratio of any two of the functions a_{jk} for $j, k = 1, \cdots, n$ must be independent y^{n+1}. When $\Omega = 0$, the functions a_{jk} are independent of y^{n+1}. Hence we have:

A necessary and sufficient condition that a V_{n+1} admit a family of hypersurfaces with indeterminate lines of curvature is that its fundamental form be reducible to

$$\varphi = A a_{ij} dy^i dy^j + B(dy^{n+1})^2 \qquad (i, j = 1, \cdots, n), \tag{53.14}$$

where a_{ij} are the functions of $y^1, \cdots, y^n$, and A and B are any functions of $y^1 \cdots, y^{n+1}$; according as A involves y^{n+1} or not, Ω is different from or equal to zero.

As a corollary we have:

If a space admits a family of hypersurfaces with indeterminate lines of curvature, their orthogonal trajectories determine a conformal correspondence between them.

If a V_{n+1} is conformal to an S_{n+1}, and the coördinates y^α are chosen so that

$$\varphi = \varrho \sum_\alpha e_\alpha (dy^\alpha)^2, \tag{53.15}$$

the conditions of the above theorems are satisfied by any of the coördinate hypersurfaces. If ϱ does not involve y^{n+1} then from (53.13) it is seen that $\Omega = 0$ for the hypersurfaces $y^{n+1} = \text{const}$. Since any S_n in the S_{n+1} to which V_{n+1} is conformal can be chosen as a hypersurface $y^{n+1} = \text{const.}$, we have:

If a V_{n+1} is conformal to an S_{n+1}, the hypersurface of V_{n+1} corresponding to any S_n in the S_{n+1} has indeterminate lines of curvature.

Another way of stating this result is that in such a V_{n+1} at each point and in each direction there is a V_n with indeterminate lines of curvature. In order that a V_{n+1} for $n > 2$ may possess the latter property, it is necessary that equations (53.6) be satisfied by every orthogonal ennuple in V_{n+1}. In § 37 we saw that in this case V_{n+1} must be conformal to an S_{n+1}. Hence we have the theorem of Schouten*:

* 1921, 2, p. 86.

A necessary and sufficient condition that at each point and in each direction of a V_{n+1} for $n > 2$ there is a hypersurface with indeterminate lines of curvature is that the V_{n+1} be conformal to an S_{n+1}.

54. Totally geodesic varieties in a space. If all the geodesics of a V_n are geodesics of an enveloping V_{n+1}, the former is called a *totally geodesic hypersurface* of V_{n+1}. These hypersurfaces are an evident generalization of the planes of euclidean 3-space.

From (44.5) we have that a necessary and sufficient condition that a V_n be a totally geodesic hypersurface is that $1/R = 0$ and from (44.3) that

$$\Omega_{ij} = 0 \qquad (i, j = 1, \cdots, n). \tag{54.1}$$

Then from § 45 we have:

*The lines of curvature of a totally geodesic hypersurface are indeterminate.**

Since Ω, as defined by (53.1), is zero, the first theorem of § 53 applies to totally geodesic hypersurfaces.

From (54.1) and (51.20) we have

$$\omega_{hk} = 0 \qquad (h, k = 1, \ldots, n). \tag{54.2}$$

In consequence of (51.17) we have from (51.15) See App. 19

$$\xi^{\alpha}{}_{,\beta} = 0. \tag{54.3}$$

Hence we have:

The normals to a totally geodesic hypersurface are parallel in the enveloping space.

We shall not write down the systems of differential equations determining a space admitting a totally geodesic hypersurface,† but will consider the case when there are ∞^1 such hypersurfaces. From the second theorem of § 53 we have:

A necessary and sufficient condition that a V_{n+1} admit a family of totally geodesic hypersurfaces is that its fundamental form be reducible to

$$\varphi = a_{ij}\, dy^i\, dy^j + B (dy^{n+1})^2 \qquad (i, j = 1, \cdots, n), \tag{54.4}$$

* Cf. *Ricci*, 1903, 2, p. 412.

† Cf. *Ricci*, 1903, 2, p. 414.

*where a_{ij} are independent of y^{n+1} and B is any function of the y's.**

As a corollary we have:

If a space admits ∞^1 totally geodesic hypersurfaces, their orthogonal trajectories determine an isometric correspondence between them.†

When all the geodesics of a V_n in a V_m for $m > n+1$ are geodesics of V_m, we say that V_n is *totally geodesic*. Since the matrix $\|\xi_{\sigma|}{}^{\alpha}\|$ is of rank $m-n$, it follows from (48.5) that

A necessary and sufficient condition that a V_n immersed in a V_m be totally geodesic is that

(54.5) $$\Omega_{\sigma|ij} = 0 \quad (\sigma = n+1, \cdots, m;\ i, j = 1, \cdots, n).$$

From (48.3) we have:

If any vector in a totally geodesic sub-space of a V_m is transported parallel to itself along a curve, it moves parallel to itself also in V_m.

Also from the results of § 52 we have:

A totally geodesic sub-space of a V_m is a minimal variety of V_m.

Exercises.

1. A minimal surface in any V_m is characterized by the property that its lines of length zero form a conjugate system, and it is a surface of translation in the sense of Ex. 4, p. 175. *Bompiani*, 1919, 2, p. 841

2. The equations

$$\begin{aligned} x &= \varphi - u\varphi' + \psi' + \varphi_0 - u_0\varphi_0' + \psi_0', \\ iy &= \varphi - u\varphi' - \psi' - \varphi_0 + u_0\varphi_0' + \psi_0', \\ z &= \psi - u\psi' - \varphi' + \psi_0 - u_0\psi_0' - \varphi_0', \\ it &= -\psi + u\psi' - \varphi' + \psi_0 - u_0\psi_0' + \varphi_0', \end{aligned}$$

where φ and ψ are arbitrary functions of u, and φ_0 and ψ_0 of u_0, and where primes denote differentiation with respect to the argument, define a minimal surface in euclidean 4-space. *Eisenhart*, 1912, 1, p. 224.

3. If $f(x+iy)$ is an analytic function and

$$f(x+iy) = u(x, y) + iv(x, y),$$

the equations

$$x = x, \quad y = y, \quad z = u(x, y), \quad t = v(x, y)$$

define a minimal surface in euclidean 4-space. *Kommerell*, 1905, 2, p. 586.

---

* This theorem for $n = 2$ is due to *Hadamard*, 1901, 2, p. 40; for any n *Ricci*, 1903, 2, p. 412, derived the result for the case $a_{ij} = 0$ $(i \neq j)$; cf. also *Bompiani*, 1924, 5, p. 122.

† Cf. *Bompiani*, 1924, 5, p. 122.

4. A necessary and sufficient condition that all the lines of curvature of a V_n be indeterminate at a point is that at the point

$$\Omega_{\sigma|ij} = \varrho_\sigma g_{ij} \qquad (\sigma = n+1, \cdots, m).$$

Such a point is called an *umbilical point.* Show that at an umbilical point the osculating geodesic variety is determined by V_n and the mean curvature normal.
Struik, 1922, 8, p. 106.

5. When the lines of curvature of a hypersurface of a space of constant curvature K_{n+1} are indeterminate, the hypersurface has constant curvature K_n, and

$$K_n = e\frac{\Omega^2}{n^2} + K_{n+1}.$$

6. A necessary and sufficient condition that a V_{n+1} for $n > 2$ admit at each point and in each direction a V_n with indeterminate lines of curvature and that Ω be the same constant for all the V_n's is that V_{n+1} have constant Riemannian curvature.
Schouten, 1924, 1, p. 181.

7. A necessary and sufficient condition that for an orthogonal ennuple $\lambda_{h|}{}^i$ the congruence $\lambda_{n|}{}^i$ be normal to a family of hypersurfaces with indeterminate lines of curvature is that

$$\gamma_{nhk} = 0 \qquad (h, k = 1, \cdots, n-1,;\ h \neq k),$$

$$e_1\gamma_{n11} = e_2\gamma_{n22} = \cdots = e_{n-1}\gamma_{n\,n-1\,n-1}.$$

8. When a V_n admits p independent fields of parallel vectors, the congruence of curves of each field are the orthogonal trajectories of a family of totally geodesic hypersurfaces.

9. A necessary and sufficient condition that a V_n be totally geodesic in a V_m is that

$$\frac{\partial^2 y^\alpha}{\partial x^i \partial x^j} - \left\{ {k \atop ij} \right\}_g \frac{\partial y^\alpha}{\partial x^k} + \left\{ {\alpha \atop \beta\gamma} \right\}_a \frac{\partial y^\beta}{\partial x^i}\frac{\partial y^\gamma}{\partial x^j} = 0 \qquad \left({\alpha, \beta, \gamma = 1, \cdots, m; \atop i, j, k = 1, \cdots, n} \right).$$

10. When a V_m admits ∞^1 totally geodesic sub-spaces V_n, they determine a V_{n+1} of which they are totally geodesic hypersurfaces.
Bompiani, 1924, 5, p. 123.

11. When a V_m admits ∞^1 totally geodesic sub-spaces V_n, the tangents to their orthogonal trajectories at points of the same V_n are parallel with respect to V_m.

12. Show by means of (47.4) that a necessary and sufficient condition that the sub-spaces $y^\sigma =$ const. for $\sigma = n+1, \cdots, m$ of a V_m with the fundamental form (42.2) be totally geodesic is that

$$\left\{ {\sigma \atop ij} \right\}_a = 0, \qquad \overline{\left\{ {k \atop ij} \right\}} = \left\{ {k \atop ij} \right\}_a \qquad \left({i, j, k = 1, \cdots, n; \atop \sigma = n+1, \cdots, m} \right),$$

where the Chrisioffel symbols $\overline{\left\{ {k \atop ij} \right\}}$ are formed with respect to $a_{ij}dy^i dy^j$ $(i, j = 1, \cdots, n)$.

13. When the fundamental form of a V_m is

$$\varphi = a_{ij}\, dy^i\, dy^j + a_{\sigma\tau}\, dy^\sigma\, dy^\tau \qquad \begin{pmatrix} i, j = 1, \cdots, n; \\ \sigma, \tau = n+1, \cdots, m \end{pmatrix},$$

where the functions a_{ij} are independent of $y^{n+1}, \cdots, y^m$, the sub-spaces $y^\sigma =$ const. are totally geodesic in V_m. *Bompiani*, 1924, 5, p. 124.

14. When two totally geodesic sub-spaces of a V_m intersect, the variety of intersection is totally geodesic in V_m. *Struik*, 1922, 8, p. 97.

15. If the order of the osculating geodesic varieties at points of a V_n in a V_m is less than m (§ 49), then for each value $q \leqq m - \tau$ there are an infinity of sub-spaces V_{n+q} with respect to which V_n is totally geodesic.
Struik, 1922, 8, p. 113.

16. If a V_n lies in a V_m for $m > n+1$, for each value of $q \leqq m - n - 1$ there are an infinity of sub-spaces V_{n+q} of V_m in which the curves of a given congruence in V_n are geodesics. *Struik*, 1922, 8, p. 113.

17. If a V_n lies in a V_m for $m > n+1$, for each value of $q \leqq m - n - 1$ there is an infinity of sub-spaces V_{n+q} with respect to which V_n is a minimal variety. *Struik*, 1922, 8, p. 114.

CHAPTER V

Sub-spaces of a flat space

55. The class of a space V_n. In § 10 it was shown that a necessary and sufficient condition that there exist for a space V_n a coördinate system in terms of which the components of the fundamental tensor are constants is that all the components of the Riemann tensor in any coördinate system be zero. We have called such a space a flat space and have denoted by S_n a flat space of n dimensions (§ 26). For an S_m there exist real coördinates z^α in terms of which the fundamental form is

$$\varphi = \sum_\alpha c_\alpha (dz^\alpha)^2 \qquad (\alpha = 1, \cdots, m), \tag{55.1}$$

where the c's are plus or minus one according to the character of the space. There are other real coördinate systems in terms of which φ assumes the form (55.1), but the number of positive c's and of negative c's is the same for all of these systems. In particular, when all of the c's are plus one, S_m is a euclidean space of m dimensions and the z's are cartesian coördinates. When φ for any S_m assumes the form (55.1), we call the coördinates *cartesian*.

In order that a space V_n with the fundamental form

$$\varphi = g_{ij}\, dx^i\, dx^j \tag{55.2}$$

be a real sub-space of S_m, it is necessary and sufficient that the system of equations (cf. § 16)

$$\sum_\alpha c_\alpha \frac{\partial z^\alpha}{\partial x^i} \frac{\partial z^\alpha}{\partial x^j} = g_{ij} \tag{55.3}$$

admit m independent real solutions

$$z^\alpha = f^\alpha (x^1, \cdots, x^n) \qquad (\alpha = 1, \cdots, m). \tag{55.4}$$

The signs of the c's in (55.3) depend upon the character of the form (55.2). In fact, from the theory of matrices* and (55.3), it follows that the determinant $g = |g_{ij}|$ is equal to the sum of terms each of which is the square of a determinant of order n of the matrix $\left\| \frac{\partial z^\alpha}{\partial x^i} \right\|$ with a plus or minus sign according as the corresponding determinant of the nth order of the determinant of (55.1) is plus or minus one. Consequently, if g is negative, all of the c's cannot be positive, that is, V_n cannot be immersed in a real euclidean space.

The coördinates x^i can be chosen so that at any point of V_n the form (55.2) involves only squared terms with plus and minus signs. Since n of the z's can be identified with x's at the point, we have that (55.1) at the point must have at least as many positive and as many negative c's as there are positive and negative terms in the reduced form of (55.2) at the point. Thus, for example, one of Einstein's postulates concerning the space-time continuum V_4 of general relativity is that at each point the fundamental form is reducible to $-(dx^1)^2 - (dx^2)^2 - (dx^3)^2 + (dx^4)^2$. Consequently, for a flat space in which V_4 can be immersed one of the c's must be positive and three negative.

If (55.2) is a positive definite form and we take all the c's equal to $+1$ in (55.3), we have $n(n+1)/2$ equations for the determination of the z's. If we take $m = n(n+1)/2$, we have a system of equations which admits in general real solutions in accordance with the theory of partial differential equations. Thus a V_n with a positive definite form can be immersed in general in a euclidean space of $n(n+1)/2$ dimensions.** Similar results hold when (55.2) is not positive definite and the c's have been chosen in accordance with the preceding observations.

We have just seen that in general a V_n can be immersed in a flat space of $n(n+1)/2$ dimensions. However, it may be immersible in a flat space of a lower order. If the lowest order is $n+p$, we say that V_n is of *class* p.†

Consider, for example, the space-time continuum V_4 outside a symmetric mass m with the Schwarzschild form‡

* § 31.

† *Ricci*, 1898, 2, p. 75; also, *Struik*, 1922, 8, p. 99.

‡ Cf. Ex. 6, p. 93.

** Cf. *Janet*, 1926, 7 and *Cartan*, 1927, 7.

$$(55.5)\quad \varphi = \left(1-\frac{2m}{r}\right)dt^2 - \frac{1}{1-\frac{2m}{r}}dr^2 - r^2(d\theta^2+\sin^2\theta\, d\varphi^2),$$

where $r>2m$. If we put

$$z^1 = \sqrt{\frac{r-2m}{r}}\cos t, \qquad z^2 = \sqrt{\frac{r-2m}{r}}\sin t, \qquad z^3 = f(r),$$

$$z^4 = r\sin\theta\cos\varphi, \qquad z^5 = r\sin\theta\sin\varphi, \qquad z^6 = r\cos\theta,$$

where $f(r)$ is such that

$$\left(\frac{df}{dr}\right)^2 = \frac{1}{r-2m}\left(\frac{m^2}{r^3}+2m\right),$$

then (55.5) becomes

$$\varphi = (dz^1)^2 + (dz^2)^2 - (dz^3)^2 - (dz^4)^2 - (dz^5)^2 - (dz^6)^2.$$

Hence the given V_4 can be immersed in a flat space of six dimensions, and consequently its class is two or one. From the results at the end of § 59, it follows that $p=2$.*

56. A space V_n of class $p>1$. If V_n with the fundamental form (55.2) is of class $p(>1)$, the enveloping flat space S_{n+p} has the fundamental form (55.1) in which $\alpha = 1, \cdots, n+p$. Let $\eta_{\sigma|}{}^{\alpha}$ denote the components of p mutually orthogonal unit vectors normal to V_n; then we have

$$(56.1)\qquad \sum_\alpha c_\alpha (\eta_{\sigma|}{}^{\alpha})^2 = e_\sigma, \qquad \sum_\alpha c_\alpha \eta_{\sigma|}{}^{\alpha}\eta_{\tau|}{}^{\alpha} = 0$$
$$(\sigma, \tau = n+1, \cdots, n+p;\ \sigma \neq \tau).†$$

The equations for this case analogous to (47.4) and (47.9) are

$$(56.2)\qquad z^{\alpha}{}_{,ij} = \sum_\sigma e_\sigma b_{\sigma|ij}\eta_{\sigma|}{}^{\alpha}$$

and

$$(56.3)\qquad \eta_{\sigma|}{}^{\alpha}{}_{,j} = -b_{\sigma|lj}g^{lm}z^{\alpha}{}_{,m} + \sum_\tau e_\tau \nu_{\tau\sigma|j}\eta_{\tau|}{}^{\alpha}$$
$$(\sigma, \tau = n+1, \cdots, n+p),$$

---

* Cf. *Kasner*, 1921, 6, p. 130.

† In this and the next section, unless stated otherwise, Greek indices take the values $1, \cdots, n+p$ and Latin $1, \cdots, n$.

where for each σ and τ the quantities $\nu_{\tau\sigma|j}$ are the covariant components of a vector, subject to the conditions

$$\nu_{\tau\sigma|j} + \nu_{\sigma\tau|j} = 0, \qquad \nu_{\sigma\sigma|j} = 0. \tag{56.4}$$

The conditions of integrability of (56.2) are reducible to [cf. (47.11) and (47.12)]

$$R_{ijkl} = \sum_{\sigma} e_\sigma (b_{\sigma|ik}\, b_{\sigma|jl} - b_{\sigma|il}\, b_{\sigma|jk}) \tag{56.5}$$

and

$$b_{\sigma|ij,k} - b_{\sigma|ik,j} = \sum_{\tau} e_\tau (\nu_{\tau\sigma|k}\, b_{\tau|ij} - \nu_{\tau\sigma|j}\, b_{\tau|ik}), \tag{56.6}$$

since the components of the Riemann tensor for S_m are zero. By means of these equations the conditions of integrability of (56.3) reduce to

$$\begin{aligned} \nu_{\tau\sigma|j,k} - \nu_{\tau\sigma|k,j} + \sum_{\varrho}^{n+1,\ldots,n+p} e_\varrho (\nu_{\varrho\tau|j}\, \nu_{\varrho\sigma|k} - \nu_{\varrho\tau|k}\, \nu_{\varrho\sigma|j}) \\ + g^{lm} (b_{\tau|lj}\, b_{\sigma|mk} - b_{\tau|lk}\, b_{\sigma|mj}) = 0.* \end{aligned} \tag{56.7}$$

See App. 20 Conversely, if we have a symmetric tensor g_{ij}, p symmetric tensors $b_{\sigma|ij}$ and $p(p-1)/2$ vectors $\nu_{\sigma\tau|i} = (-\nu_{\tau\sigma|i})$ satisfying (56.5), (56.6) and (56.7), the conditions of integrability of (56.2), (56.3) and (56.4) are satisfied. If we put

$$\sum_{\alpha} c_\alpha \overset{\alpha}{z}_{,i} \overset{\alpha}{z}_{,j} - g_{ij} = A_{ij}, \qquad \sum_{\alpha} c_\alpha \overset{\alpha}{\eta}_{\sigma|} \overset{\alpha}{z}_{,i} = B_{\sigma|i},$$

$$\sum_{\alpha} c_\alpha \overset{\alpha}{\eta}_{\sigma|} \overset{\alpha}{\eta}_{\tau|} - e_{\sigma\tau} = C_{\sigma\tau}, \quad e_{\sigma\sigma} = e_\sigma, \quad e_{\sigma\tau} = 0 \quad (\sigma \neq \tau),$$

then A_{ij} are the components of a tensor, $B_{\sigma|i}$ of p vectors and $C_{\sigma\tau}$ are invariants in V_n. If we differentiate these equations with respect to x^k and make use of (56.2) and (56.3), we find $C_{\sigma\tau}$ constant and the first derivatives of A_{ij} and $B_{\sigma|i}$ equal to expressions linear and homogeneous in these functions. In like manner the derivatives of any order are linear functions of A_{ij} and $B_{\sigma|i}$ and of the derivatives of lower order. Hence if we choose a set of solutions of (56.2) and (56.3), whose initial values satisfy

* Cf. *Ricci*, 1898, 2, p. 90.

$$(56.8)\qquad \begin{aligned} &\sum_\alpha c_\alpha z^\alpha_{,i} z^\alpha_{,j} = g_{ij}, && \sum_\alpha c_\alpha \eta_{\sigma|}{}^\alpha z^\alpha_{,i} = 0, \\ &\sum_\alpha c_\alpha (\eta_{\sigma|}{}^\alpha)^2 = e_\sigma, && \sum_\alpha c_\alpha \eta_{\sigma|}{}^\alpha \eta_{\tau|}{}^\alpha = 0 \qquad (\sigma \neq \tau), \end{aligned}$$

these conditions will be satisfied by the functions for all values of the x's. Since there are $(n+p)(n+p+1)/2$ of these conditions on the $(n+p)^2$ functions $z^\alpha_{,i}$ and $\eta_{\sigma|}{}^\alpha$, the desired solution involves $(n+p)(n+p-1)/2$ arbitrary constants in addition to $n+p$ additive arbitrary constants, arising from the determination of the z's by the integrals

$$(56.9)\qquad z^\alpha = \int z^\alpha_{,i}\, dx^i.$$

These results obtain for an arbitrary choice of the c's in (56.8). These can be chosen so that for a domain of the x's the set of solutions are real. In fact, if the coördinates x^i are chosen so that at a given point P we have $g_{ij} = 0$ $(i \neq j)$, and we make the choice

$$c_i = g_{ii} \quad (i = 1, \cdots, n), \qquad c_\sigma = e_\sigma \quad (\sigma = n+1, \cdots, n+p),$$

where the e's appear in (56.5), (56.6) and (56.7), the conditions (56.8) are satisfied by the values

$$z^\alpha_{,i} = \delta^\alpha_i, \qquad \eta_{\sigma|}{}^\alpha = \delta^\alpha_\sigma$$

at P, and thus for a domain in the neighborhood of P we have real solutions, and consequently a real S_{n+p} enveloping the V_n with the fundamental tensor g_{ij}.

As previously seen, the desired type of solution of equations (56.2) and (56.3) involve $(n+p)(n+p+1)/2$ arbitrary constants. We give an interpretation of the significance of these constants by observing that, if z^α and $\eta_{\sigma|}{}^\alpha$ are a set of solutions, so also are

$$(56.10)\qquad \bar{z}^\alpha = a^\alpha{}_\beta z^\beta + b^\alpha,$$

$$(56.11)\qquad \bar{\eta}_{\sigma|}{}^\alpha = a^\alpha{}_\beta \eta_{\sigma|}{}^\beta \qquad \left(\begin{aligned} \alpha, \beta &= 1, \cdots, n+p; \\ \sigma &= n+1, \cdots, n+p \end{aligned}\right),$$

where the a's and b's are constants, and that the conditions (56.8) are satisfied, if the constants $a^\alpha{}_\beta$ satisfy

$$(56.12) \qquad \sum_\alpha c_\alpha (a^\alpha{}_\beta)^2 = c_\beta, \quad \sum_\alpha c_\alpha a^\alpha{}_\beta a^\alpha{}_\gamma = 0 \qquad (\beta \neq \gamma).$$

Because of these $(n+p)(n+p+1)/2$ conditions, $(n+p)(n+p-1)/2$ of the constants $a^\alpha{}_\beta$ and all of the b's are arbitrary. Hence the general solution is obtained from a particular solution by means of (56.10) and (56.11). From (56.10) and (56.12) it follows that

$$\varphi = \sum_\alpha c_\alpha (d\bar{z}^\alpha)^2 = \sum_\alpha c_\alpha (dz^\alpha)^2,$$

and consequently equations (56.10) and (56.12) define in cartesian coördinates the most general motion (§ 27) of the S_{n+p} into itself. Generalizing the ideas of motions in euclidean space, we say that the a's determine a *rotation* and the b's a *translation*. Thus we have that different sets of solutions of (56.2) and (56.3) define V_n's which are superposable by a motion in S_{n+p}. Hence the foregoing results may be formulated as follows:

In order that a symmetric tensor g_{ij}, p symmetric tensors $b_{\sigma|ij}$ and $p(p-1)/2$ vectors $\nu_{\sigma\tau|i}$ $(= -\nu_{\tau\sigma|i})$ for $i, j = 1, \cdots, n$, $\sigma, \tau = n+1, \cdots, n+p$ determine a V_n with g_{ij} as fundamental tensor immersed in a real S_{n+p}, it is necessary and sufficient that these quantities satisfy equations (56.5), (56.6) *and* (56.7); *the fundamental form of S_{n+p} is determined by the first of* (56.8), *and V_n is determined to within a motion in S_{n+p}.*

From the definition of the class of a V_n it follows that equations (55.3) admit solutions (55.4) when $\alpha = n+p$. Evidently the equations (55.3) for $\alpha = n+p+r$ admit solutions of the type (55.4), and if r of these solutions are not constants, the given V_n is a sub-space of an S_{n+p+r}. For the cases when $r = 0$ and $r \neq 0$ the geometric properties of V_n depending entirely upon its fundamental form, that is, the *intrinsic* properties, are the same. But this is not true for geometrical properties depending upon the enveloping space. We are familiar with this idea in the case of surfaces of euclidean 3-space and those of euclidean 4-space.

See App. 21

57. Evolutes of a V_n in an S_{n+p}. The coördinates of a point on the normal of components $\eta_{\sigma|}{}^\alpha$ to a V_n in an S_{n+p} are given by

$$(57.1) \qquad z_{\sigma|}{}^\alpha = z^\alpha + \varrho \eta_{\sigma|}{}^\alpha.$$

In order that the point with these coördinates shall undergo a displacement tangential to the normal, when the corresponding point in V_n is displaced in V_n, it is necessary that

$$\frac{dz_{\sigma|}{}^{\alpha}}{\eta_{\sigma|}{}^{\alpha}} = \lambda \qquad (\alpha = 1, \cdots, n+p),$$

where λ is an invariant, or, in consequence of (57.1),

$$(z^{\alpha}{}_{,i} + \varrho\eta_{\sigma|}{}^{\alpha}{}_{,i})\, dx^i + \eta_{\sigma|}{}^{\alpha}(d\varrho - \lambda) = 0.$$

If this equation is multiplied by $c_\alpha \eta_{\sigma|}{}^{\alpha}$ and summed for α, in consequence of (56.1) and (56.3), we find that $\lambda = d\varrho$, so that these equations become

$$(z^{\alpha}{}_{,i} + \varrho\eta_{\sigma|}{}^{\alpha}{}_{,i})\, dx^i = 0. \tag{57.2}$$

If (57.2) be multiplied by $c_\alpha \eta_{\tau|}{}^{\alpha}$ $(\tau \neq \sigma)$ and summed for α, and also by $c_\alpha z^{\alpha}{}_{,j}$ and summed for α, we obtain the respective sets of equations

$$\nu_{\tau\sigma|i}\, dx^i = 0 \quad (\tau = n+1, \cdots, n+p) \tag{57.3}$$

and

$$(g_{ij} - \varrho b_{\sigma|ij})\, dx^i = 0 \qquad (i, j = 1, \cdots, n). \tag{57.4}$$

Conversely, for a displacement in V_n satisfying (57.3) and (57.4) the conditions (57.2) are satisfied.

As in § 50 we say that the congruences defined by (57.4) consist of the *lines of curvature* of V_n for the normal $\eta_{\sigma|}{}^{\alpha}$, and the roots of the determinant equation

$$|g_{ij} - \varrho b_{\sigma|ij}| = 0 \tag{57.5}$$

are the corresponding *principal radii of normal curvature* for the vector $\eta_{\sigma|}{}^{\alpha}$. When, in particular, the directions of a line of curvature satisfy (57.3), we say that the V_n defined by (57.1) is an *evolute* of the given V_n.

Suppose that for a root ϱ_1, of (57.5) the direction determined by (57.4) satisfies (57.3) (or one of the directions, if ϱ_1 is a multiple root). If we take these curves for the curves of parameter x^1, from (57.1) and (57.2) we have

$$z_{\sigma|}{}^{\alpha}{}_{,1} = \eta_{\sigma|}{}^{\alpha} \frac{\partial \varrho_1}{\partial x^1}, \qquad z_{\sigma|}{}^{\alpha}{}_{,r} = z^{\alpha}{}_{,r} + \eta_{\sigma|}{}^{\alpha}{}_{,r}\, \varrho_1 + \frac{\partial \varrho_1}{\partial x^r} \eta_{\sigma|}{}^{\alpha}$$
$$(r = 2, \cdots, n).$$

If $g_{\sigma|ij}$ denote the components of the fundamental tensor of the corresponding evolute $V_{\sigma|}$, we have

$$g_{\sigma|11} = e_\sigma \left(\frac{\partial \varrho_1}{\partial x^1}\right)^2, \qquad g_{\sigma|1r} = e_\sigma \frac{\partial \varrho_1}{\partial x^1} \frac{\partial \varrho_1}{\partial x^r}.$$

Consequently the fundamental form of $V_{\sigma|}$ can be written

(57.6) $$e_\sigma\, d\varrho_1^2 + \bar{g}_{\sigma|rs}\, dx^r\, dx^s \qquad (r, s = 2, \cdots, n),$$

where $\bar{g}_{\sigma|rs}$ are determinate functions. Hence the varieties $\varrho_1 = \text{const.}$ in $V_{\sigma|}$ are geodesically parallel (§ 19) and have for orthogonal trajectories the curves of parameter ϱ_1; these are geodesics in $V_{\sigma|}$.*

Conversely, let a $\bar{V}_n$ in an S_{n+p} be referred to a family of geodesically parallel hypersurfaces whose orthogonal geodesics are not null curves, and take the latter for curves of parameter x^1; then the fundamental form of $\bar{V}_n$ is

(57.7) $$e_1 (dx^1)^2 + \bar{g}_{rs}\, dx^r\, dx^s \qquad (r, s = 2, \cdots, n).$$

Let z_1^α be the coördinates in S_{n+p} of points of the $\bar{V}_n$, then $\frac{\partial z_1^\alpha}{\partial x^1}$ are the components in the z's of the tangents to the curves of parameter x^1. Moreover, from (57.7) it follows that

(57.8) $$\sum_\alpha c_\alpha \left(\frac{\partial z_1^\alpha}{\partial x^1}\right)^2 = e_1, \qquad \sum_\alpha c_\alpha \frac{\partial z_1^\alpha}{\partial x^1} \frac{\partial z_1^\alpha}{\partial x^r} = 0,$$

and from the first of these we have

(57.9) $$\sum_\alpha c_\alpha \frac{\partial z_1^\alpha}{\partial x^1} \frac{\partial^2 z_1^\alpha}{\partial x^1\, \partial x^i} = 0 \qquad (i = 1, \cdots, n).$$

---

* This is a generalization of a well-known result concerning the evolute of a surface in euclidean 3-space; cf. 1909, 1, p. 181.

If we put

$$z^\alpha = z_1^\alpha - (x^1 + a)\frac{\partial z_1^\alpha}{\partial x^1}, \tag{57.10}$$

where a is a constant, we have by differentiation

$$\frac{\partial z^\alpha}{\partial x^1} = -(x^1+a)\frac{\partial^2 z_1^\alpha}{\partial x^{1\,2}}, \qquad \frac{\partial z^\alpha}{\partial x^r} = \frac{\partial z_1^\alpha}{\partial x^r} - (x^1+a)\frac{\partial^2 z_1^\alpha}{\partial x^1 \partial x^r}$$
$$(r = 2, \cdots, n).$$

In consequence of the second of (57.8) and (57.9) we have

$$\sum_\alpha c_\alpha \frac{\partial z^\alpha}{\partial x^i}\frac{\partial z_1^\alpha}{\partial x^1} = 0 \qquad (i = 1, \cdots, n),$$

whatever be a. Hence equations (57.10) define a family of V_n's of which the given $\overline{V}_n$ is an evolute.

58. A subspace V_n of a V_m immersed in an S_{m+p}. In this section we shall derive the equations of § 47 from the equations of § 56 by considering V_m as immersed in an S_{m+p}. We let the coördinates of V_n be x^i, those of V_m be y^λ and of S_{m+p} be z^α.* There are $m - n + p$ mutually orthogonal non-null vectors in S_{m+p} normal to V_n; we denote by $\eta_{\sigma|}{}^\alpha$ for $\sigma = n+1, \cdots, m$ the components in the z's of these vectors which lie in V_m and by $\eta_{\varrho|}{}^\alpha$ for $\varrho = m+1, \cdots, m+p$ the components of the vectors normal to both V_m and V_n at points of the latter.

From the equations

$$z^\alpha{}_{,i} = \frac{\partial z^\alpha}{\partial y^\mu} y^\mu{}_{,i} \tag{58.1}$$

we have by covariant differentiation with respect to x^j and the fundamental form of V_n

$$z^\alpha{}_{,ij} = \frac{\partial z^\alpha}{\partial y^\mu} y^\mu{}_{,ij} + \frac{\partial^2 z^\alpha}{\partial y^\mu \partial y^\nu} y^\mu{}_{,i}\, y^\nu{}_{,j}. \tag{58.2}$$

For V_m immersed in S_{m+p} we have equations of the form (56.2), namely

* In this section $i, j = 1, \cdots, n$; $\lambda, \mu, \nu = 1, \cdots, m$; $\alpha = 1, \cdots, m+p$.

(58.3) $$\frac{\partial^2 z^\alpha}{\partial y^\mu \partial y^\nu} - \left\{ {\lambda \atop \mu\,\nu} \right\}_a \frac{\partial z^\alpha}{\partial y^\lambda} = \sum_\varrho e_\varrho \bar{b}_{\varrho|\mu\nu} \eta_{\varrho|}{}^\alpha \quad (\varrho = m+1, \cdots, m+p),$$

where $\left\{ {\lambda \atop \mu\,\nu} \right\}_a$ are formed with respect to the fundamental form of V_m, namely

(58.4) $$a_{\mu\nu}\, dy^\mu\, dy^\nu .$$

In like manner for V_n immersed in S_{m+p} we have

(58.5) $$z^\alpha{}_{,ij} = \sum_\sigma e_\sigma b_{\sigma|ij} \eta_{\sigma|}{}^\alpha + \sum_\varrho e_\varrho b_{\varrho|ij} \eta_{\varrho|}{}^a \begin{pmatrix} \sigma = n+1, \cdots, m; \\ \varrho = m+1, \cdots, m+p \end{pmatrix}.$$

If $\xi_{\sigma|}{}^\lambda$ are the components in the y's of the vectors $\eta_{\sigma|}{}^\alpha$ in the z's, we have

(58.6) $$\eta_{\sigma|}{}^\alpha = \xi_{\sigma|}{}^\lambda \frac{\partial z^\alpha}{\partial y^\lambda}.$$

Substituting the expressions for $\dfrac{\partial^2 z}{\partial y^\mu \partial y^\nu}$ from (58.3) in (58.2) and for $\eta_{\sigma|}{}^\alpha$ from (58.6) in (58.5) and subtracting the resulting equations, we have

$$\frac{\partial z^\alpha}{\partial y^\lambda}\left(y^\lambda{}_{,ij} + \left\{ {\lambda \atop \mu\,\nu} \right\}_a y^\mu{}_{,i}\, y^\nu{}_{,j} - \sum_\sigma e_\sigma b_{\sigma|ij} \xi_{\sigma|}{}^\lambda \right)$$
$$+ \sum_\varrho e_\varrho (\bar{b}_{\varrho|\mu\nu} y^\mu{}_{,i}\, y^\nu{}_{,j} - b_{\varrho|ij}) \eta_{\varrho|}{}^\alpha = 0.$$

Since the determinant of the quantities $\dfrac{\partial z^\alpha}{\partial y^\lambda}$ and $\eta_{\varrho|}{}^\alpha$ is not zero, we have

(58.7) $$y^\lambda{}_{,ij} = -\left\{ {\lambda \atop \mu\,\nu} \right\}_a y^\mu{}_{,i}\, y^\nu{}_{,j} + \sum_\sigma e_\sigma b_{\sigma|ij} \xi_{\sigma|}{}^\lambda,$$

(58.8) $$b_{\varrho|ij} = \bar{b}_{\varrho|\mu\nu} y^\mu{}_{,i}\, y^\nu{}_{,j} \quad \begin{pmatrix} \sigma = n+1, \cdots, m; \\ \varrho = m+1, \cdots, m+p \end{pmatrix}.$$

If we differentiate equations (58.6) with respect to x^j, we have in consequence of (58.3)

$$\eta_{\sigma|}{}^\alpha{}_{,j} = \left(\xi_{\sigma|}{}^\lambda{}_{,j} + \left\{ {\lambda \atop \mu\,\nu} \right\}_a \xi_{\sigma|}{}^\nu y^\mu{}_{,j} \right) \frac{\partial z^\alpha}{\partial y^\lambda} + \sum_\varrho e_\varrho \bar{b}_{\varrho|\mu\nu} \xi_{\sigma|}{}^\nu y^\mu{}_{,j} \eta_{\varrho|}{}^\alpha .$$

From (56.3) and (58.6) we have

$$(58.9)\quad \eta_{\sigma|}{}^{\alpha}{}_{,j} = -b_{\sigma|lj}\, g^{lm} \frac{\partial z^{\alpha}}{\partial y^{\lambda}} y^{\lambda}{}_{,m} + \sum_{\tau} e_{\tau}\, \nu_{\tau\sigma|j}\, \xi_{\tau|}{}^{\lambda} \frac{\partial z^{\alpha}}{\partial y^{\lambda}} + \sum_{\varrho} e_{\varrho}\, \nu_{\varrho\sigma|j}\, \eta_{\varrho|}{}^{\alpha}.$$

Subtracting these equations and proceeding as above, we obtain

$$(58.10)\quad \xi_{\sigma|}{}^{\lambda}{}_{,j} = -b_{\sigma|lj}\, g^{lm} y^{\lambda}{}_{,m} - \left\{ {\lambda \atop \mu\,\nu} \right\}_a y^{\mu}{}_{,j}\, \xi_{\sigma|}{}^{\nu} + \sum_{\tau} e_{\tau}\, \nu_{\tau\sigma|j}\, \xi_{\tau|}{}^{\lambda},$$

$$(58.11)\quad \nu_{\varrho\sigma|j} = \bar{b}_{\varrho|\mu\nu}\, y^{\mu}{}_{,j}\, \xi_{\sigma|}{}^{\nu}. \quad \begin{pmatrix} \sigma, \tau = n+1, \cdots, m; \\ \varrho = m+1, \cdots, m+p \end{pmatrix}.$$

Equations (58.7) and (58.10) are of the form (47.4) and (47.9) respectively, where

$$(58.12)\quad b_{\sigma|ij} = \Omega_{\sigma|ij}, \qquad \nu_{\tau\sigma|j} = \mu_{\tau\sigma|j} \qquad (\sigma,\tau = n+1 \cdots, m).$$

If the expressions for $b_{\varrho|ij}$, $\nu_{\varrho\sigma|j}$, $b_{\sigma|ij}$ and $\nu_{\tau\sigma|j}$ from (58.8), (58.11) and (58.12) are substituted (56.5), (56.6) and (56.7), and we remark that from (56.5) and (58.3) it follows that the components of the Riemann tensor of V_m are given by

$$\bar{R}_{\lambda\mu\nu\pi} = \sum_{\varrho} e_{\varrho} (\bar{b}_{\varrho|\lambda\nu}\, \bar{b}_{\varrho|\mu\pi} - \bar{b}_{\varrho|\lambda\pi}\, \bar{b}_{\varrho|\mu\nu}),$$

the resulting equations are reducible to (47.11), (47.12) and (47.13) respectively.

59. Spaces V_n of class one. When V_n is of class 1, we have in place of (56.2) and (56.3)

$$(59.1)\quad z^{\alpha}{}_{,ij} = e\, b_{ij}\, \eta^{\alpha}$$

and

$$(59.2)\quad \eta^{\alpha}{}_{,j} = -b_{lj}\, g^{lm} z^{\alpha}{}_{,m}.$$

The conditions of integrability of these equations are

$$(59.3)\quad R_{ijkl} = e(b_{ik}\, b_{jl} - b_{il}\, b_{jk})$$

and

$$b_{ij,k} - b_{ik,j} = 0, \tag{59.4}$$

in place of (56.5) and (56.6).

From considerations similar to those of the preceding sections we obtain the theorem:

In order that

$$\varphi = g_{ij}\, dx^i\, dx^j, \qquad \psi = b_{ij}\, dx^i\, dx^j \tag{59.5}$$

be the first and second fundamental forms of a space V_n *immersed in a real* S_{n+1}, *it is necessary and sufficient that* (59.3) *and* (59.4) *be satisfied; then* V_n *is determined to within a motion in* S_{n+1}.

See App. 22

The roots of the determinant equation

$$|R\, b_{ij} - g_{ij}| = 0 \tag{59.6}$$

are the principal radii of normal curvature of V_n in the S_{n+1}, and the congruences of curves defined by

$$(R_h\, b_{ij} - g_{ij})\, \lambda_{h|}{}^i = 0 \tag{59.7}$$

are the lines of curvature of V_n (§ 45). Since (57.3) are satisfied identically in this case, it follows that the normals to V_n along a line of curvature are tangents to a curve, and that these curves lie in the sheets of the evolute of V_n, just as in the case of surfaces of euclidean 3-space.

If the elementary divisors of (59.6) are simple, there are n families of lines of curvature, whose directions at any point are mutually orthogonal and are not null directions. At any point P the coördinate system can be chosen so that these are the coördinate directions. Hence at P in this coördinate system we have

$$\begin{aligned} g_{ii} = e_i, \quad g_{ij} = 0\ (i \neq j); \qquad & \lambda_{h|}{}^i = 0\ (h \neq i), \quad \lambda_{h|}{}^h = 1; \\ b_{ij} = 0\ (i \neq j), \qquad & \frac{1}{R_h} = e_h\, b_{hh}. \end{aligned} \tag{59.8}$$

If we denote by r_{hk} the Riemannian curvature for the orientation determined by $\lambda_{h|}{}^i$ and $\lambda_{k|}{}^i$, we have from (25.9), (59.3) and (59.8)

$$r_{hk} = e\frac{1}{R_h R_k}. \tag{59.9}$$

Since these quantities are invariants, we have the theorem:

When a V_n *is of class one and the elementary divisors of* (59.6) *are simple, the Riemannian curvature at a point for the orientation determined by the directions of two lines of curvature at the point is numerically equal to the product of the corresponding normal curvatures; the sign is determined by the character of the normal to* V_n *in the enveloping* S_{n+1}.*

From (59.3) we have for the components of the Ricci tensor

$$R_{jk} = eg^{il}(b_{ik}b_{jl} - b_{il}b_{jk}). \tag{59.10}$$

Hence at a point for the coördinates such that (59.8) hold we have

$$R_{jj} = -e b_{jj} \sum_{i \neq j}^{1,\cdots,n} e_i b_{ii}, \qquad R_{jk} = 0 \qquad (j \neq k). \tag{59.11}$$

From these equations and (34.4), we have:

When a V_n *is of class one and the elementary divisors of* (59.6) *are simple, the Ricci principal directions coincide with the directions of the lines of curvature.*†

We seek now under what conditions $R_{jj} = 0$ for $j = 1, \cdots, n$ in (59.11). These conditions are satisfied, if one of the b's, say b_{11}, is not zero, and all the others vanish. Suppose now that $p(>1)$ of the b's do not vanish, say $b_{11}, \cdots, b_{pp}$, and that the others vanish. Then we must have

$$\sum_{i \neq j}^{1,\cdots,p} e_i b_{ii} = 0 \qquad (j = 1, \cdots, p). \tag{59.12}$$

Subtracting two of these equations for $j = r, s$, we get $e_r b_{rr} = e_s b_{ss}$. Since this must be true for $r, s = 1, \cdots, p \, (r \neq s)$, it follows from (59.12) that all of these b's are zero contrary to hypothesis. Hence at most one of the b's can be different from zero, and then from (59.3) it follows that the components of the Riemann tensor are zero. Since this situation must hold at every point of V_n, the latter is of class zero and not of class one. Hence we have:

---

* This is a generalization of the theorem of Gauss for surfaces in euclidean 3-space, cf. 1909, 1, pp. 120, 155.

† *Schouten* and *Struik*, 1921, 3, p. 214.

*There are no spaces of class one for which all the components of the Ricci tensor are zero.**

See App. 23

60. Applicability of hypersurfaces of a flat space. For a space V_2 with the fundamental form

$$\varphi = g_{ij}\, dx^i\, dx^j \tag{60.1}$$

equations (59.3) and (59.4) for $e = 1$; $i, j = 1, 2$ are the Gauss and Codazzi equations of a surface in euclidean 3-space. The problem of finding other surfaces applicable to the given V_2, that is, with the same fundamental form (60.1) is the problem of finding sets of functions b_{ij} satisfying these equations. It is of the generality of a partial differential equation of the second order,† and thus there are many surfaces applicable to a given surface. When $n > 2$, this is not the case. In fact, it will be shown that

If a V_n is of class one, real quantities b_{ij} are determined by (59.3) *to within sign, if one of the determinants of the third order of the b's is not zero.*‡

Since by hypothesis the rank of the determinant $b = |b_{ij}|$ is at least three, the coördinates can be chosen so that at a point P we have $b_{rr} \neq 0$, $b_{rs} = 0$ for $r, s = 1, 2, 3$ and $r \neq s$, and thus

$$B \equiv |b_{rs}| \neq 0 \qquad (r, s = 1, 2, 3). \tag{60.2}$$

We consider first the possibility of two sets of solutions b_{ij} and $\overline{b}_{ij}$ of (59.3) for $e = 1$ and $\overline{e} = -1$ respectively. If B_{rs} denotes the cofactor of b_{rs} in B and similarly $\overline{B}_{rs}$ for the determinant $\overline{B} = |\overline{b}_{rs}|$, it follows from (59.3) that $\overline{B}_{rs} = -B_{rs}$. Since

$$|B_{rs}| = B^2, \tag{60.3}$$

we have that $\overline{B}^2 = -B^2$. Consequently if equations (59.3) and (59.4) admit a set of real solutions for $e = 1$ or -1, the solutions are imaginary for $e = -1$ or 1. Accordingly as we are concerned only with real solutions, e must be the same for both sets of solutions and consequently $\overline{B}_{rs} = B_{rs}$ and from (60.3) $\overline{B}^2 = B^2$.

---

* *Kasner*, 1921, 7, p. 126; also *Schouten* and *Struik*, 1921, 3, p. 215.

† Cf. 1909, 1, p. 331.

‡ Cf. *Killing*, 1885, 1, pp. 236–237; also *Bianchi*, 1902, 1, p. 465.

Hence

$$\bar{B} = \pm B. \tag{60.4}$$

Since the cofactor of B_{rs} in (60.3) is $b_{rs}B$,* we have in consequence of (60.4)

$$\bar{b}_{rs} = \pm b_{rs} \qquad (r, s = 1, 2, 3). \tag{60.5}$$

From the equality

$$\bar{b}_{rr}\bar{b}_{st} - \bar{b}_{rs}\bar{b}_{rt} = b_{rr}b_{st} - b_{rs}b_{rt} \quad (r, s = 1, 2, 3;\ t = 1, \cdots, n), \tag{60.6}$$

from (60.5) and $b_{rr} \neq 0$, $b_{rs} = 0$ $(r \neq s)$, we have

$$\bar{b}_{rt} = \pm b_{rt} \quad (r = 1, 2, 3;\ t = 1, \cdots, n). \tag{60.7}$$

Again if in (60.6) we take $r = 1, 2, 3$ and $s, t = 1, \cdots, n$, we obtain

$$\bar{b}_{st} = \pm b_{st} \qquad (s, t = 1, \cdots, n)$$

and the theorem is proved.

From (59.6) it is seen that the case where the rank of the determinant is less than three is that for which $n-2$ of the roots of (59.6) are infinite, and from (59.9) that the Riemannian curvature is zero for all but one of the orientations determined by the lines of curvature of V_n. Hence the preceding result may be stated as follows:

A hypersurface in an S_{n+1} for $n > 2$ is indeformable, if more than two of its principal radii of curvature are finite, or, in other words, if the Riemannian curvature determined by more than one pair of directions of the lines of curvature is not zero.

It should be observed that, although the functions b_{ij} are determined to within sign by (59.3) for $n > 2$, except in the cases indicated, the conditions (59.4) must be satisfied also, in order that the space be of class one.†

61. Spaces of constant curvature which are hypersurfaces of a flat space. In a flat space with the fundamental form

$$\varphi = \sum_{\alpha} c_{\alpha}(dz^{\alpha})^2 \qquad (\alpha = 1, \cdots, n+1) \tag{61.1}$$

* Cf. *Bôcher*, 1907, 1, p. 33.

† Cf. *Sbrana*, 1909, 3.

the hypersurfaces defined by

$$\sum_\alpha c_\alpha (z^\alpha)^2 = eR^2, \tag{61.2}$$

where e is plus or minus one, and R is an arbitrary constant, will be called the *fundamental hyperquadrics* of the space. When all the c's are positive, that is, when the space is euclidean, there is only one family of real hyperquadrics; in this case $e = 1$ and the hyperquadrics are hyperspheres. In all other cases (except when all the c's are negative which case we exclude) there are two families of such real hyperquadrics, corresponding to $e = 1$ and $e = -1$ and arbitrary values of R. When the hyperquadrics are subjected to a translation in the S_{n+1}, we get in place of (61.2) the equation

$$\sum_\alpha c_\alpha (z^\alpha - b^\alpha)^2 = eR^2, \tag{61.3}$$

where the b's are the constants defining the translation (§ 56). We shall show that the hyperquadrics are spaces V_n of constant curvature; we take their equations in the form (61.2).*

Assuming that the z's are functions of x^i for $i = 1, \cdots, n$ so that (61.2) holds, we have from (61.2) by differentiation

$$\begin{aligned} &\sum_\alpha c_\alpha z^\alpha z^\alpha{}_{,i} = 0, \\ &\sum_\alpha c_\alpha z^\alpha z^\alpha{}_{,ij} = -\sum c_\alpha z^\alpha{}_{,i} z^\alpha{}_{,j} = -g_{ij}. \end{aligned} \tag{61.4}$$

From the first of (61.4) and (61.2) it follows that the components η^α of the unit vector normal to V_n are given by

$$\eta^\alpha = \frac{z^\alpha}{R}. \tag{61.5}$$

When the expressions for $z^\alpha{}_{,ij}$ from (59.1) are substituted in the second of (61.4), we find

$$b_{ij} = -\frac{1}{R} g_{ij}. \tag{61.6}$$

* From the results of § 56 it follows that there is no loss of generality in so doing.

Now (59.4) are satisfied identically and (59.3) become

(61.7) $$R_{ijkl} = K_0 (g_{ik} g_{jl} - g_{il} g_{jk}),$$

where

(61.8) $$K_0 = \frac{e}{R^2},$$

and (59.1) become

(61.9) $$z^{\alpha}{}_{,ij} = -e g_{ij} \frac{z^{\alpha}}{R^2} = -K_0 g_{ij} z^{\alpha}.$$

Hence:

The fundamental hyperquadrics of a flat space are spaces of constant Riemannian curvature.

We shall prove the converse theorem:

The fundamental hyperquadrics are the only hypersurfaces of constant Riemannian curvature $\neq 0$ *of a flat space.*

In fact, if V_n is any space of constant curvature, a solution of (59.3) and (59.4) is given by (61.6), where R and e are determined by (61.8). Moreover, by the arguments of § 60 this is the only solution, to within algebraic sign, if $g = |g_{ij}| \neq 0$. Equations (59.2) become

$$\frac{\partial \eta^{\alpha}}{\partial x^j} = \frac{1}{R} \frac{\partial z^{\alpha}}{\partial x^j},$$

of which the integral is (61.5), if we neglect additive constants, that is, a translation in S_{n+1}. Then (61.2) follows from (56.1) and the theorem is proved.

If z^{α} are one set of solutions of (61.9), the equations

(61.10) $$\bar{z}^{\alpha} = a^{\alpha}{}_{\beta} z^{\beta},$$

where the a's are constants define other sets of solutions. In order that (61.2) may be satisfied, these constants must satisfy the conditions

(61.11) $$\sum_{\alpha} c_{\alpha} (a^{\alpha}{}_{\beta})^2 = c_{\beta}, \qquad \sum_{\alpha} c_{\alpha} a^{\alpha}{}_{\beta} a^{\alpha}{}_{\gamma} = 0 \qquad (\beta \neq \gamma).$$

There are $n+1$ and $n(n+1)/2$ of the conditions respectively, and consequently $n(n+1)/2$ of the $(n+1)^2$ constants $a^{\alpha}{}_{\beta}$ are arbitrary.

We shall show that (61.10) and (61.11) define the most general solution of (61.2) and (61.9). In fact, if we put

(61.12) $$z^\alpha_{,i} = p^\alpha_i$$

and write (61.9) in the form

(61.13) $$p^\alpha_{i,j} = -K_0\, g_{ij}\, z^\alpha,$$

the system of equations (61.12) and (61.13) is completely integrable in the $(n+1)$ functions z^α and the $n(n+1)$ functions p^α_i. Our problem consists in the determination of the solutions of these equations satisfying (61.2) and the conditions

(61.14) $$\sum_\alpha c_\alpha\, p^\alpha_i\, p^\alpha_j = g_{ij}, \qquad \sum_\alpha c_\alpha\, z^\alpha\, p^\alpha_i = 0,$$

that is, $(n+1)(n+2)/2$ conditions, so that the desired solution involves $n(n+1)/2$ arbitrary constants, as was to be proved.

For any set of the a's satisfying (61.11) equations (61.10) define a motion (§ 27) of the hyperquadric into itself; from the point of view of the enveloping S_{n+1}, this is a rotation (§ 56) about the origin of the cartesian coördinates of S_{n+1}. Since the quantities p^α_i determine a direction at a point in V_n, the number of arbitrary a's is just sufficient for the determination of a motion which carries a point P into a desired point Q, and an orthogonal ennuple in V_n at P into a chosen orthogonal ennuple at Q. This result is in keeping with the last theorem of § 27.

62. Coördinates of Weierstrass. Motion in a space of constant curvature. In the preceding section the z's have been interpreted as the cartesian coördinates of a flat space S_{n+1} in which a given space V_n of constant Riemannian curvature is immersed. If we are concerned only with intrinsic properties of the V_n, that is, those depending only on its fundamental form, we may adopt another point of view and treat the z's as a particular type of coördinates, $n+1$ in number, in terms of which the equations for a space V_n of constant curvature assume a form advantageous to the consideration of certain problems. Thus we may state the results of the preceding section as follows:

For a space V_n of constant Riemannian curvature K_0, there exist sets of $n+1$ real coördinates z^α satisfying the condition

(62.1) $$\sum_\alpha c_\alpha\,(z^\alpha)^2 = \frac{1}{K_0} \qquad (\alpha = 1, \cdots, n+1),$$

in terms of which the fundamental form of V_n may be written

$$\varphi = \sum_\alpha c_\alpha (dz^\alpha)^2, \tag{62.2}$$

where the c's are plus or minus one according to the character of the fundamental form; when one such system is known, others are given by

$$\bar{z}^\alpha = a^\alpha{}_\beta z^\beta, \tag{62.3}$$

where the a's are constants subject to the conditions

$$\sum_\alpha c_\alpha (a^\alpha{}_\beta)^2 = c_\beta, \qquad \sum_\alpha c_\alpha a^\alpha{}_\beta a^\alpha{}_\gamma = 0 \tag{62.4}$$

$$(\alpha, \beta, \gamma = 1, \cdots, n+1;\ \beta \neq \gamma).$$

When the V_n is defined in terms of any set of coördinates x^i $(i = 1, \cdots, n)$, the determination of the z's reduces to the solution of equations (61.9), (62.1) and

$$\sum_\alpha c_\alpha z^\alpha{}_{,i} z^\alpha{}_{,j} = g_{ij}. \tag{62.5}$$

From (61.4) it follows that a set of $n+1$ quantities η^α such that

$$\sum_\alpha c_\alpha z^\alpha \eta^\alpha = 0 \tag{62.6}$$

are the components in the z's of a vector in V_n; the components λ^i of the same vector in the x's are given by

$$\eta^\alpha = \lambda^i z^\alpha{}_{,i}. \tag{62.7}$$

If $\eta_{1|}^\alpha$ and $\eta_{2|}^\alpha$ are the components of two of these vectors, it follows from (62.7) and (62.5) that

$$\begin{aligned} \sum_\alpha c_\alpha (\eta_{1|}^\alpha)^2 &= g_{ij} \lambda_{1|}^i \lambda_{1|}^j, \\ \sum_\alpha c_\alpha \eta_{1|}^\alpha \eta_{2|}^\alpha &= g_{ij} \lambda_{1|}^i \lambda_{2|}^j. \end{aligned} \tag{62.8}$$

Consequently, the angle between two unit vectors is given by

$$\cos\theta = \sum_\alpha c_\alpha \eta_{1|}^\alpha \eta_{2|}^\alpha. \tag{62.9}$$

An equation of the form

$$a_1 z^1 + a_2 z^2 + \cdots + a_{n+1} z^{n+1} = 0$$

in which the a's are constants defines a hypersurface of the V_n, which we shall show is a space of constant curvature. In fact, by a transformation (62.3) of the z's this can be reduced to the form $z^{n+1} = 0$. Then from (62.1) and (62.2) we see that the hypersurface $z^{n+1} = 0$ has the same constant curvature as the enveloping V_n.

When all the c's in (62.2) are positive, it follows from (62.2) that the fundamental form of the V_n is positive definite, and from (62.1) that K_0 is positive for the space to be real. When $c_i = 1$ for $i = 1, \cdots, n$, $c_{n+1} = -1$ and $K_0 = -1/R^2$, if we solve (62.1) for z^{n+1} and substitute in (62.2), it becomes

$$(62.10)\quad \varphi = \frac{1}{\sum_i z^{i^2} + R^2}\left[R^2 \sum_i (dz^i)^2 + \sum_{i,j} (z^i\, dz^j - z^j\, dz^i)^2\right] \qquad (i, j = 1, \cdots, n),$$

from which it is found that φ is positive definite. In consequence of the first theorem of § 27, when a V_n has a positive definite form, it is possible to choose a set of $n+1$ coördinates in terms of which the fundamental form is (62.2) with all the c's plus one or all but one plus one, according as the curvature of V_n is positive or negative.

If we put $z^\alpha = y^\alpha/R$, the y's are the coördinates which Bianchi has called the point coördinates of Weierstrass, since they are a generalization of coördinates used by Weierstrass in non-euclidean geometries of two dimensions.* In deriving these results for spaces of constant curvature with positive definite fundamental forms, Bianchi used a different point of view, which seems less direct than the foregoing. We generalize his notation so as to apply to spaces of constant curvature with any type of fundamental form and call the z's the *point coördinates of Weierstrass* and the components η^α the *vector components of Weierstrass.*

In the above theorem equations (62.3) and (62.4) have been interpreted as a transformation of coördinates of Weierstass into

* Cf. *Bianchi*, 1902, 1, pp. 407, 434–444.

coördinates of the same kind. They serve also as a basis for the equations of motion of a space of constant curvature V_n into itself in terms of general coördinates. In fact, we have seen (§ 27) that the portion of V_n in the neighborhood of a point P can be applied to the portion in the neighborhood of another point $\overline{P}$. Consequently, there exist coördinate systems x^i and $\overline{x}^i$ in V_n such that the fundamental forms of V_n in the two coördinate systems are

$$\varphi = g_{ij}\, dx^i\, dx^j = \overline{g}_{ij}\, d\overline{x}^i\, d\overline{x}^j, \tag{62.11}$$

where any $\overline{g}_{ij}$ is the same function of the $\overline{x}$'s as g_{ij} is of the x's, and the coördinates $\overline{x}^i$ have the same values at $\overline{P}$ as the corresponding x^i at P. If z^α denote a particular set of solutions of (61.9) satisfying (62.1), evidently the same functions of the $\overline{x}$'s are a solution of the corresponding equations (61.9) in the $\overline{x}$'s. When these expressions for $\overline{z}^\alpha$ and z^α are substituted in (62.3), we have the $\overline{x}$'s defined as functions of the x's and $n(n+1)/2$ parameters, and thus we have in general coördinates the equations of the continuous group of motions of V_n into itself.

63. Equations of geodesics in a space of constant curvature in terms of coördinates of Weierstrass. For a non-minimal geodesic in a space V_n of coördinates x^i we have (§ 17)

$$\frac{d^2 x^i}{ds^2} = -\left\{ {i \atop jk} \right\} \frac{dx^j}{ds} \frac{dx^k}{ds}. \tag{63.1}$$

If V_n is a hyperquadric (61.2) of a flat space, we have in consequence of (63.1) and (61.9)

$$\begin{aligned} \frac{d^2 z^\alpha}{ds^2} &= \frac{\partial z^\alpha_{\ ,i}}{\partial x^j} \frac{dx^i}{ds} \frac{dx^j}{ds} + z^\alpha_{\ ,i} \frac{d^2 x^i}{ds^2} \\ &= z^\alpha_{\ ,ij} \frac{dx^i}{ds} \frac{dx^j}{ds} = -K_0\, g_{ij} \frac{dx^i}{ds} \frac{dx^j}{ds} z^\alpha. \end{aligned}$$

Because of (61.8) and

$$g_{ij} \frac{dx^i}{ds} \frac{dx^j}{ds} = e_1, \tag{63.2}$$

the above equations reduce to

$$\frac{d^2 z^\alpha}{d s^2} = -\frac{e e_1}{R^2} z^\alpha. \tag{63.3}$$

There are two cases to be considered according as $e e_1$ is $+1$ or -1.

1°. $e e_1 = +1$. In this case the integrals of (63.3) are

$$z^\alpha = z_0^\alpha \cos\frac{s}{R} + R\eta_0^\alpha \sin\frac{s}{R}, \tag{63.4}$$

where z_0^α are the coördinates at the point $s = 0$, and η_0^α are the components in the z's of the unit vector tangent to the geodesic at $s = 0$, as is seen from the equations

$$\frac{d z^\alpha}{d s} = \left(-z_0^\alpha \sin\frac{s}{R} + R\eta_0^\alpha \cos\frac{s}{R}\right)\frac{1}{R}. \tag{63.5}$$

Since the expressions (63.4) must satisfy (61.2) for all values of s, we must have

$$\sum_\alpha c_\alpha (z_0^\alpha)^2 = e R^2, \quad \sum_\alpha c_\alpha (\eta_0^\alpha)^2 = e_1, \quad \sum_\alpha c_\alpha \eta_0^\alpha z_0^\alpha = 0, \tag{63.6}$$

which are in agreement with the preceding observations and results.

From (63.5) it follows that the functions η^α, defined by

$$R\eta^\alpha = -z_0^\alpha \sin\frac{s}{R} + R\eta_0^\alpha \cos\frac{s}{R}, \tag{63.7}$$

are the components of the unit vector tangent to the geodesic at the point of coördinates z^α. From (63.4) and (63.7) we have

$$\begin{aligned} z_0^\alpha &= z^\alpha \cos\frac{s}{R} - R\eta^\alpha \sin\frac{s}{R}, \\ R\eta_0^\alpha &= z^\alpha \sin\frac{s}{R} + R\eta^\alpha \cos\frac{s}{R}, \end{aligned} \tag{63.8}$$

which reveals the reciprocal character of these formulas.

From (63.4) and (63.6) we have

$$\sum_\alpha c_\alpha (z^\alpha - z_0^\alpha)^2 = 4 e R^2 \sin^2\frac{s}{2R},$$

and consequently the distance in the enveloping space between two points whose geodesic distance is s is $2R\sin\frac{s}{2R}$. From this it follows that two points coincide whose geodesic distance is $2\pi R$; this is seen also from (63.4). Hence:

In a space of constant curvature the geodesics for which $ee_1 = 1$, *where* e *is defined by* (61.2) *and* e_1 *by* (63.2), *are closed curves of length* $2\pi R$.

From (63.7) we have

$$\sum_\alpha c_\alpha \eta^\alpha \eta_0^\alpha = e\cos\frac{s}{R}. \tag{63.9}$$

Consequently the angle, as determined by the metric of the enveloping space, between the tangents is s/R or $\pi - s/R$, according as e is $+1$ or -1, whereas from the definition of parallelism these tangents are parallel with respect to the curve in the metric of the given V_n (§ 21).

2°. $ee_1 = -1$. The integrals of (63.3) are

$$z^\alpha = z_0^\alpha \cosh\frac{s}{R} + R\eta_0^\alpha \sinh\frac{s}{R}. \tag{63.10}$$

The components of the unit vector tangent to the geodesic at the point of coördinates z^α are given by

$$R\eta^\alpha = z_0^\alpha \sinh\frac{s}{R} + R\eta_0^\alpha \cosh\frac{s}{R}. \tag{63.11}$$

Since

$$\sum_\alpha c_\alpha (z^\alpha - z_0^\alpha)^2 = 2eR^2\left(1 - \cosh\frac{s}{R}\right) = -4eR^2\sinh^2\frac{s}{2R},$$

we have that the distance in the enveloping space of two points, whose geodesic distance is s, is $2R\sinh\frac{s}{2R}$. Moreover, since

$$\sum_\alpha c_\alpha \eta^\alpha \eta_0^\alpha = e\cosh\frac{s}{R},$$

we see that in calling the left-hand member the cosine of the angle between the tangents (§ 16) the term cosine is a mere notation.

When, in particular, the fundamental forms of the spaces of constant curvature are positive definite, we have $e_1 = 1$, and consequently the cases 1° and 2° apply respectively to spaces of positive and negative constant curvature.*

When the fundamental form of V_n is not definite, there remains for consideration the case of minimal geodesics. If in accordance with the observations following equations (17.11) we choose the parameter t so that the equations of the geodesics are of the form (63.1) with s replaced by t, equations (63.3) assume the form $\frac{d^2 z^\alpha}{dt^2} = 0$, and consequently in the coördinates of Weierstrass the equations of the minimal geodesics are

$$z^\alpha = \eta_0^\alpha t + z_0^\alpha. \tag{63.12}$$

Accordingly the components in the z's of the tangent vector are the same at all points of a minimal geodesic.

64. Equations of a space V_n immersed in a V_m of constant curvature. As an application of the results of §§ 58 and 61 we establish the equations of a sub-space V_n of a space V_m of constant curvature in terms of the coördinates of Weierstrass, making use of the notation of these sections and observing that $p = 1$ in § 58.

From (61.5), (61.6) and (61.8) we have

$$\eta_{m+1|}{}^\alpha = \frac{z^\alpha}{R}, \qquad \bar{b}_{m+1|\mu\nu} = -\frac{1}{R} a_{\mu\nu}, \qquad K_0 e = \frac{1}{R^2}, \tag{64.1}$$

where $a_{\mu\nu}\, dy^\mu\, dy^\nu$ is the fundamental form of V_m.†

From (58.8), (58.11) and (64.1) we have

$$\begin{aligned} b_{m+1|ij} &= -\frac{1}{R} a_{\mu\nu}\, y^\mu{}_{,i}\, y^\nu{}_{,j} = -\frac{1}{R} g_{ij}, \\ \nu_{m+1\sigma|j} &= -\frac{1}{R} a_{\mu\nu}\, y^\mu{}_{,j}\, \xi_{\sigma|}{}^\nu = 0, \qquad (\sigma = n+1, \cdots, m). \end{aligned} \tag{64.2}$$

* Cf. *Bianchi*, 1902, 1, pp. 434–440, where these results are obtained from a different point of view.

† In this section $\alpha = 1, \cdots, m+1$; $\lambda, \mu, \nu = 1, \cdots, m$ and Latin indices take the values $1, \cdots, n$.

Because of these results and (58.12) equations (58.5) and (58.9) reduce to

$$(64.3)\quad z^{\alpha}_{\ ,ij} = \sum_{\sigma} e_{\sigma}\, \Omega_{\sigma|ij}\, \eta_{\sigma|}^{\ \alpha} - K_0\, g_{ij}\, z^{\alpha},$$

$$(64.4)\quad \eta_{\sigma|}^{\ \alpha}{}_{,j} = -\Omega_{\sigma|lj}\, g^{lm}\, z^{\alpha}_{\ ,m} + \sum_{\tau} e_{\tau}\, \mu_{\tau\sigma|j}\, \eta_{\tau|}^{\ \alpha}$$

$$(\sigma, \tau = n+1, \cdots, m).$$

Proceeding as in § 47, we find that the conditions of integrability of (64.3) and (64.4) are

$$(64.5)\quad R_{ijkl} = \sum_{\sigma} e_{\sigma} (\Omega_{\sigma|ik}\, \Omega_{\sigma|jl} - \Omega_{\sigma|il}\, \Omega_{\sigma|jk}) + K_0 (g_{ik}\, g_{jl} - g_{il}\, g_{jk}),$$

$$(64.6)\quad \Omega_{\sigma|ij,k} - \Omega_{\sigma|ik,j} = \sum_{\tau} e_{\tau} (\mu_{\tau\sigma|k}\, \Omega_{\tau|ij} - \mu_{\tau\sigma|j}\, \Omega_{\tau|ik}),$$

$$(64.7)\quad \begin{aligned} &\mu_{\tau\sigma|j,k} - \mu_{\tau\sigma|k,j} + \sum_{\varrho} e_{\varrho} (\mu_{\varrho\tau|j}\, \mu_{\varrho\sigma|k} - \mu_{\varrho\tau|k}\, \mu_{\varrho\sigma|j}) \\ &\quad + g^{lh} (\Omega_{\tau|lj}\, \Omega_{\sigma|hk} - \Omega_{\tau|lk}\, \Omega_{\sigma|hj}) = 0 \quad (\varrho, \sigma, \tau = n+1, \cdots, m). \end{aligned}$$

These equations follow directly from (56.5), (56.6) and (56.7), if we make use of (64.2) and (58.12).* In this case we can show as in § 56 that, if we take the equations preceding (56.8) for $\sigma, \tau = n+1, \cdots, m$ and

$$(64.8)\quad \sum_{\alpha} c_{\alpha}\, \eta_{\sigma|}^{\ \alpha}\, z^{\alpha} = D_{\sigma}, \quad \sum_{\alpha} c_{\alpha}\, z^{\alpha}\, z^{\alpha}_{\ ,i} = E_i, \quad \sum_{\alpha} c_{\alpha} (z^{\alpha})^2 - \frac{1}{K_0} = F$$

and choose initial values so that A_{ij}, $B_{\sigma|i}$, $C_{\sigma\tau}$, D_{σ}, E_i and F vanish, then they vanish for all values of the x's.†

There are $(m+1)(m+2)/2$ of these equations of condition on the $(m+1)^2$ functions $z^{\alpha}_{\ ,i}$, $\eta_{\sigma|}^{\ \alpha}$, z^{α}. Hence a solution of (64.3) and (64.4) satisfying these conditions involves $m(m+1)/2$ arbitrary constants. We may account for these arbitrary constants by observing that, if z^{α} and $\eta_{\sigma|}^{\ \alpha}$ are a set of solutions of (64.3) and (64.4), so also are $\bar{z}^{\alpha}$ given by (61.10) and $\bar{\eta}_{\sigma|}^{\ \alpha}$ given by

* They follow also from (47.11), (47.12) and (47.14), if we note that

$$\bar{R}_{\lambda\mu\nu\pi} = K_0 (a_{\lambda\nu}\, a_{\mu\pi} - a_{\lambda\pi}\, a_{\mu\nu}).$$

† Equations (64.8) are merely forms of (56.8) for $\eta_{m+1|}^{\ \alpha}$ given by (64.1).

$$\bar{\eta}_{\sigma|}{}^{\alpha} = a^{\alpha}{}_{\beta}\, \eta_{\sigma|}{}^{\beta},$$

where the a's are subject to the conditions (61.11), when $\alpha = m+1$. Recalling the intrepretation of (61.10), we have the theorem:

When a symmetric tensor g_{ij}, $(m-n)(m-n-1)/2$ vectors $\mu_{\tau\sigma|i}(= -\mu_{\sigma\tau|i})$ and $m-n$ tensors $\Omega_{\sigma|ij}$ satisfy equations (64.5), (64.6) *and* (64.7), *in which K_0 is a constant, the tensor g_{ij} is the fundamental tensor of a space V_n immersed in a space V_m of curvature K_0; V_n is determined to within a motion in V_m.*

When $m = n+1$, that is, when V_n is a hypersurface of a space of constant curvature, we have in place of (64.3) and (64.4)

$$z^{\alpha}{}_{,ij} = e\Omega_{ij}\,\eta^{\alpha} - K_0\, g_{ij}\, z^{\alpha}, \tag{64.9}$$

$$\eta^{\alpha}{}_{,j} = -\Omega_{lj}\, g^{lm}\, z^{\alpha}{}_{,m}, \tag{64.10}$$

where the functions z^{α} and η^{α} are in the relations

$$\sum_{\alpha} c_{\alpha}\,(z^{\alpha})^2 = \frac{1}{K_0}, \qquad \sum_{\alpha} c_{\alpha}\,(\eta^{\alpha})^2 = e, \tag{64.11}$$

the c's being plus or minus one, such that the equations

$$\sum_{\alpha} c_{\alpha}\, z^{\alpha}{}_{,i}\, z^{\alpha}{}_{,j} = g_{ij} \tag{64.12}$$

admit solutions z^{α} which are real functions of the x's. The conditions of integrability are

$$R_{ijkl} = e(\Omega_{ik}\,\Omega_{jl} - \Omega_{il}\,\Omega_{jk}) + K_0\,(g_{ik}\, g_{jl} - g_{il}\, g_{jk}), \tag{64.13}$$

$$\Omega_{ij,k} - \Omega_{ik,j} = 0.^{*} \tag{64.14}$$

When two tensors g_{ij} and Ω_{ij} satisfy these conditions, there exists a V_n immersed in a space V_{n+1} of curvature K_0, which is determined to within a motion in the space.

The arguments applied in § 60 to equations (59.3) apply in like manner to (64.13) with the result:

A hypersurface of a space V_n of constant Riemannian curvature for $n > 3$ is indeformable in the V_n, if more than two of the principal radii of curvature are finite.

---

* Cf. (43.20) and (43.21).

We establish the following theorem which evidently is a generalization of the results of §§ 57 and 59 for spaces of class one: *A necessary and sufficient condition that the geodesics of a space of constant curvature V_{n+1} normal to a hypersurface V_n of V_{n+1} along a curve C of V_n be tangent to a curve in V_{n+1} is that C be a line of curvature of V_n.**

We establish this theorem by means of the results of § 63, where z_0^α and η_0^α denote respectively the coördinates of a point of V_n and the components of the normal to V_n at the point, this normal lying in V_{n+1}.

We consider first the case when the coördinates of points on the geodesics normal to V_n along a curve of the latter are expressible in the form [cf. (63.4)]

$$z^\alpha = z_0^\alpha \cos\frac{w}{R} + R\eta_0^\alpha \sin\frac{w}{R}, \tag{64.15}$$

where z_0^α, η_0^α and w are functions of x^i which are functions of s for the curve. Now

$$\begin{aligned}\frac{dz^\alpha}{ds} = {} & \frac{dz_0^\alpha}{ds}\cos\frac{w}{R} + R\frac{d\eta_0^\alpha}{ds}\sin\frac{w}{R} \\ & + \left[-z_0^\alpha \sin\frac{w}{R} + R\eta_0^\alpha\cos\frac{w}{R}\right]\frac{1}{R}\frac{dw}{ds}.\end{aligned} \tag{64.16}$$

In order that the point of coördinates z^α be displaced tangentially to the geodesic at the point, we must have as follows from (64.16) and (63.7)

$$\frac{dz_0^\alpha}{ds}\cos\frac{w}{R} + R\frac{d\eta_0^\alpha}{ds}\sin\frac{w}{R} = \varrho\left(-z_0^\alpha\sin\frac{w}{R} + R\eta_0^\alpha\cos\frac{w}{R}\right),$$

where ϱ is a factor of proportionality. If we multiply by $c^\alpha\eta_0^\alpha$, sum for α and make use of (63.6) and $\sum_\alpha c^\alpha z_0{}^\alpha{}_{,i}\eta_0^\alpha = 0$, we find that $\varrho = 0$. Hence we have

$$\left(z_0{}^\alpha{}_{,i} + R\eta_0{}^\alpha{}_{,i}\tan\frac{w}{R}\right)\frac{dx^i}{ds} = 0. \tag{64.17}$$

* Cf. *Bianchi*, 1902, 1, pp. 488–491.

When this equation is multiplied by $c^\alpha z_0{}^\alpha{}_{,j}$ and summed for α, we obtain, in consequence of (64.10) and (64.12),

$$\left(g_{ij} - R\Omega_{ij} \tan \frac{w}{R}\right) \frac{dx^i}{ds} = 0. \tag{64.18}$$

Comparing this equation with (45.1), we see that the curves possessing the desired property are the lines of curvature of V_n, and if R_i denote the principal radii of normal curvature, the quantities w_i are given by

$$\tan \frac{w_i}{R} = \frac{R_i}{R}. \tag{64.19}$$

App. 24 replaces the last sentence of this paragraph

Conversely, when (64.18) are satisfied, we have

$$c_\alpha z_0{}^\alpha{}_{,j} \left(z_0{}^\alpha{}_{,i} + R\eta_0{}^\alpha{}_{,i} \tan \frac{w}{R}\right) \frac{dx^i}{ds} = 0.$$

Also the equations

$$c_\alpha \eta_0{}^\alpha \left(z_0{}^\alpha{}_{,i} + R\eta_0{}^\alpha{}_{,i} \tan \frac{w}{R}\right) \frac{dx^i}{ds} = 0$$

are satisfied identically. Since the determinant of these equations is different from zero, equations (64.17) follow, and consequently the theorem is proved for the case (64.15).

Proceeding in like manner with the second case of § 63, we obtain similar results. In place of (64.19) we have

$$\tanh \frac{w_i}{R} = \frac{R_i}{R},$$

from which it follows that w_i is real or imaginary according as R_i is less or greater than R.

65. Spaces V_n conformal to an S_n. In § 28 we established by direct processes the conditions in tensor form that a space V_n be conformal to an S_n. In this section we show that such a V_n can be immersed in an S_{n+2} and make use of the results of § 56 to obtain the conditions obtained in § 28.

If V_n is conformal to an S_n, there exists a coördinate system x^i for which the fundamental form of V_n is

$$\varphi = \psi^2 \sum_i c_i (dx^i)^2 \qquad (i = 1, \cdots, n), \tag{65.1}$$

where the c's are plus or minus one and ψ is a function of the x's. If we put

$$(65.2)\qquad \begin{aligned} z^i &= \psi x^i, \qquad z^{n+1} = \psi\left(\sum_i c_i (x^i)^2 - \frac{1}{4}\right), \\ z^{n+2} &= \psi\left(\sum_i c_i (x^i)^2 + \frac{1}{4}\right), \end{aligned}$$

we have from (65.1)

$$(65.3)\qquad \varphi = \sum_\alpha c_\alpha (dz^\alpha)^2 \qquad (\alpha = 1, \cdots, n+2),$$

where

$$(65.4)\qquad c_{n+1} = 1, \qquad c_{n+2} = -1,$$

and from (65.2)

$$(65.5)\qquad \sum_\alpha c_\alpha (z^\alpha)^2 = 0.$$

If we call (65.5) the *fundamental hypercone* of the S_{n+2} with the fundamental form (65.3), we have that V_n is immersible in an S_{n+2} and is in fact a hypersurface of the fundamental hypercone (65.5).

Conversely, equation (65.5) and any equation $F(z^1, \cdots, z^{n+2}) = 0$ not homogeneous in the z's define a hypersurface of the hypercone. If in the equation $F = 0$, we substitute the expressions (65.2), we find the function ψ of the x's in terms of which (65.3) is reducible to (65.1). Hence we have the following theorem which is a generalization of a theorem due to Brinkmann:*

Any V_n which is conformal to an S_n is a hypersurface of the fundamental hypercone of a certain S_{n+2}, and any hypersurface of the fundamental hypercone of an S_{n+2} which is not a hypercone with the same vertex is conformal to an S_n.

In terms of any coördinates x^i in V_n we have from (65.3)

$$(65.6)\qquad \sum_\alpha c_\alpha \overset{\alpha}{z}_{,i} \overset{\alpha}{z}_{,j} = g_{ij}.$$

Differentiating (65.5) covariantly with respect to x^i and x^j and the fundamental form of V_n, we have in consequence of (65.6)

$$(65.7)\qquad \begin{aligned} \sum_\alpha c_\alpha \overset{\alpha}{z} \overset{\alpha}{z}_{,i} &= 0, \\ \sum_\alpha c_\alpha \overset{\alpha}{z} \overset{\alpha}{z}_{,ij} &= -g_{ij}. \end{aligned}$$

* 1923, 7, p. 1.

As in § 56 we denote by $\eta_{\sigma|}{}^{\alpha}$ for $\sigma = 1, 2$ the components of two mutually orthogonal unit vectors in S_{n+2} normal to V_n. From the first of (65.7) it follows that $z^{\alpha} = r\,\eta_{1|}{}^{\alpha} + t\,\eta_{2|}{}^{\alpha}$. Substituting in (65.5) we find that $r^2 e_1 + t^2 e_2 = 0$. Hence e_1 and e_2 differ in sign. Without loss of generality we take $e_1 = -e_2 = 1$, so that

$$(65.8) \qquad \sum_{\alpha} c_{\alpha} (\eta_{1|}{}^{\alpha})^2 = 1, \qquad \sum_{\alpha} c_{\alpha} (\eta_{2|}{}^{\alpha})^2 = -1$$

and then

$$(65.9) \qquad z^{\alpha} = r(\eta_{1|}{}^{\alpha} + \eta_{2|}{}^{\alpha}),$$

where r is an invariant. From the conditions

$$\sum_{\alpha} c_{\alpha} z^{\alpha}{}_{,i}\, \eta_{\sigma|}{}^{\alpha} = 0,$$

(65.9) and (65.8) we have

$$(65.10) \quad \sum_{\alpha} c_{\alpha}\, \eta_{2|}{}^{\alpha}\, \eta_{1|}{}^{\alpha}{}_{,i} = -\sum_{\alpha} c_{\alpha}\, \eta_{1|}{}^{\alpha}\, \eta_{2|}{}^{\alpha}{}_{,i} = \frac{\partial \log r}{\partial x^i}.$$

Hence from (56.3) we have

$$(65.11) \qquad \nu_{21|i} = -\nu_{12|i} = \frac{\partial \log r}{\partial x^i},$$

so that

$$(65.12) \quad \eta_{\sigma|}{}^{\alpha}{}_{,i} = -b_{\sigma|li}\, g^{lm} z^{\alpha}{}_{,m} - \eta_{\tau|}{}^{\alpha} \frac{\partial \log r}{\partial x^i} \quad (\sigma, \tau = 1, 2;\ \sigma \neq \tau).$$

From the second of (65.7) and from (65.9), (56.2) and (56.1) we have

$$(65.13) \qquad b_{2|ij} = -\left(b_{1|ij} + \frac{1}{r}\, g_{ij}\right).$$

In this case equations (56.5) reduce to

$$(65.14) \qquad \begin{aligned} R_{ijkl} = {} & \frac{1}{r}(b_{1|il}\, g_{jk} + b_{1|jk}\, g_{il} - b_{1|ik}\, g_{jl} - b_{1|jl}\, g_{ik}) \\ & + \frac{1}{r^2}(g_{il}\, g_{jk} - g_{ik}\, g_{jl}); \end{aligned}$$

in place of (56.6) we have

$$(65.15)\quad b_{1|ij,k} - b_{1|ik,j} = \frac{\partial \log r}{\partial x^k}\left(b_{1|ij} + \frac{1}{r} g_{ij}\right) - \frac{\partial \log r}{\partial x^j}\left(b_{1|ik} + \frac{1}{r} g_{ik}\right),$$

and (56.7) are satisfied identically.

By means of (65.9) and (65.13) equations (56.2) can be written

$$(65.16)\qquad z^{\alpha}_{\ ,ij} = \frac{1}{r}(b_{1|ij}\, z^{\alpha} + g_{ij}\, \eta_{2|}{}^{\alpha}),$$

and (65.12) for $\sigma = 2$ becomes

$$(65.17)\quad \eta_{2|}{}^{\alpha}{}_{,i} = b_{1|li}\, g^{lm}\, z^{\alpha}_{\ ,m} + \frac{1}{r} z^{\alpha}_{\ ,i} + \left(\eta_{2|}{}^{\alpha} - \frac{z^{\alpha}}{r}\right)\frac{\partial \log r}{\partial x^i}.$$

When (65.14) and (65.15) are satisfied, equations (65.16), (65.17) and

$$\frac{\partial z^{\alpha}}{\partial x^i} = z^{\alpha}_{\ ,i}$$

form a completely integrable system in z^{α}, $z^{\alpha}_{\ ,i}$ and $\eta_{2|}{}^{\alpha}$. As in § 56 it can be shown that if the initial values of the quantities are chosen so that

$$(65.18)\qquad \begin{aligned} &\sum_{\alpha} c_{\alpha}\, z^{\alpha}_{\ ,i}\, z^{\alpha}_{\ ,j} = g_{ij}, &&\sum_{\alpha} c_{\alpha}\, z^{\alpha}\, z^{\alpha}_{\ ,i} = 0,\\ &\sum_{\alpha} c_{\alpha}\, z^{\alpha}_{\ ,i}\, \eta_{2|}{}^{\alpha} = 0, &&\sum_{\alpha} c_{\alpha}\, (z^{\alpha})^2 = 0, \end{aligned}$$

these equations will be satisfied by all values of the x's. Hence two tensors g_{ij}, $b_{1|ij}$ and an invariant r in the relations (65.14) and (65.15) determine a V_n with the fundamental tensor g_{ij} which is conformal with an S_n.

If we put

$$(65.19)\qquad d_{ij} = \frac{1}{r} b_{1|ij} + \frac{1}{2r^2} g_{ij},$$

equations (65.14) and (65.15) become

$$(65.20)\qquad R_{ijkl} = g_{jk}\, d_{il} + g_{il}\, d_{jk} - g_{ik}\, d_{jl} - g_{jl}\, d_{ik}$$

and

$$(65.21)\qquad d_{ij,k} - d_{ik,j} = 0.$$

From (65.20) we have for the components of the Ricci tensor

$$R_{jk} = g_{jk}\, g^{il}\, d_{il} + (n-2)\, d_{jk} \tag{65.22}$$

and consequently

$$R = 2(n-1) g^{ij} d_{ij}.$$

Hence from (65.22) we have

$$d_{jk} = \frac{1}{n-2} R_{jk} - \frac{1}{2(n-1)(n-2)} R\, g_{jk}. \tag{65.23}$$

When these expressions for the d's are substituted in (65.20) and (65.21), we get equations (28.17) and (28.19) respectively.

Exercises

1. Determine the conditions which the functions t_ϱ^σ must satisfy, in order that for the normal of components

$$\bar{\eta}_{\varrho|}^{\;\alpha} = t_\varrho^\sigma\, \eta_{\sigma|}^{\;\alpha} \qquad (\sigma = n+1, \cdots, n+p)$$

the conditions (57.3) are satisfied identically.

2. Show that in a euclidean space of $n\,(>3)$ dimensions there are no hypersurfaces of constant negative curvature. *Bianchi*, 1902, 1, p. 485.

3. When the fundamental form of a space of constant curvature K_0 is definite, the hypersurfaces of constant curvature K are such that $K > K_0$. *Levy*, 1925, 1.

4. A necessary and sufficient condition that a hypersurface of a space of constant curvature be of constant curvature is that the lines of curvature of the hypersurface be indeterminate. *Levy*, 1925, 1.

5. Show that, if in (27.4) the b's are given the values zero and the c's are chosen so that $\sum_i e_i c_i = 1/4$, the fundamental form is reducible to

$$\varphi = R^2 \frac{\sum e_i (dx^i)^2}{\left(\sum e_i x^{i^2} + \frac{e}{4}\right)^2},$$

on replacing x^i by Rx^i and K_0 by e/R^2.

6. When in (61.2) we put $c_i = e_i$ and $c_{n+1} = e$, this equation is satisfied by

$$z^i = R \frac{x^i}{\sum_i e_i x^{i^2} + \frac{e}{4}}, \qquad z^{n+1} = R \frac{\sum_i e_i x^{i^2} - \frac{e}{4}}{\sum_i e_i x^{i^2} + \frac{e}{4}},$$

and in terms of the x's the fundamental form is that of Ex. 5.

7. When a hypersurface V_n of a space of constant curvature admits n congruences of lines of curvature, their tangents are Ricci principal directions for V_n.

8. When in the equations of § 65 we put $z^{n+2} = R$ or $z^{n+1} = R$, we have the case of spaces V_n conformal to spaces of constant curvature.

9. The *third fundamental form* of a hypersurface of an S_{n+1} is given by

$$\psi = \sum_{\alpha} c_\alpha (d\eta^\alpha)^2 = b_{ik}\, b_{jl}\, g^{kl}\, dx^i\, dx^j.$$

10. When V_n is a hypersurface of a space of constant curvature, from (64.10) it follows that

$$\psi = \sum_{\alpha} c_\alpha (d\eta^\alpha)^2 = \Omega_{ik}\, \Omega_{jl}\, g^{kl}\, dx^i\, dx^j.$$

Bianchi calls this the third fundamental form of the hypersurface.

Bianchi, 1902, 1, p. 488.

11. When in (27.4) $a = 0$, $c_i = 0$, $b_j = 0\ (j = 1, \cdots, n-1)$, this equation becomes $K_0 = -4 e_n b_n^2$ and the fundamental form of the space of constant curvature K_0 is

$$\varphi = \frac{e_1 (dx^1)^2 + \cdots + e_n (dx^n)^2}{4 b_n^2\, x^{n^2}}.$$

12. When the fundamental form of a space of constant curvature is

$$\varphi = \frac{e_1 (dx^1)^2 + \cdots + e_n (dx^n)^2}{e_n\, x^{n^2}},$$

the function

$$U = \frac{e_n}{x^n} [e_1 (x^1 - a^1)^2 + \cdots + e_{n-1} (x^{n-1} - a^{n-1})^2 + e_n\, x^{n^2}],$$

where the a's are arbitrary constants, in such that $\Delta_1 U = U^2$. Hence (§ 19) the finite equations of the geodesics are

$$x^j - a^j = \frac{1}{2} e_j\, e_n\, b_j\, x^n\, U \qquad (j = 1, \cdots, n-1),$$

where the b's are arbitrary constants. *Bianchi,* 1902, 1, p. 422.

13. If U in Ex. 12 be replaced by $\frac{2}{b} e^s$, where $e_n b^2 = e_1 b_1^2 + \cdots + e_{n-1} b_{n-1}^2$, the equations of the geodesics can be written

$$x^n = \frac{1}{b \cosh s}, \qquad x^j = c^j + e_j\, e_n \frac{b_j}{b^2} \tanh s \quad (j = 1, \cdots, n-1),$$

where the c's are arbitrary constants. *Bianchi,* 1902, 1, p. 422.

14. For a given set of values of c^j in Ex. 13, the geodesics of V_n lie in the hypersurface

$$\sum_j^{1,\ldots,n-1} e_j (x^j - c^j)^2 + e_n\, x^{n^2} = \frac{e_n}{b^2},$$

and are geodesics of this hypersurface (§ 24). *Beltrami,* 1868, 1, p. 234.

15. Show by means of the theorem of Beltrami (§ 40) that the hypersurfaces in Ex. 14 have constant Riemannian curvature.

16. The determination of n-tuply orthogonal systems of hypersurfaces in a space of constant curvature K_0 reduces to the solution of the system of equations (cf. § 37)

$$\frac{\partial H_i}{\partial x^j} = H_j \beta_{ji}, \qquad \frac{\partial \beta_{ij}}{\partial x^k} - \beta_{ik} \beta_{kj} = 0 \qquad (i, j, k \neq),$$

$$e_i \frac{\partial \beta_{ji}}{\partial x^j} + e_j \frac{\partial \beta_{ij}}{\partial x^i} + \sum_k e_i\, e_j\, e_k\, \beta_{ki}\, \beta_{kj} + K_0\, e_i\, e_j\, H_i\, H_j = 0,$$

where $\beta_{ii} = 0$.

17. When a space of constant curvature K_0 is referred to an n-tuply orthogonal system of hypersurfaces $x^i =$ const. and the fundamental tensor has the form (37.1), the functions $\eta_{i|}{}^{\alpha}$, defined by

$$\frac{\partial z^{\alpha}}{\partial x^i} = \eta_{i|}{}^{\alpha} H_i \qquad (i = 1, \cdots, n),$$

where z^{α} are coördinates of Weierstrass, satisfy the equations

$$\sum_{\alpha} c_{\alpha}\, z^{\alpha}\, \eta_{i|}{}^{\alpha} = 0, \qquad \sum_{\alpha} c_{\alpha} (\eta_{i|}{}^{\alpha})^2 = e_i, \qquad \sum_{\alpha} c_{\alpha}\, \eta_{i|}{}^{\alpha}\, \eta_{j|}{}^{\alpha} = 0 \qquad (i \neq j).$$

Show that (cf. § 37)

$$\frac{\partial \eta_{i|}{}^{\alpha}}{\partial x^j} = \eta_{j|}{}^{\alpha} \beta_{ij}, \qquad \frac{\partial \eta_{i|}{}^{\alpha}}{\partial x^i} = -e_i \sum_k e_k\, \eta_{k|}{}^{\alpha}\, \beta_{ki} - e_i\, K_0\, H_i\, z^{\alpha}.$$

Bianchi, 1924, 3, p. 651.

CHAPTER VI

Groups of motions

66. Properties of continuous groups. For a V_n expressed in terms of coördinates x^i the equations

$$\bar{x}^i = f^i(x^1, x^2, \cdots, x^n; a) \qquad (i = 1, \cdots, n), \tag{66.1}$$

where a is a parameter, define for each value of a a point transformation of V_n. If the functions f^i are such that the combination of two such transformations is one of the transformations (66.1), and if also the identity transformation and the inverse of every transformation is in the set, then these transformations are said to form a *one-parameter continuous group of transformations.** In this case the $\bar{x}$'s considered as functions of a satisfy a system of differential equations of the form

$$\frac{d\bar{x}^i}{da} = \psi(a)\, \xi^i(\bar{x}^1, \bar{x}^2, \cdots, \bar{x}^n).\dagger \tag{66.2}$$

If a_0 is the value of a for the identity transformation and if we put $t = \int_{a_0}^{a} \psi(a)\, da$, the equations (66.2) become

$$\frac{d\bar{x}^i}{dt} = \xi^i(\bar{x}^1, \cdots, \bar{x}^n), \tag{66.3}$$

and the identity is given by $t = 0$. If the functions ξ^i are assumed to be regular in the domain of x^i, the integrals of (66.3) can be written in the form

$$\bar{x}^i = x^i + \xi^i(x)\, t + \xi^j \frac{\partial \xi^i}{\partial x^j} \frac{t^2}{2} + \cdots. \tag{66.4}$$

---

* The restriction that the identity and the inverse of every transformation be in the group is not made in the general definition of a group as given by *Lie*, 1893, 3, p. 368. However, the above definition is in keeping with that generally in vogue today, and the groups of the less restricted type are called *semi-groups*.

† Cf. *Lie*, 1893, 3, p. 371; also *Bianchi*, 1918, 4, p. 63.

If we introduce the notation

$$(66.5) \qquad Xf = \xi^i \frac{\partial f}{\partial x^i},$$

and indicate by $X^r f$ the result of performing the operation X on f r times in succession, equations (66.4) can be written

$$(66.6) \qquad \overline{x}^i = x^i + tXx^i + \frac{t^2}{2} X^2 x^i + \cdots + \frac{t^r}{r!} X^r x^i + \cdots.$$

Moreover any function $F(x^1, \cdots, x^n)$ regular in the domain of x^i is expressible in the form

$$(66.7) \quad F(\overline{x}^1, \cdots, \overline{x}^n) = F(x^1, \cdots, x^n) + tXF + \cdots + \frac{t^r}{r!} X^r F + \cdots.$$

When in (66.4) we replace t by the infinitesimal δt, we obtain, on neglecting terms of higher order,

$$(66.8) \qquad \overline{x}^i = x^i + \xi^i \delta t.$$

This is the *infinitesimal transformation* of the group and from (66.8) we have that the x's undergo the infinitesimal change

$$(66.9) \qquad \delta x^i = \xi^i \delta t.$$

Moreover from (66.7) we have that the change of any function F is given by

$$(66.10) \qquad \delta F = XF \cdot \delta t.$$

The equations (66.8) are uniquely defined by the form of Xf which Lie* calls the *symbol* of the infinitesimal transformation of the group. The equations (66.4) of the group are then determined; the group is said to be *generated* by Xf. It is understood that Xf and $a\,Xf$, where a is any constant, generate the same group. We shall at times refer to Xf as the *generator* of the group.

Equations (66.3) define a congruence of curves in V_n, the *paths* of the group, each of which is described by a point as the latter undergoes the continuous transformation of the group.

* 1893, 3, p. 390; *Bianchi*, 1918, 4, p. 67.

From (66.3) it is seen that ξ^i are the contravariant components of a vector; we call them the *contravariant components* of the infinitesimal transformation. From § 2 it follows that there exists a transformation of coördinates $x'^1 = x^1$, $x'^j = \varphi^j(x^1, \cdots, x^n)$ for $j = 2, \cdots, n$ so that in the new system the components $\xi'^j = 0$ for $j = 2, \cdots, n$. If we effect the further change defined by App. 25 replaces lines 3-8

$$x''^1 = \int \frac{dx'^1}{\xi'^1}, \quad x''^j = x'^j \qquad (j = 2, \cdots, n),$$

it follows from (66.5) that in this system $Xf = \dfrac{\partial f}{\partial x''^1}$. Hence:

The coördinates of a V_n can be chosen so that the contravariant components of the infinitesimal transformation of a one-parameter group are

(66.11) $$\xi^1 = 1, \qquad \xi^j = 0 \qquad (j = 2, \cdots, n).$$

In this coördinate system the finite equations of the group are

(66.12) $$\bar{x}^1 = x^1 + t, \qquad \bar{x}^j = x^j,$$

as follows from (66.4). As an immediate consequence we have that a one parameter group containing the identity contains also the inverse of every transformation of the group.

When the equations of a transformation involve r essential parameters, thus

(66.13) $$\bar{x}^i = f^i(x^1, \cdots, x^n;\ a^1, \cdots, a^r) \qquad (i = 1, \cdots, n),$$

and these transformations possess the property referred to in connection with (66.1), they are said to form a group G_r. We say that r infinitesimal transformations

(66.14) $$X_\alpha f = \xi_{\alpha|}^{\ i} \frac{\partial f}{\partial x^i} \qquad (\alpha = 1, \cdots, r)$$

are *linearly independent*, when there do not exist constants c^α for which

(66.15) $$c^\alpha \xi_{\alpha|}^{\ i} = 0.$$

Suppose that r linearly independent infinitesimal transformations satisfy the conditions (cf. § 23)

$$(66.16) \qquad (X_\alpha, X_\beta) f = c_{\alpha\beta}^{\ \ \gamma} X_\gamma f \qquad (\alpha, \beta, \gamma = 1, \cdots, r),$$

where the c's are constants, called the *constants of composition* of the group, and are subject to the conditions

$$(66.17) \qquad \begin{aligned} c_{\alpha\beta}^{\ \ \gamma} + c_{\beta\alpha}^{\ \ \gamma} &= 0, \\ c_{\alpha\beta}^{\ \ \gamma} c_{\gamma\delta}^{\ \ \varepsilon} + c_{\beta\delta}^{\ \ \gamma} c_{\gamma\alpha}^{\ \ \varepsilon} + c_{\delta\alpha}^{\ \ \gamma} c_{\gamma\beta}^{\ \ \varepsilon} &= 0 \\ (\alpha, \beta, \gamma, \delta, \varepsilon &= 1, \cdots, r). \end{aligned}$$

It can be shown* that the $X_\alpha f$ generate a group G_r consisting of all the groups G_1 generated by the infinitesimal transformations

$$(66.18) \qquad a^\alpha X_\alpha f,$$

where the a's are arbitrary constants; and conversely every group G_r can be generated by r linearly independent infinitesimal transformations (66.14) satisfying (66.16) and (66.17).

If the components ξ^i of an infinitesimal transformation are regular in the neighborhood of a point P_0 of coördinates x_0^i, and they are expressed in the form

$$(66.19) \qquad \xi^i = \xi_0^i + \alpha^i_{\ j}(x^j - x_0^j) + \alpha^i_{\ jk}(x^j - x_0^j)(x^k - x_0^k) + \cdots,$$

we say that the transformation is of *order zero* at P_0, when not all of the ξ_0^i's are zero; that is of *order one* when all the ξ_0^i's are zero but not all the $\alpha^i_{\ j}$'s, and so on.

Consider the matrix

$$(66.20) \qquad M = \left\| \begin{matrix} \xi_{1|}^{\ 1}, & \cdots, & \xi_{1|}^{\ n} \\ \cdot & \cdot & \cdot \\ \cdot & \cdot & \cdot \\ \xi_{r|}^{\ 1}, & \cdots, & \xi_{r|}^{\ n} \end{matrix} \right\|$$

of the components of the generators of a G_r. If the rank of M, when the x's are replaced by the x_0's, is r_0, then in the equations

$$a^\alpha \xi_{\alpha|}^{\ i}(x_0) = 0$$

* *Lie*, 1893, 3, pp. 391, 396· *Bianchi*, 1918, 4, pp. 97, 98.

$r - \tau_0$ of the a's can be chosen arbitrarily and the others expressed in terms of them. Hence there are $r - \tau_0$ linearly independent transformations (66.18) of order greater than zero at P_0, and τ_0 linearly independent transformations of order zero.

From equations (66.4) it is seen that if an infinitesimal transformation is of order >0 at P_0, the finite equations of the group generated by it leave P_0 fixed. The $r - \tau_0$ infinitesimal generators of order >0 generate a $G_{r-\tau_0}$, which is the sub-group of G_r leaving P_0 fixed; it is called the sub-group of *stability* of P_0.*

67. Transitive and intransitive groups. Invariant varieties. A group is said to be *transitive*, when by means of its transformations any point can be transformed into any other point; otherwise it is *intransitive*. For example, the group of motions in euclidean space of three dimensions is transitive, whereas the group of rotations about a point is intransitive. From the finite equations (66.13) of a G_r it follows that for a transitive group $r \geqq n$.

For an intransitive group G_r there are subspaces V_m of V_n such that any point of a V_m is transformable only into points of V_m; otherwise by a combination of transformations a given point could be transformed into any other point of V_n. Such a V_m is called an *invariant variety* for G_r.

If we consider any point P_0 of V_n and as in § 66 denote by τ_0 the rank of the matrix M for P_0, there are τ_0 independent infinitesimal transformations which transform P_0 into nearby points and any linear combination of the form (66.18) for $\alpha = 1, \cdots, \tau_0$ possesses this property. Hence the paths of these transformations determine a V_{τ_0} into points of which P_0 is transformable, and the sub-group of stability of P_0 is of order $r - \tau_0$. If G_r is transitive, $\tau_0 = n$, since V_{τ_0} is the same as V_n by the above definition of a transitive group. If G_r is intransitive, V_{τ_0} is a sub-space of V_n. It is the invariant variety of lowest order containing P_0 and is called the *minimum invariant variety* for P_0. If T denotes the transformation by means of which P_0 is transformed into a point P' of V_{τ_0}, T^{-1} its inverse and $\overline{T}$ any transformation of stability of P_0, then

$$T\overline{T}T^{-1}(P') = P'.$$

---

* *Bianchi*, 1918, 4, p. 147.

Since all these transformations $T\bar{T}T^{-1}$ are distinct, the group of stability of P' is of at least the same order as for P_0, and by reversing the process we have that it is of the same order. Consequently V_{τ_0} is the minimum invariant variety for any point of it. App. 26 replaces this sentence Accordingly the equations of V_{τ_0} are obtained by equating to zero all the determinants of M of order τ_0+1. Moreover, the sub-group of G_r generated by the τ_0 infinitesimal transformations referred to at the beginning of this paragraph is a transitive group for V_{τ_0}.

From the foregoing considerations it follows that, if the equations obtained by equating to zero all the determinants of the same order of M are consistent, they define an invariant variety with respect to G_r. From this we have

According as the generic rank of the matrix M in the ξ's is n or less, G_r is transitive or intransitive.

If the rank of M (66.20) in the x's is $q(<n)$, and P is a point for the coördinates of which M is of rank q, then the minimum invariant variety for P is a V_q. But if for the coördinates of P the rank is $r<q$, then the minimum invariant variety for P is a V_r and is obtained by equating to zero all the determinants of M of order $r+1$.

If the rank of M is $q(<n)$, then all of the equations

$$X_\alpha f = 0 \tag{67.1}$$

are expressible in terms of q of them. In consequence of this result and of equations (66.16) it follows from the theorem of § 23 that equations (67.1) form a complete system and admit $n-q$ independent solutions $\varphi_1, \cdots, \varphi_{n-q}$. From (66.10) it follows that any solution of equations (67.1) is an invariant for G_r, and conversely any invariant is a solution of (67.1). Hence every invariant of G_r is a function of $\varphi_1, \cdots, \varphi_{n-q}$. From these considerations we see that the equations

$$\varphi_\beta(x^1, \cdots, x^n) = \varphi_\beta(x_0^1, \cdots, x_0^n) \quad (\beta = 1, \cdots, n-q) \tag{67.2}$$

define the minimum variety for the ordinary point P_0 of coördinates x_0^i.

Let V_m be an invariant variety for a G_r, defined by the equations

$$x^i = \varphi^i(y^1, \cdots, y^m). \tag{67.3}$$

Since the paths of the transformations must be in V_m, we must have

$$\xi_{\sigma|}{}^{i} = \eta_{\sigma|}{}^{\alpha} \frac{\partial x^i}{\partial y^\alpha} \qquad \begin{pmatrix} \sigma = 1, \cdots, r; \; \alpha = 1, \cdots, m; \\ i = 1, \cdots, n \end{pmatrix}, \tag{67.4}$$

where the η's are functions of the y's. Now

$$X_\sigma f = \xi_{\sigma|}{}^{i} \frac{\partial f}{\partial x^i} = \eta_{\sigma|}{}^{\alpha} \frac{\partial f}{\partial y^\alpha} \equiv Y_\sigma f. \tag{67.5}$$

Hence the Y's are the generators of a group Γ in V_m which is said to be *induced* by G_r.

If Γ is of order less than r, there exist relations of the form

$$c^\sigma \eta_{\sigma|}{}^{\alpha} = 0 \tag{67.6}$$

and from (67.4)

$$c^\sigma \xi_{\sigma|}{}^{i} = 0 \tag{67.7}$$

at points of V_m. In this case the transformation of G_r of components

$$\xi^i = c^\sigma \xi_{\sigma|}{}^{i} \tag{67.8}$$

leaves V_m point-wise invariant. Conversely, if (67.8) leaves V_m point-wise invariant, then (67.7) must hold at points of V_m, and since the Jacobian matrix $\left\| \frac{\partial x^i}{\partial y^\alpha} \right\|$ is of rank m, (67.6) must hold. Hence:

*If V_m is an invariant variety for a G_r and a sub-group G_p of G_r leaves V_m point-wise invariant, the group induced on V_m by G_r is a G_{r-p}; and conversely.**

From the definition of minimum variety it follows that the group induced in such a variety is transitive, whereas for any other invariant variety it is intransitive.

68. Infinitesimal transformations which preserve geodesics. If a V_n with the fundamental form

$$\varphi = g_{ij}\, dx^i\, dx^j \tag{68.1}$$

---

* Cf. *Bianchi*, 1918, 4, p. 165.

is subjected to an infinitesimal transformation defined by (66.8), then from (66.8), (66.9) and (66.10) we have

$$(68.2)\quad \delta\, dx^i = d\,\delta x^i = \frac{\partial \xi^i}{\partial x^j} dx^j\, \delta t, \qquad \delta g_{ij} = \frac{\partial g_{ij}}{\partial x^k} \xi^k\, \delta t,$$

and consequently from (68.1)

$$(68.3)\quad \delta\varphi = h_{ij}\, dx^i\, dx^j\, \delta t,$$

where

$$(68.4)\quad h_{ij} = \xi^k \frac{\partial g_{ij}}{\partial x^k} + g_{ik} \frac{\partial \xi^k}{\partial x^j} + g_{jk} \frac{\partial \xi^k}{\partial x^i} = g_{ik}\, \xi^k{}_{,j} + g_{jk}\, \xi^k{}_{,i}.$$

From (68.3) it follows that the fundamental tensor of the transform $\overline{V}_n$ is given by

$$(68.5)\quad \overline{g}_{ij} = g_{ij} + h_{ij}\, \delta t.$$

For infinitesimal transformations which preserve geodesics we have equations of the form (40.6) and (40.8), in which ψ_i is replaced by $\psi_{,i}\,\delta t$, where $\psi_{,i}$ is the gradient of a function ψ. From the latter and (68.5) we obtain

$$(68.6)\quad h_{ij,k} = 2 g_{ij}\, \psi_{,k} + g_{jk}\, \psi_{,i} + g_{ik}\, \psi_{,j}.$$

From (68.5) we have (cf. § 6)

$$\overline{g} = g(1 + g^{ij} h_{ij}\, \delta t),$$

and from equations analogous to (40.7)

$$(68.7)\quad \psi_{,k} = \frac{1}{2(n+1)} g^{ij} h_{ij,k}.$$

Since (68.4) can be written in the form

$$(68.8)\quad h_{ij} = \xi_{i,j} + \xi_{j,i},$$

equations (68.7) become

$$(68.9)\quad \psi_{,k} = \frac{1}{n+1} g^{ij}\, \xi_{i,jk}.$$

From (68.6) we have

$$(68.10)\quad h_{ij,k} + h_{ik,j} - h_{jk,i} = 2(g_{ij}\, \psi_{,k} + g_{ik}\, \psi_{,j}).$$

Substituting in this equation from (68.8) and making use of Ricci identities (§ 11) of the form

$$(68.11) \qquad \xi_{i,jk} - \xi_{i,kj} = \xi_m R^m{}_{ijk}$$

and of the identity (8.11), we obtain

$$(68.12) \qquad \xi_{i,jk} = -\xi_m R^m{}_{kij} + g_{ij}\psi_{,k} + g_{ik}\psi_{,j}.$$

From these equations and (68.9) we must have

$$\xi_m R^m{}_{kij} g^{ij} = 0,$$

which is identically satisfied, since $R^m{}_{kij}$ is skew-symmetric in i and j.
The conditions of integrability of (68.12) are [cf. (11.15)]

$$(68.13) \qquad \begin{aligned} \xi_m(R^m{}_{kij,l} - R^m{}_{lij,k}) + \xi_{m,l} R^m{}_{kij} - \xi_{m,k} R^m{}_{lij} + \xi_{i,m} R^m{}_{jkl} \\ + \xi_{m,j} R^m{}_{ikl} + g_{il}\psi_{,jk} - g_{ik}\psi_{,jl} = 0. \end{aligned}$$

Multiplying by g^{il} and summing for i and l, we have

$$(68.14) \quad \psi_{,jk} = \frac{1}{n-1}(\xi_m R^m{}_{j,k} + \xi_{m,k} R^m{}_j + \xi_{m,j} R^m{}_k + g^{il}\xi_m R^m{}_{kji,l}),$$

where $R^m{}_j = g^{mk} R_{kj}$.*
Since $\psi_{,jk}$ must be symmetric in j and k, we have from (68.14)

$$\xi_m[R^m{}_{j,k} - R^m{}_{k,j} + g^{il}(R^m{}_{kji,l} - R^m{}_{jki,l})] = 0.$$

When the expressions (68.14) are substituted in (68.13), we obtain equations of condition linear in ξ_i and $\xi_{i,j}$ for $i, j = 1, \cdots, n$. In addition, the conditions of integrability of (68.14) are linear in ξ_i, $\xi_{i,j}$ and $\psi_{,i}$. From these equations we obtain by continued differentiation other equations linear in ξ_i, $\xi_{i,j}$ and $\psi_{,i}$. All of these equations must be algebraically consistent, if the given V_n is to admit infinitesimal transformations preserving geodesics.

---

* For, $g^{il}\xi_{i,m} R^m{}_{jkl} = g^{il}\xi_{i,m} g^{mt} R_{tjkl} = g^{il}\xi_{i,m} g^{mt} R_{lkjt} = \xi_{m,i} g^{il} R^m{}_{kjl}$ by changing the dummy indices.

When V_n is of constant curvature $K_0 (\neq 0)$, equations (68.13) reduce in consequence of (40.12) to

$$g_{il}[K_0(\xi_{j,k} + \xi_{k,j}) + \psi_{,jk}] - g_{ik}[K_0(\xi_{j,l} + \xi_{l,j}) + \psi_{,jl}] = 0,$$

from which follows, for $n \neq 1$,

$$K_0(\xi_{i,j} + \xi_{j,i}) + \psi_{,ij} = 0. \tag{68.15}$$

From the second of (40.13), where now $A_{ij} = K_0 g_{ij} - \psi_{,ij} \delta t$, and from (68.5) we have $\overline{R}_{hijl} = K_0(\overline{g}_{hj}\overline{g}_{il} - \overline{g}_{hl}\overline{g}_{ij})$, which is in accordance with the theorem of Beltrami (§ 40). In this case equations (68.12) reduce to

$$\xi_{i,jk} = K_0(g_{ik}\xi_j - g_{jk}\xi_i) + g_{ij}\psi_{,k} + g_{ik}\psi_{,j}. \tag{68.16}$$

Differentiating (68.15) covariantly with respect to x^k and substituting from (68.16), we find that ψ must satisfy the equations

$$\psi_{,ijk} + K_0(2g_{ij}\psi_{,k} + g_{ik}\psi_{,j} + g_{jk}\psi_{,i}) = 0. \tag{68.17}$$

The conditions of integrability (40.17) of these equations are satisfied identically.

If we put

$$\xi_i = \overline{\xi}_i - \frac{1}{2K_0}\psi_{,i}, \tag{68.18}$$

where ψ is any solution of (68.17), equations (68.15) and (68.16) reduce respectively to

$$\begin{aligned} \overline{\xi}_{i,j} + \overline{\xi}_{j,i} &= 0, \\ \overline{\xi}_{i,jk} &= K_0(g_{ik}\overline{\xi}_j - g_{jk}\overline{\xi}_i). \end{aligned} \tag{68.19}$$

In § 71 it will be shown that these equations admit $n(n+1)/2$ independent solutions. Hence for each solution of (68.17) there are $n(n+1)/2$ independent infinitesimal transformations of a V_n of constant curvature preserving geodesics.

69. Infinitesimal conformal transformations. From (68.5) and (68.8) we have the $\overline{V}_n$ resulting from an infinitesimal transformation of a V_n is conformal with V_n, when

$$h_{ij} = \xi_{i,j} + \xi_{j,i} = \psi g_{ij}, \tag{69.1}$$

where ψ is an invariant. The case where $\psi = 0$ will be treated in the next and subsequent sections.

A necessary and sufficient condition that the paths of two transformations ξ^i and $\bar{\xi}^i$ be the same is that $\bar{\xi}^i = \varrho\, \xi^i$. From (69.1) and analogous equations in the $\bar{\xi}$'s we have in this case

$$\varrho_{,j}\, \xi_i + \varrho_{,i}\, \xi_j = (\bar{\psi} - \varrho\, \psi) g_{ij}.$$

Consider first the case when $\bar{\psi} - \varrho\, \psi = 0$. One of the ξ's must be different from zero, say ξ_1. When we take $i = j = 1$, we get $\varrho_{,1} = 0$; and when we take $i = 1$, $j \neq 1$, we get $\varrho_{,j} = 0$. Hence ϱ is a constant and the two transformations are the same. When $\bar{\psi} - \varrho\, \psi \neq 0$, it follows from the above equations that the rank of the determinant $|g_{ij}|$ is not greater than 2. Hence we have the theorem of Fubini:*

Two infinitesimal conformal transformations of a V_n for $n > 2$ cannot have the same paths.

From (69.1) we have

$$h_{ij,k} + h_{ik,j} - h_{jk,i} = g_{ij}\, \psi_{,k} + g_{ik}\, \psi_{,j} - g_{jk}\, \psi_{,i}.$$

Proceeding with this equation in a manner similar to that followed in the case of (68.10), we get

$$\xi_{i,jk} = -\xi_m R^m_{\cdot kij} + \frac{1}{2}(g_{ij}\, \psi_{,k} + g_{ik}\, \psi_{,j} - g_{jk}\, \psi_{,i}). \tag{69.2}$$

The conditions of integrability of these equations are

$$\begin{aligned} &\xi_m (R^m_{\cdot kij,l} - R^m_{\cdot lij,k}) + \xi_{m,l} R^m_{\cdot kij} - \xi_{m,k} R^m_{\cdot lij} + \xi_{i,m} R^m_{\cdot jkl} \\ &\quad + \xi_{m,j} R^m_{\cdot ikl} + \frac{1}{2}(g_{il}\, \psi_{,jk} - g_{ik}\, \psi_{,jl} + g_{jk}\, \psi_{,il} - g_{jl}\, \psi_{,ik}) = 0. \end{aligned} \tag{69.3}$$

If these equations be multiplied by g^{il} and be summed for i and l, we get

* 1903, 3. p. 410.

$$g^{il}\xi_m R^m{}_{kij,l} - \xi_m R^m{}_{j,k} - \xi_{m,k} R^m{}_j - \xi_{m,j} R^m{}_k + \frac{1}{2}(n-2)\psi_{,jk} \tag{69.4}$$
$$+ \frac{1}{2} g_{jk}\Delta_2\psi = 0,^*$$

where $\Delta_2\psi$ is defined by (14.3). Multiplying by g^{jk} and summing for j and k, we have

$$\Delta_2\psi = \frac{2}{n-1}(\xi_m R^{mi}{}_{,i} + \xi_{m,i} R^{mi}), \tag{69.5}$$

where $R^{mi} = g^{ml} R^i_l = g^{ml} g^{ri} R_{lr}$. Substituting this expression for $\Delta_2\psi$ in (69.4), we have $\psi_{,jk}$ expressed linearly in terms of ξ_i and $\xi_{i,j}$ for $i, j = 1, \cdots, n$, and the general procedure to be applied to this case is similar to that applied to (68.12), (68.13) and (68.14).

When V_n is a space of constant curvature $K_0 \neq 0$, equations (69.3) reduce in consequence of (40.12) and (69.1) to

$$K_0\psi(g_{il}g_{jk} - g_{ik}g_{jl}) \tag{69.6}$$
$$+ \frac{1}{2}(g_{il}\psi_{,jk} - g_{ik}\psi_{,jl} + g_{jk}\psi_{,il} - g_{jl}\psi_{,ik}) = 0,$$

and (69.4) to

$$2K_0(n-1)g_{jk}\psi + (n-2)\psi_{,jk} + g_{jk}\Delta_2\psi = 0. \tag{69.7}$$

Multiplying by g^{jk} and summing for j and k, we have $\Delta_2\psi + K_0 n\psi = 0$, by means of which (69.7) reduces for $n > 2$ to

$$\psi_{,jk} + K_0 g_{jk}\psi = 0. \tag{69.8}$$

When ψ is a solution of these equations, equations (69.6) are satisfied identically. Moreover, the conditions of integrability of (69.8) are satisfied. If we have any solution of (69.8), equations (69.1) may be written by means of (69.8) in the form (68.15). If in this equation and (69.2) we make the substitution (68.18), we obtain (68.19). Consequently for each solution of (69.8) there are $n(n+1)/2$ independent infinitesimal conformal transformations of a V_n of constant curvature.

See App. 27 Let G_r be an intransitive group of conformal transformations of a V_n and take for the hypersurfaces $x^1 =$ const. ∞^1 invariant

* Cf. footnote p. 229.

varieties, x^1 being the parameter of the orthogonal trajectories of these invariant varieties; also we take hypersurfaces formed by these trajectories for $x^j = \text{const.}$ where $j = 2, \cdots, n$. It is assumed that the orthogonal trajectories are not null curves; hence we have

$$g_{11} \neq 0, \qquad g_{1j} = 0 \qquad (j = 2, \cdots, n). \tag{69.9}$$

Since the hypersurfaces $x^1 = \text{const.}$ are invariant varieties it follows from (69.9) that $\xi^1_{\sigma|} = 0$. When in the equations (69.1), in which h_{ij} is given by (68.4), we take $i = 1,\ j \neq 1$, we get $g_{jk} \dfrac{\partial \xi^k_{\sigma|}}{\partial x^1} = 0$. Hence $\xi^j_{\sigma|}$ for $j = 2, \cdots, n$ are independent of x^1. Consequently the coördinates can be chosen so as to involve $n-1$ variables, and the group transforms conformally into itself not only V_n, but also each of the V_{n-1}'s. For $i = j = 1$, the equation of condition is $\xi^k_{\sigma|} \dfrac{\partial g_{11}}{\partial x^k} = \psi g_{11}$. Since $\psi \neq 0$ and $g_{11} \neq 0$ by hypothesis, the V_{n-1}'s do not admit a sub-group of stability and are minimum invariant varieties, if the rank of M (66.20) is $n-1$.

If the V_{n-1}'s are not the minimum invariant varieties, we may proceed with any of them as we did with V_n, and reduce the group to one operating on $n-2$ variables; and so on. Hence we have:*

If a group G of conformal transformations of a V_n admits minimum invariant varieties of order m, the group may be reduced by means of a transformation of variables to a group on m variables with only m linearly independent transformations.

70. Infinitesimal motions. The equations of Killing. When, as remarked in § 27, a space V_n is of such a character that there exist two systems of coördinates, x^i and $\bar{x}^i$, for which the corresponding coefficients g_{ij} and $\bar{g}_{ij}$ of the fundamental forms are the same functions of x^i and $\bar{x}^i$ respectively and the equations of transformation of the two sets of coördinates involve one or more parameters, these equations may be interpreted as defining a continuous motion of the space into itself. In § 27 it was shown that any space of constant curvature admits a continuous group of motions of $n(n+1)/2$ parameters, and that spaces of constant

---

* *Fubini*, 1903, 3, p. 405.

curvature are the only ones admitting a group with $n(n+1)/2$ parameters. Also it was pointed out that the method of Christoffel (§ 10) could be used to determine whether a given space admits a group of motions. In the remainder of this chapter we apply the Lie theory to this problem.

We remark that if a V_n admits a group of motions, the fundamental form (68.1) of V_n must remain invariant for every infinitesimal transformation of the group, which accordingly determines an infinitesimal motion of V_n into itself.

From (68.4) and (68.5) it follows that the contravariant components ξ^i of an infinitesimal motion must satisfy the equations

$$\xi^k \frac{\partial g_{ij}}{\partial x^k} + g_{ik}\frac{\partial \xi^k}{\partial x^j} + g_{jk}\frac{\partial \xi^k}{\partial x^i} = 0, \tag{70.1}$$

which by (68.8) are equivalent to

$$\xi_{i,j} + \xi_{j,i} = 0. \tag{70.2}$$

These equations of condition were first obtained by Killing* and are known as the *equations of Killing.*

From (66.9) we have that the magnitude of the infinitesimal displacement in a motion is given by

$$(\delta s)^2 = e g_{ij}\,\xi^i \xi^j (\delta t)^2, \tag{70.3}$$

and consequently in order that there may be a non-null motion, we must have

$$g_{ij}\,\xi^i \xi^j \neq 0. \tag{70.4}$$

We shall show conversely that if equations (70.1) are consistent and admit a solution satisfying (70.4), these ξ's determine the infinitesimal generator of a group G_1 of motions of V_n. To this end we assume that the coördinates are chosen so that the ξ's have the values (66.11). Then equations (70.1) reduce to

$$\frac{\partial g_{ij}}{\partial x^1} = 0 \qquad (i, j = 1, \cdots, n). \tag{70.5}$$

* 1892, 1, p. 167.

Hence the g's are independent of x^1 and consequently the fundamental form is transformed into itself by the finite equations (66.12) of the group. Hence:

When a space admits an infinitesimal motion, it admits the finite continuous group G_1 of motions generated by the infinitesimal motion.

Conversely, when (70.5) are satisfied, a solution of (70.1) is given by (66.11). Therefore:

A necessary and sufficient condition that a V_n admits an infinitesimal motion is that there exist a coördinate system in terms of which all of the g's do not involve one of the coördinates, say x^1; then the curves of parameter x^1 are the paths of the infinitesimal motion and also of the finite motion.

From the foregoing considerations and those of § 68 it follows that lengths are preserved in a motion and that geodesics go into geodesics. We shall show directly that angles are preserved. The angle between two directions defined by $d_1 x^i$ and $d_2 x^i$ is given by [cf. § 13.4)]

$$\cos\alpha = \frac{g_{ij}\, d_1 x^i\, d_2 x^j}{\sqrt{(e_1\, g_{ij}\, d_1 x^i\, d_1 x^j)\,(e_2\, g_{kl}\, d_2 x^k\, d_2 x^l)}}\,.$$

In consequence of (68.2) and (70.1) we have

$$\delta\,(g_{ij}\, d_1 x^i\, d_2 x^j) = \left(\xi^k \frac{\partial g_{ij}}{\partial x^k} + g_{ik}\frac{\partial \xi^k}{\partial x^j} + g_{jk}\frac{\partial \xi^k}{\partial x^i}\right) d_1 x^i\, d_2 x^j\, \delta t = 0,$$

and therefore $\delta \cos\alpha = 0$. Hence:

When a V_n undergoes a motion into itself, lengths and angles are preserved and geodesics go into geodesics.

By considerations similar to those at the beginning of § 69 for $\psi = \overline{\psi} = 0$ we have:

Two motions of a V_n cannot have the same paths.

We shall prove the following theorem:

If a space V_n admits an intransitive group G_r of motions, and a hypersurface V_{n-1} is an invariant variety, the hypersurfaces geodesically parallel to it are invariant varieties.

Let the family of geodesically parallel hypersurfaces be the spaces $x^1 =$ const. and choose for the parameter x^1 the distance from the given V_{n-1} measured along the normal geodesics. Then

$$g_{11} = e_1, \qquad g_{1j} = 0 \qquad (j \neq 1).$$

Since $x^1 = 0$ is an invariant variety, it follows from (66.9) that $\xi_{\sigma|}^{\ 1} = 0$ for $x^1 = 0$ and $\sigma = 1, \cdots, r$. From (70.1) for $i = j = 1$, we have $\dfrac{\partial \xi_{\sigma|}^{\ 1}}{\partial x^1} = 0$ and consequently $\xi_{\sigma|}^{\ 1} = 0$ for all values of x^1 and the theorem is proved.

Suppose now that the minimum invariant varieties of a group G_r of motions are hypersurfaces. We take them as the hypersurfaces $x^1 =$ const. and choose the other coördinates so that (69.9) hold. Then $\xi_{\sigma|}^{\ 1} = 0$. Equations (70.1) for $i = j = 1$ reduce to $\xi_{\sigma|}^{\ k} \dfrac{\partial g_{11}}{\partial x^k} = 0$ $(k = 2, \cdots, n)$. Since the rank of M (66.20) is $n-1$, g_{11} is a function of x^1 alone. Hence:

If the minimum invariant varieties of a G_r of motions are hypersurfaces, they are geodesically parallel.

Exercises

1. Determine the solution of equations (68.13) and (68.14) when the space is flat and the coördinates are cartesian.

2. Show that a V_3 with the fundamental form

$$\varphi = e_1 (dx^1)^2 + X_1 [e_2 (dx^2)^2 + e_3 (dx^3)^2],$$

where X_1 is an arbitrary function of x^1 alone, admits the intransitive group G_3 of motions of which the generators are

$$\frac{\partial}{\partial x^2}; \quad \frac{\partial}{\partial x^3}; \quad e_3 x^3 \frac{\partial}{\partial x^2} - e_2 x^2 \frac{\partial}{\partial x^3}.$$

Bianchi, 1918, 4, p. 545.

3. Show that a V_4 with the fundamental form

$$\varphi = e_1 (dx^1)^2 + X_1 [e_2 (dx^2)^2 + e_3 (dx^3)^2 + e_4 (dx^4)^2],$$

where X_1 is an arbitrary function of x^1 alone, admits the intransitive group G_6 of motions of which the generators are

$$\frac{\partial}{\partial x^2}; \quad \frac{\partial}{\partial x^3}; \quad \frac{\partial}{\partial x^4};$$

$$e_3 x^3 \frac{\partial}{\partial x^2} - e_2 x^2 \frac{\partial}{\partial x^3}; \quad e_4 x^4 \frac{\partial}{\partial x^3} - e_3 x^3 \frac{\partial}{\partial x^4}; \quad e_2 x^2 \frac{\partial}{\partial x^4} - e_4 x^4 \frac{\partial}{\partial x^2}.$$

Fubini, 1904, 4, p. 64.

4. Show that a V_4 with the fundamental form

$$\varphi = X_4 [(dx^1)^2 + e^{2x^1} (dx^2)^2 + e^{2x^1} (dx^3)^2] + (dx^4)^2,$$

where X_4 is an arbitrary function of x^4 alone, admits the intransitive group G_6 of motions of which the generators are

$$\frac{\partial}{\partial x^2};\quad \frac{\partial}{\partial x^3};\quad x^3\frac{\partial}{\partial x^2}-x^2\frac{\partial}{\partial x^3};\quad -\frac{\partial}{\partial x^1}+x^2\frac{\partial}{\partial x^2}+x^3\frac{\partial}{\partial x^3};$$

$$-x^\alpha\frac{\partial}{\partial x^1}+\frac{1}{2}[(x^\alpha)^2-(x^\beta)^2-e^{-2x^1}]\frac{\partial}{\partial x^\alpha}+x^\alpha x^\beta\frac{\partial}{\partial x^\beta}\qquad \begin{pmatrix}\alpha,\beta=2,3;\\ \alpha\neq\beta\end{pmatrix}.$$

Fubini, 1904, 4, p. 64.

5. If ξ_i are the components of a motion and λ^i the components of the unit vector tangent to a non-minimal geodesic, then $\xi_i\lambda^i$ is constant along the geodesic.

71. Conditions of integrability of the equations of Killing. Spaces of constant curvature. From equations (69.2) we have, on putting $\psi = 0$,

(71.1) $$\xi_{i,jk} = -\xi_m R^m{}_{kij},$$

and from (69.3) we have as the conditions of integrability of these equations

(71.2) $$\begin{aligned}\xi_m(R^m{}_{kij,l}-R^m{}_{lij,k})+\xi_{m,l}R^m{}_{kij}-\xi_{m,k}R^m{}_{lij}\\ +\xi_{i,m}R^m{}_{jkl}+\xi_{m,j}R^m{}_{ikl}=0.\end{aligned}$$

From these equations and (70.2) we have:

If $\xi_{\sigma|i}$ for $\sigma = 1, \cdots, r$ are the components of infinitesimal motions of a V_n, so also are $a^\sigma\xi_{\sigma|i}$, where the a's are arbitrary constants.

We establish also the following theorem:

If $X_\sigma f$ for $\sigma = 1, \cdots, r$ are the generators of infinitesimal motions of a V_n, so also are $(X_\sigma, X_\tau)f$ for $\sigma, \tau = 1, \cdots, r$ $(\sigma \neq \tau)$.

Consider the case where $\sigma = 1$, $\tau = 2$. If ξ^i are the components of $(X_1, X_2)f$, then

$$\xi^i = \xi_{1|}{}^k\frac{\partial\xi_{2|}{}^i}{\partial x^k}-\xi_{2|}{}^k\frac{\partial\xi_{1|}{}^i}{\partial x^k} = \xi_{1|}{}^k\xi_{2|}{}^i{}_{,k}-\xi_{2|}{}^k\xi_{1|}{}^i{}_{,k},$$

from which by means of (70.2) we have

$$\xi_i = -\xi_{1|}{}^k\xi_{2|k,i}+\xi_{2|}{}^k\xi_{1|k,i}.$$

In consequence of (71.1) we have

$$\xi_{i,j} = (-\xi_{1|}{}^k{}_{,j}\xi_{2|k,i}+\xi_{2|}{}^k{}_{,j}\xi_{1|k,i})+\xi_{1|}{}^m\xi_{2|}{}^k(R_{kijm}-R_{mijk}).$$

Because of (8.10) it follows that $\xi_{i,j}+\xi_{j,i}=0$ which was to be proved.

From (71.1) it follows that the second and higher derivatives of ξ_i are expressible linearly and homogeneously in terms of ξ_i and their first derivatives. Hence (§ 66):

The transformations of a group of motions are of order zero or one at any point of V_n.

We observe that (70.2) are the conditions (§ 39) that the equations of the geodesics of V_n admit the linear first integral $\xi_i \frac{dx^i}{ds} = \text{const.}$ Hence:

When a V_n admits a group G_r of motions, the equations of the geodesics of V_n admit r linearly independent first integrals, and conversely.

We have seen that the second and higher derivatives of the ξ's are expressible linearly and homogeneously in terms of the ξ's and their first derivatives. These $n(n+1)$ quantities must satisfy the $n(n+1)/2$ linearly independent conditions (70.2), and consequently the general solution of (70.1) admits at most $n(n+1)/2$ arbitrary constants. Hence the complete group of motions of a V_n involves at most $n(n+1)/2$ parameters. In § 27 it was shown from other considerations that a space of constant curvature admits a continuous group of motions of $n(n+1)/2$ parameters, and that this is a characteristic property of such spaces. We shall establish this result from the present point of view, and observe that the condition is that equations (71.2) must be satisfied identically, when the conditions (70.2) are imposed.

As a first consequence we have

$$R^m{}_{kij,l} - R^m{}_{lij,k} = 0, \tag{71.3}$$

and since the other terms of (71.2) can be written in the form

$$\xi_{m,p}(\delta_l^p R^m{}_{kij} - \delta_k^p R^m{}_{lij} + \delta_j^p R^m{}_{ikl} - \delta_i^p R^m{}_{jkl}) = 0,$$

on taking account of (70.2), we have the conditions

$$\delta_l^p R^m{}_{kij} - \delta_l^m R^p{}_{kij} - \delta_k^p R^m{}_{lij} + \delta_k^m R^p{}_{lij} + \delta_j^p R^m{}_{ikl} - \delta_j^m R^p{}_{ikl}$$
$$- \delta_i^p R^m{}_{jkl} + \delta_i^m R^p{}_{jkl} = 0 \qquad (p \neq m,\ k \neq l).$$

Contracting for l and p and making use of (8.11), we get

$$R^m{}_{kij} = \frac{1}{n-1}(\delta_j^m R_{ik} - \delta_i^m R_{jk}),$$

which are equivalent to (40.21), the conditions that V_n have constant curvature. In view of the preceding theorem we have:

A group of motions of a V_n *has at most* $n(n+1)/2$ *parameters, and this number only in case* V_n *has constant curvature.*

A space of constant curvature is characterized by the property that the equations of its geodesics admit $n(n+1)/2$ *linearly independent linear first integrals.*

72. Infinitesimal translations. In § 23 we saw that when a V_n admits a field of parallel vectors, the curves to which the vectors are tangent form a normal geodesic congruence and that any two V_{n-1}'s orthogonal to the congruence can be brought into coincidence with one another by a motion in which each point describes the same distance, that is, by a *translation.* We observe that (23.15) satisfies the conditions of the second theorem of § 70 and that in this case

$$g_{ij}\,\xi^i\,\xi^j = \text{const.}, \tag{72.1}$$

which from (70.3) is seen to be a necessary and sufficient condition that an infinitesimal motion be a translation.

If the coördinates are chosen so that the components of the infinitesimal translation have the components (66.11), then (70.5) must hold and from (72.1) it follows that g_{11} must be constant. From this result and the second theorem of § 19 we have:*

The paths of a motion are geodesics, when, and only when, the motion is a translation.

The spaces for which the tangents to the paths form a field of parallel vectors are only a particular type of spaces admitting translations. The following theorem gives a geometrical characterization of the general case:

A necessary and sufficient condition that a field of unit vectors be such that the vectors at points of any non-minimal geodesic whatever make a constant angle with the geodesic is that the vectors be tangent to the paths of a translation.

Let C be any geodesic along which the coördinates are expressed in terms of the arc. The cosine of the angle at each point of C between the vector ξ^i and C is $\xi_j \dfrac{dx^j}{ds}$. For this to be constant we must have

* *Bianchi*, 1918, 4, p. 499.

$$\frac{dx^k}{ds}\left(\xi_j \frac{dx^j}{ds}\right)_{,k} = \frac{dx^k}{ds}\left[\xi_{j,k}\frac{dx^j}{ds} + \xi_j\left(\frac{dx^j}{ds}\right)_{,k}\right]$$
$$= \xi_{j,k}\frac{dx^j}{ds}\frac{dx^k}{ds} = 0.$$

Since this condition must be satisfied for every C, we must have (70.2), and since the vector is a unit vector, the theorem is proved.

73. Geometrical properties of the paths of a motion. If ξ_i are the components of an infinitesimal motion, not a translation, and we put

(73.1) $$\xi_i = e^{-\psi}\lambda_{n|i},$$

where $\lambda_{n|i}$ are the components of the corresponding unit vector-field, and associate with $\lambda_{n|i}$ $n-1$ other unit vectors forming an orthogonal ennuple with it, equations (70.2) become in consequence of (30.2)

(73.2) $$\sum_{l,m} e_l e_m (\gamma_{nlm} + \gamma_{nml})\lambda_{l|i}\lambda_{m|j} - \psi_{,j}\lambda_{n|i} - \psi_{,i}\lambda_{n|j} = 0.$$

Multiplying by $\lambda_{p|}{}^{i}\lambda_{q|}{}^{j}$ for $p, q = 1, \cdots, n-1$ and summing for i and j, we get

(73.3) $$\gamma_{npq} + \gamma_{nqp} = 0 \qquad (p, q = 1, \cdots, n-1).$$

If (73.2) be multiplied by $\lambda_{n|}{}^{i}$ and summed for i, we have

(73.4) $$\sum_m e_m \gamma_{mnn}\lambda_{m|j} + e_n\psi_{,j} + \psi_{,i}\lambda_{n|}{}^{i}\lambda_{n|j} = 0.$$

If we multiply (73.4) by $\lambda_{n|}{}^{j}$ and sum for j, we get

(73.5) $$\lambda_{n|}{}^{i}\psi_{,i} = 0,$$

and consequently (73.4) reduces to

$$(73.6) \qquad \psi_{,j} = -e_n \sum_m e_m \gamma_{mnn} \lambda_{m|j}.$$

Conversely, when (73.3) and (73.6) are satisfied, so also are (73.2) and (73.5). Hence (73.3) and (73.6) constitute a necessary and sufficient condition that the congruence of curves $\lambda_{n|i}$ be the paths of a motion.

When $p \neq q$, equations (73.3) are the condition that the congruences $\lambda_{p|i}$ for $p = 1, \cdots, n-1$ be canonical with respect to $\lambda_{n|i}$ (Cf. § 38). From (30.16) we have $\lambda_{n|i} \mu_l^{\,i} = \gamma_{lnl}$ $(l \neq n)$. Hence equations (73.3) for $p = q$ are necessary and sufficient conditions that the curves of the congruences $\lambda_{l|}^{\,i}$ for $l \neq n$ be geodesics, or that their principal normals be orthogonal to the paths. Moreover, from (30.14) and (73.6) we have $\lambda_{n|}^{\,k} \lambda_{n|j,k} = e_n \psi_{,j}$; consequently the principal normals to C are normal to a family of surfaces $\psi =$ const. Hence we have the following theorem of Ricci:*

In order that a congruence C of curves be the paths of a motion, not a translation, it is necessary and sufficient that (1) *any $n-1$ mutually orthogonal congruences orthogonal to C be canonical with respect to C*; (2) *the curves of any congruence orthogonal to C be geodesics or their principal normals be orthogonal to the curves of C at corresponding points*; (3) *the principal normals to the curves of C form a normal congruence.*

From (73.6) and (35.9) we have:

When the paths of a motion, not a translation, form a normal congruence, the hypersurfaces orthogonal to the paths are isothermic.

When the paths C are geodesics, and consequently the motion is a translation, equations (73.6) are satisfied identically in consequence of (30.15). Hence:

In order that a congruence of geodesics be the paths of a translation, it is necessary and sufficient that conditions (1) *and* (2) *of the above theorem be satisfied.*

74. Spaces V_2 which admit a group of motions. We consider first the case of a group of motions G_1 of a V_2, take the components in the form (66.11), and choose the curves of param-

* 1899, 1, p. 79; also *Ricci* and *Levi-Civita*, 1901, 1, pp. 173, 608.

eter x^2 orthogonal to the paths. Then $g_{12} = 0$, and from (70.1) we find that g_{11} and g_{22} are independent of x^1, so that by a suitable choice of x^2, we have

$$\varphi = g_{11}(dx^1)^2 + e_2(dx^2)^2, \tag{74.1}$$

that is, V_2 is applicable to a surface of revolution, if φ is definite.

In order to determine whether a V_2 can admit more than one motion, we consider the equations of Killing for the form (74.1). They reduce to

$$\xi^2 \frac{\partial g_{11}}{\partial x^2} + 2 g_{11} \frac{\partial \xi^1}{\partial x^1} = 0, \qquad g_{11} \frac{\partial \xi^1}{\partial x^2} + e_2 \frac{\partial \xi^2}{\partial x^1} = 0, \quad \frac{\partial \xi^2}{\partial x^2} = 0.$$

From the third of these equations, we have $\xi^2 = X_1$, where X_1 is a function of x^1 alone. Indicating by primes derivatives with respect to the arguments, from the first two we have

$$\frac{\partial \xi^1}{\partial x^1} = -X_1 \frac{\partial \log \sqrt{g_{11}}}{\partial x^2}, \qquad \frac{\partial \xi^1}{\partial x^2} = -\frac{e_2}{g_{11}} X_1', \tag{74.2}$$

of which the condition of consistency is

$$g_{11} \frac{\partial^2 \log \sqrt{g_{11}}}{\partial x^{2^2}} = e_2 \frac{X_1''}{X_1} = c, \tag{74.3}$$

where c is a constant, since the first and second terms of this equation are independent of x^1 and x^2 respectively. Equating to zero the derivatives of the first term with respect to x^2, we find from the resulting equation that $\frac{1}{\sqrt{g_{11}}} \frac{\partial^2 \sqrt{g_{11}}}{\partial x^{2^2}} = k$, where k is a constant. Then from (15.8) we have $R_{2112} = g_{11} k$, that is, V_2 is of constant curvature. For a given V_2 the constant c in (74.3) is determined, and the general solution of $X_1'' = c e_2 X_1$ involves two arbitrary constants. Another is introduced in the determination of ξ^1 from (74.2). Hence the general group is a G_3, and since the rank of M (66.20) is two, the group is transitive. Thus we have the theorem, well-known for the case where φ is definite:*

* 1909, 1, pp. 323, 326; *Bianchi*, 1902, 1, p. 508.

The fundamental form of any surface admitting a continuous deformation is reducible to (74.1), *where* g_{11} *is independent of* x^1. *and the group involves one parameter, unless the surface is of constant curvature; in the latter case the complete group is a* G_3.

In order to determine whether V_2 can admit a sub-group G_2 of motions, we have that (66.16) must hold for $\alpha, \beta, \gamma = 1, 2$. There are two cases to consider, according as the constants of composition are zero or not. In the former case we have

$$(74.4) \qquad (X_1, X_2) f = 0,$$

called the *Abelian* case, and in the latter linear combinations with constant coefficients of $X_1 f$ and $X_2 f$ can be made so that

$$(74.5) \qquad (X_1, X_2) f = X_1 f.^*$$

We choose the paths for the coördinate lines, which is possible in consequence of the fourth theorem of § 70. Then $\xi_{1|}{}^2 = \xi_{2|}{}^1 = 0$.

For the case (74.4) we have that $\xi_{1|}{}^1$ is a function of x^1 alone and $\xi_{2|}{}^2$ of x^2 alone. Hence the coördinates can be chosen so that $\xi_{1|}{}^1 = \xi_{2|}{}^2 = 1$, that is,

$$(74.6) \qquad X_1 f = \frac{\partial f}{\partial x^1}, \quad X_2 f = \frac{\partial f}{\partial x^2}.$$

From the equations of Killing (70.1) it follows that g_{ij} are constants, and consequently V_2 is an S_2.

For the case (74.5) we have

$$\frac{\partial \xi_{2|}{}^2}{\partial x^1} = 0, \quad \frac{\partial \log \xi_{1|}{}^1}{\partial x^2} = -1/\xi_{2|}{}^2.$$

Hence the coördinates can be chosen so that $\xi_{2|}{}^2 = 1$, $\xi_{1|}{}^1 = e^{-x^2}$. From (70.1) we have

$$\frac{\partial g_{ij}}{\partial x^2} = 0, \quad \frac{\partial g_{11}}{\partial x^1} = 0, \quad \frac{\partial g_{12}}{\partial x^1} = g_{11}, \quad \frac{\partial g_{22}}{\partial x^1} = 2 g_{12},$$

and consequently

$$(74.7) \qquad g_{11} = a, \quad g_{12} = a x^1 + b, \quad g_{22} = a (x^1)^2 + 2 b x^1 + c,$$

* *Bianchi*, 1918, 4, p. 235.

where a, b and c are constants. The generators are

$$(74.8) \qquad X_1 f = e^{-x^2} \frac{\partial f}{\partial x^1}, \quad X_2 f = \frac{\partial f}{\partial x^2}.*$$

In this case the curvature of V_2 is $a/(b^2 - ac)$†.

75. Intransitive groups of motions. Since the group induced by a G_r upon a minimum variety is transitive (§ 67), the problem of finding the groups of motions of a V_n is reduced to the problem of transitive groups by means of the following theorem due to Fubini:‡

If a space V_n admits an intransitive group of motions G_r, which is the complete§ group for V_n or one of its sub-groups, the group induced on any minimum variety V_{n-k} has r parameters, and the finite equations of G_r are reducible by a suitable choice of coördinates to those of a transitive group on $n-k$ variables.

We recall from § 67 that the order $n-k$ of the minimum varieties is the rank of the matrix M (66.20), that there passes one of these varieties V_{n-k} through every point, and that if the induced group on any V_{n-k} is not of order r, there exists a sub-group G_σ of G_r leaving this V_{n-k} point-wise invariant. Let P_0 be a ordinary point of V_n, V^0_{n-k} the minimum invariant variety through P_0 and P be any point of V_n not in V^0_{n-k}. Consider now the V_{n-k+1} consisting of an infinity of invariant V_{n-k}'s including V^0_{n-k} and the one through P; this evidently is an invariant variety of G_r, and in particular of the sub-group G_σ leaving V^0_{n-k} point-wise invariant. In V_{n-k+1} draw the geodesics of V_{n-k+1} normal to V^0_{n-k}. Any motion of G_r induces a motion in V_{n-k+1} which sends geodesics into geodesics, preserves angles and distances (§ 70). In particular, any transformation G_σ holds the points of V^0_{n-k} fixed and consequently all the points of the geodesics fixed, and in particular P. Hence G_σ consists of the identity and thus the first part of the theorem is proved.

In order to prove the second part, we consider a hypersurface

* Cf. *Bianchi*, 1918, 4, p. 510.

† Cf. 1909, 1, p. 155.

‡ 1903, 4, p. 40; also *Bianchi*, 1918, 4, p. 514.

§ By the *complete* group we mean the group with the maximum number of parameters which satisfies the conditions of the problem.

V^0_{n-1} of V_n consisting of invariant V_{n-k}'s. It is an invariant variety of G_r and the induced group of V^0_{n-1} contains r parameters, by the first part of the theorem. Consider V_n referred to V^0_{n-1} and the hypersurfaces geodesically parallel to it (§ 19) as the spaces $x^1 = \text{const.}$, x^1 being the distance measured from V^0_{n-1} along a geodesic normal to it, in which case the latter is the hypersurface $x^1 = 0$. For each motion of G_r each of the hypersurfaces $x^1 = \text{const.}$ moves into itself just as V^0_{n-1} does. Suppose further that the other coördinates $x^2, \cdots, x^n$ are chosen in any manner whatever so that the normal geodesics are the curves $x^2 = \text{const.}, \cdots, x^n = \text{const.}$ Since the geodesics are interchanged among themselves in a motion, it follows that the coördinates $x'^2, \cdots, x'^n$ of a point on one V_{n-1} into which a point of coördinates $x^2, \cdots, x^n$ goes are the same for any other V_{n-1}, and consequently the finite equations of any motion are of the form

$$x'^1 = x^1, \qquad x'^j = \varphi^j(x^2, \cdots, x^n) \quad (j = 2, \cdots, n).$$

Thus for the space V^0_{n-1} we have shown that the finite equations can be put in the form stated in the theorem. If $k > 1$, we take V^0_{n-1} in place of V_n in the above process and reduce the equations to those in $n-2$ variables and so on, which proves the theorem. See App. 28

76. Spaces V_3 admitting a G_2 of motions. Complete groups of motions of order $n(n+1)/2-1$. A group G_2 of a V_3 is intransitive and from the fourth theorem of § 70 it follows that the minimum invariant varieties are V_2's. From § 75 we have that the induced group on these varieties is a G_3 and from § 74 that their curvature is constant. From the last theorem of § 70 it follows also that they are geodesically parallel, and that if they be taken for the surfaces $x^3 = \text{const.}$, then $\xi_{\sigma|}{}^3 = 0$ for $\sigma = 1, 2$ and $\xi_{\sigma|}{}^i$ for $i = 1, 2$ are independent of x^3. We take for the curves $x^1 = \text{const.}$, $x^2 = \text{const.}$ the geodesics orthogonal to one of the surfaces $x^3 = \text{const.}$, at points of the paths, and write the fundamental form

$$\varphi = g_{ij}\, dx^i\, dx^j + e_3 (dx^3)^2 \qquad (i, j = 1, 2). \tag{76.1}$$

For this particular surface the infinitesimal transformations are given by (74.6) and (74.8), and from the preceding observations these are the generators for V_3.

In order that equations (70.1) be satisfied by the transformations (74.6), it is necessary and sufficient that g_{ij} be functions of x^3 alone, subject only to the condition $g_{11}\, g_{22} - g_{12}{}^2 \neq 0$. In order that equations (70.1) be satisfied by the transformations (74.8), it is necessary and sufficient that

$$(76.2) \qquad g_{11} = \alpha, \quad g_{12} = \alpha x^1 + \beta, \quad g_{22} = \alpha (x^1)^2 + 2\beta x^1 + \gamma,$$

where α, β, γ are arbitrary functions of x^3 such that $\alpha\gamma - \beta^2 \neq 0$. In the former case the curvature of the surfaces $x^3 = \text{const.}$ is zero, and in the latter $\alpha/(\beta^2 - \alpha\gamma)$ (cf. § 74).*

By means of these results we shall show that a V_3 cannot admit a complete group G_5 of motions. The group cannot be intransitive, otherwise a family of surfaces (the minimum varieties) would admit a G_5, which is impossible since $5 > (2 \cdot 3)/2$. Hence the group must be transitive, and the sub-group of stability (§§ 66, 67) of any point P_0 is of order $5 - 3 = 2$. If there were such a G_2, the points at a constant geodesic distance from P_0 would constitute a minimum invariant variety, and thus we should have a family of geodesically parallel invariant varieties. This is the case just considered, and from the form of the transformations (74.6) and (74.8) it follows that all the transformations of such a G_2 are of order zero (§ 66), and consequently there cannot be an invariant point.†

We are now in a position to prove the following theorem due to Fubini.‡

A V_n for $n > 2$ cannot admit a complete group of motions of order $n(n+1)/2 - 1$.

We prove this theorem by induction, assuming it to hold for a V_{n-1}. If a V_n admits a G_r with $r = n(n+1)/2 - 1$, it must be transitive; otherwise by the theorem of § 75, a variety of order $n-1$, or less, would admit a group of this order, which is impossible since a V_{n-1} can admit at most a group of order $n(n-1)/2$. If G_r is transitive, there is a sub-group of order $r_1 = r - n = n(n-1)/2 - 1$ leaving a point P_0 fixed, which is a group of

* Cf. *Bianchi*, 1918, 4, p. 542.

† Cf. *Bianchi*, 1918, 4, p. 540.

‡ 1903, 4, p. 54.

motions of $\infty^1 V_{n-1}$'s, the loci of points at constant geodesic distance from P_0. But this is contrary to the assumption that the theorem holds for a V_{n-1}. Since we have shown that the theorem holds for a V_3, the proof is complete.

77. Simply transitive groups as groups of motions. When for a group G_n in n variables the matrix M(66.20) is of rank n, the group is called *simply transitive.* We shall prove the following theorem due to Bianchi:*

Any simply transitive group in n variables is the group of motions (complete or partial) of an infinity of spaces V_n.

Let $\xi_{j|}{}^i$ for $i, j = 1, \cdots, n$ be the components of the infinitesimal transformations of the group, and denote by A_i^j the cofactor of $\xi_{j|}{}^i$ in the determinant $|\xi_{j|}{}^i|$ divided by this determinant; then

$$(77.1) \qquad A_k^j \xi_{j|}{}^i = \delta_k^i, \qquad A_j^i \xi_{k|}{}^j = \delta_k^i.$$

In order that the group may be a group of motions it is necessary and sufficient that the equations of Killing

$$(77.2) \qquad \xi_{l|}{}^k \frac{\partial g_{ij}}{\partial x^k} + g_{ih} \frac{\partial \xi_{l|}{}^h}{\partial x^j} + g_{hj} \frac{\partial \xi_{l|}{}^h}{\partial x^i} = 0$$

admit a set of solutions g_{ij} symmetric in i and j. Multiplying by A_r^l and summing for l, we have

$$(77.3) \qquad \frac{\partial g_{ij}}{\partial x^r} = g_{ih} \Gamma_{jr}^h + g_{hj} \Gamma_{ir}^h,$$

where

$$(77.4) \qquad \Gamma_{ir}^h = -\frac{\partial \xi_{l|}{}^h}{\partial x^i} A_r^l.$$

The conditions of integrability of (77.3) are

$$(77.5) \qquad g_{il} B^l{}_{jrt} + g_{lj} B^l{}_{irt} = 0,$$

---

* 1918, 4, p. 517. The method used in this section is different from that used by Bianchi. The latter considers also (pp. 522–524) the case when the group is not simply transitive.

where

$$B^l_{jrt} = \frac{\partial \Gamma^l_{jr}}{\partial x^t} - \frac{\partial \Gamma^l_{jt}}{\partial x^r} + \Gamma^m_{jr}\Gamma^l_{mt} - \Gamma^m_{jt}\Gamma^l_{mr}. \tag{77.6}$$

It is our purpose to show that the B's are zero and consequently the system is completely integrable, and thus prove the theorem.

From (77.1) and (77.4) we have

$$\Gamma^h_{ir} = \xi_{l|}{}^h \frac{\partial A^l_r}{\partial x^i}. \tag{77.7}$$

Multiplying (77.4) by $\xi_{j|}{}^r$ and summing for r, and (77.7) by A^j_h and summing for h, we get

$$\frac{\partial \xi_{j|}{}^h}{\partial x^i} = -\xi_{j|}{}^r \Gamma^h_{ir} \tag{77.8}$$

and

$$\frac{\partial A^j_r}{\partial x^i} = A^j_h \Gamma^h_{ir}. \tag{77.9}$$

Since the ξ's are components of a group, we have from (66.16)

$$\xi_{l|}{}^k \frac{\partial \xi_{m|}{}^h}{\partial x^k} - \xi_{m|}{}^k \frac{\partial \xi_{l|}{}^h}{\partial x^k} = c_{lm}{}^r \xi_{r|}{}^h.$$

Multiplying by $A^l_i A^m_j$ and summing for l and m, we have in consequence of (77.1) and (77.4)

$$\Gamma^h_{ij} - \Gamma^h_{ji} = -c_{lm}{}^r \xi_{r|}{}^h A^l_i A^m_j. \tag{77.10}$$

By means of (77.10) equations (77.6) can be written

$$B^l_{jrt} = \frac{\partial \Gamma^l_{rj}}{\partial x^t} - \frac{\partial \Gamma^l_{tj}}{\partial x^r} + c_{hk}{}^p \left[\frac{\partial}{\partial x^r}(\xi_{p|}{}^l A^h_j A^k_t) - \frac{\partial}{\partial x^t}(\xi_{p|}{}^l A^h_j A^k_r)\right] + \Gamma^m_{jr}\Gamma^l_{mt} - \Gamma^m_{jt}\Gamma^l_{mr}.$$

Substituting from (77.4) in the first two terms of the right-hand member and reducing the resulting expressions and the next two terms by means of (77.8), (77.9) and (77.7), we obtain

$$B^l{}_{jrt} = \Gamma^l_{rm}\Gamma^m_{tj} - \Gamma^l_{tm}\Gamma^m_{rj} + \Gamma^m_{jr}\Gamma^l_{mt} - \Gamma^m_{jt}\Gamma^l_{mr}$$
$$+ c_{hk}{}^p [A^h_j \xi_{p|}{}^m (\Gamma^l_{tm} A^k_r - \Gamma^l_{rm} A^k_t) + \xi_{p|}{}^l (A^k_t A^h_m \Gamma^m_{rj}$$
$$+ A^h_j A^k_m \Gamma^m_{rt} - A^k_r A^h_m \Gamma^m_{tj} - A^h_j A^k_m \Gamma^m_{tr})].$$

By repeated application of (77.10) to the right-hand member of this equation, we have, in consequence of the first of (66.17) and (77.1)

$$B^l{}_{jrt} = c_{hk}{}^p c_{ab}{}^h \xi_{p|}{}^l (A^k_t A^a_j A^b_r + A^k_r A^a_t A^b_j + A^k_j A^a_r A^b_t)$$
$$= (c_{ab}{}^h c_{hk}{}^p + c_{ka}{}^h c_{hb}{}^p + c_{bk}{}^h c_{ha}{}^p)\, \xi_{p|}{}^l A^k_t A^a_j A^b_r.$$

From the second of (66.17) it follows that the B's are zero and hence the theorem is proved. See App. 29

Exercises

1. When the paths of a motion form a normal congruence, the lines of curvature of the orthogonal hypersurfaces are indeterminate.

2. A surface admits a translation, when, and only when, it is flat.
Bianchi, 1918, 4, p. 507.

3. If a V_n admits a translation, the surface formed by an infinity of paths of translation is flat. *Bianchi*, 1918, 4, p. 501.

4. When a V_n admits a system of coördinates for which $g_{\sigma\sigma} =$ const. for $\sigma = 1, \cdots, r$ and the other g's are independent of x^σ, then the V_n admits a group G_r of translations, the curves of parameters x^σ being the paths.

5. When the paths of a translation form a normal congruence, the orthogonal hypersurfaces are totally geodesic. *Struik*, 1922, 8, p. 157.

6. If the rank of M (66.20) for an intransitive group G_r of motions of a V_n is m, then $r \leq m(m+1)/2$. *Bianchi*, 1918, 4, p. 515.

7. If a V_n admits an intransitive group G_r of motions, where $r = n(n-1)/2$, the minimum invariant varieties are a family of geodesically parallel hypersurfaces of constant curvature. *Bianchi*, 1918, 4, p. 544.

8. A group is said to be Abelian, when the constants of composition (§ 66) are zero. Show that for an Abelian group G_r, for $r \leq n$, the coördinates can be chosen so that $\xi_{\sigma|}{}^i = \delta^i_\sigma$. *Bianchi*, 1918, 4, p. 260.

9. When a V_n admits a simply transitive Abelian group of motions, it is an S_n and the group is that of translations. *Bianchi*, 1918, 4. p. 521.

10. When a V_n admits an Abelian group $G_r (r < n)$ of motions, the minimum invariant varities are of order r and are flat spaces.

11. Show that equations (71.2) can be written

$$\xi^m R_{ijkl,m} = \xi_{m,l} R^m{}_{kij} - \xi_{m,k} R^m{}_{lij} - \xi_{m,i} R^m{}_{jkl} + \xi_{m,j} R^m{}_{ikl},$$

and that, when they are multiplied by g^{jk} and summed for j and k, the resulting equations are

(1) $$\xi^m R_{il,m} = -\xi_{m,l} R^m{}_i - \xi_{m,i} R^m{}_l .$$

12. If a space admits a motion and also an orthogonal ennuple $\lambda_{h|}{}^i$ of Ricci principal directions, when equations (1) of Ex. 11 are multiplied by $\lambda_{h|}{}^i \lambda_{k|}{}^l$ and summed for i and l, the resulting equations are reducible according as $h = k$ or $h \neq k$ to the respective sets of equations (Cf. §§ 33, 34)

$$\xi^m \frac{\partial \varrho_h}{\partial x^m} = 0,$$

$$(\varrho_h - \varrho_k)\,[\xi^m \sum_p e_p \gamma_{hkp} \lambda_{p|m} + \lambda_{h|}{}^i \lambda_{k|}{}^j \xi_{i,j}] = 0,$$

where ϱ_h are Ricci principal invariants. Show that these invariants are invariant under the transformations of the group generated by ξ^i. *Ricci*, 1905, 1, p. 490.

13. When a space admits a transitive group of motions and an orthogonal ennuple of Ricci principal directions, the Ricci principal invariants are constant.

Ricci, 1905, 1, p. 491.

14. If a space admitting a motion is referred to an orthogonal ennuple $\lambda_{h|}{}^i$, the invariants a_i and b_{ij}, defined by

$$\xi^i = \sum_r e_r a_r \lambda_{r|}{}^i, \qquad \xi_{i,j} = \sum_{r,t} e_r e_t b_{rt} \lambda_{r|i} \lambda_{t|j},$$

satisfy the equations

$$\frac{\partial a_i}{\partial s_j} = b_{ij} + \sum_l e_l a_l \gamma_{ilj},$$

$$\frac{\partial b_{ij}}{\partial s_k} = \sum_r e_r (a_r \gamma_{rkji} + b_{jr} \gamma_{rik} - b_{ir} \gamma_{rjk}),$$

in consequence of (71.1) to which they are equivalent. Moreover, equations (71.2) are equivalent to

$$\sum_r e_r (a_r \gamma_{ijklr} + b_{rl} \gamma_{rkji} - b_{rk} \gamma_{rlji} + b_{ri} \gamma_{rjkl} - b_{rj} \gamma_{rikl}) = 0,$$

where

$$\gamma_{ijklr} = R_{pqst,u} \lambda_{i|}{}^p \lambda_{j|}{}^q \lambda_{k|}{}^s \lambda_{l|}{}^t \lambda_{r|}{}^u .$$

Ricci, 1905, 1, p. 489.

15. Show that the invariants γ_{ijhkl} of Ex. 14 are expressible in the form

$$\gamma_{ijhkl} = \frac{\partial \gamma_{ijhk}}{\partial s_l} + \sum_r e_r (\gamma_{ril} \gamma_{rjhk} + \gamma_{rjl} \gamma_{irhk} + \gamma_{rhl} \gamma_{ijrk} + \gamma_{rkl} \gamma_{ijhr}).$$

Ricci, 1905, 1, p. 489.

16. If a space admits an orthogonal ennuple of Ricci principal directions and the corresponding invariants γ_{ijkl} are zero when more than two of the indices are different, the space is called *regular* by Ricci. Show that when a regular space admits a motion all of the invariants γ_{ijkl} are invariant for the motion;

also that if the space admits a transitive group of motions, these invariants are constant. *Ricci*, 1905, 1, p. 491.

17. Show that the functions Γ^h_{ir} defined by (77.7), for two sets of coördinates satisfy equations of the form (8.1) in which the Christoffel symbols are replaced by Γ's; and that if $\left\{ {h \atop i\,r} \right\}$ are the Christoffel symbols formed with respect to the fundamental form of the space, then $\Gamma^h_{ir} - \left\{ {h \atop i\,r} \right\}$ are the components of a tensor.

18. If $\xi_{\lambda|}{}^i$ are the components of an orthogonal ennuple and A^j_k have the significance of § 77, the functions Γ^h_{ij}, defined by

$$\Gamma^h_{ij} = \xi_{r|}{}^h \frac{\partial A^r_i}{\partial x^j},$$

satisfy equations of the form (8.1) in two coördinate systems; also

$$\text{(1)} \qquad \frac{\partial \xi_{r|}{}^h}{\partial x^j} + \xi_{r|}{}^i \Gamma^h_{ij} = 0, \qquad \frac{\partial A^r_i}{\partial x^j} - A^r_k \Gamma^k_{ij} = 0.$$

19. Equations (1) of Ex. 18 may be interpreted as the vanishing of the first covariant derivatives of the vectors $\xi_{r|}{}^h$ and A^r_i for covariant differentiation defined by replacing the Christoffel symbols by corresponding Γ's. Show that the covariant derivatives so defined are tensors. *Eisenhart*, 1925, 12, p. 248.

Appendix 1

Replace if ν^i is a vector satisfying this condition *by* if ν^i is a vector defined by $a_{ij}\lambda_{\alpha|}{}^i\nu^j = 0$ for $\alpha = 1, \cdots, n-1$ and $\mu_i\nu^i \neq 0$.

Appendix 2

The argument following "In fact" is not valid, since one cannot form the covariant derivative of $\frac{dx^i}{ds}$. Differentiating equation (17.9) with respect to s and using (7.8) one obtains the second part of equation (17.10). Accordingly the left-hand number of (17.11) must be interpreted as meaning the quantity in parenthesis in (17.10).

Appendix 3

In this note we derive the expression for the metric tensor g_{ij} in the neighborhood of the origin O of normal coordinates x^i in terms of these coordinates and establish therefrom certain geometric results.

At O we have $g_{ij} = e_{ij}$, where

$$e_{ii} = e_i, \qquad e_{ij} = 0 \qquad (i \neq j),$$

e_i being plus or minus one as the case may be. Since at O the first derivatives of g_{ij} are equal to zero by (18.10), we have to within terms of the third and higher orders in the x's

$$g_{ij} = e_{ij} + \frac{1}{2}\left(\frac{\partial^2 g_{ij}}{\partial x^h\, \partial x^k}\right)_0 x^h x^k.$$

From the first of (18.9), (18.8), and (17.14) we have at O

$$\frac{\partial}{\partial x^k}[hj, i] + \frac{\partial}{\partial x^h}[jk, i] + \frac{\partial}{\partial x^j}[hk, i] = 0.$$

On replacing the Christoffel symbols by their expressions from (7.1), these equations become

$$\frac{\partial^2 g_{ih}}{\partial x^j\, \partial x^k} + \frac{\partial^2 g_{ij}}{\partial x^h\, \partial x^k} + \frac{\partial^2 g_{ik}}{\partial x^j\, \partial x^h} - \frac{1}{2}\left(\frac{\partial^2 g_{hj}}{\partial x^i\, \partial x^k} + \frac{\partial^2 g_{jk}}{\partial x^i\, \partial x^h} + \frac{\partial^2 g_{hk}}{\partial x^i\, \partial x^j}\right) = 0.$$

Adding to this equation the one obtained from it on interchanging i

and j, we obtain

$$\frac{1}{2}\left(\frac{\partial^2 g_{jh}}{\partial x^i\,\partial x^k}+\frac{\partial^2 g_{jk}}{\partial x^i\,\partial x^h}+\frac{\partial^2 g_{ki}}{\partial x^j\,\partial x^h}+\frac{\partial^2 g_{hi}}{\partial x^j\,\partial x^k}\right)+2\,\frac{\partial^2 g_{ij}}{\partial x^h\,\partial x^k}-\frac{\partial^2 g_{hk}}{\partial x^i\,\partial x^j}=0,$$

from which we have

$$x^h x^k\left(\frac{\partial^2 g_{jh}}{\partial x^i\,\partial x^k}+\frac{\partial^2 g_{ik}}{\partial x^j\,\partial x^h}+2\,\frac{\partial^2 g_{ij}}{\partial x^h\,\partial x^k}-\frac{\partial^2 g_{hk}}{\partial x^i\,\partial x^j}\right)=0.$$

From (8.9) we have at O

$$(R_{hijk})_0=\frac{1}{2}\left(\frac{\partial^2 g_{hk}}{\partial x^i\,\partial x^j}+\frac{\partial^2 g_{ij}}{\partial x^h\,\partial x^k}-\frac{\partial^2 g_{hj}}{\partial x^i\,\partial x^k}-\frac{\partial^2 g_{ik}}{\partial x^h\,\partial x^j}\right)_0$$

from which and the preceding equations we obtain

$$(R_{hijk})_0 x^h x^k=\frac{3}{2}\left(\frac{\partial^2 g_{ij}}{\partial x^h\,\partial x^k}\right)_0 x^h x^k.$$

Hence we have

(1) $$g_{ij}=e_{ij}+\tfrac{1}{3}(R_{hijk})_0 x^h x^k.$$

From (1) we get in the neighborhood of O by (7.1), noting the identities (8.10),

(2) $$[ij,\,k]=\tfrac{1}{3}(R_{ikjh}+R_{jkih})_0 x^h.$$

Since at O $g^{ij}=e^{ij}$, we have that to the same approximation

(3) $$\begin{Bmatrix} k \\ ij \end{Bmatrix}=e_k[ij,\,k].$$

Consider through O a geodesic whose equations in normal coordinates are (18.1)

(4) $$x^i=\xi^i s,$$

where ξ^i is a unit vector, and on it a point M for which $s=a$. At M take the geodesic whose tangent at M is the unit vector λ^i obtained by parallel displacement along OM of a given unit vector α^i at O. To within terms of higher order in s λ^i are of the form

$$\lambda^i=\alpha^i+b^i s+\tfrac{1}{2}c^i s^2.$$

In order to evaluate the constants b^i and c^i, we substitute these expressions for λ^i in equations (21.1) written

$$\frac{d\lambda^i}{ds} + \lambda^l \begin{Bmatrix} i \\ lj \end{Bmatrix} \frac{dx^j}{ds} = 0,$$

and obtain, using (3),

$$b^i + c^i s + \left(\alpha^l + b^l s + \frac{1}{2} c^l s^2\right) \frac{e_i}{3} (R_{lijh} + R_{jilh})_0 \xi^h \xi^j s = 0.$$

From this result we have, noting that R_{lijh} is skew-symmetric in the last two indices,

$$b^i = 0, \qquad c^i = -\frac{e_i}{3} \alpha^l (R_{jilh})_0 \xi^h \xi^j = \frac{e_i}{3} (R_{jihl})_0 \xi^j \xi^h \alpha^l.$$

Hence we have

(5) $$\lambda^i = \alpha^i + \tfrac{1}{6} e_i (R_{jihl})_0 \xi^j \xi^h \alpha^l a^2.$$

At M we draw the geodesic tangent to the vector of components λ^i and denote by P the point at the distance b from M. The normal coordinates y^i of P are solutions of the equations (17.8), namely

(6) $$\frac{d^2 y^i}{ds^2} + \begin{Bmatrix} i \\ jk \end{Bmatrix} \frac{dy^j}{ds} \frac{dy^k}{ds} = 0,$$

which for $s = 0$ are $\xi^i a$, the normal coordinates of M.

Since at M the quantities $\dfrac{dy^i}{ds}$ have the values λ^i, one has to within terms of higher order

(7) $$y^i = \xi^i a + \lambda^i s + \tfrac{1}{2} d^i s^2,$$

where d^i is to be determined.
From (2) and (3) we have to within terms of higher order

(8) $$\begin{Bmatrix} i \\ jk \end{Bmatrix} = \tfrac{1}{3} e_i (R_{jikh} + R_{kijh})_0 \xi^h s.$$

When the expressions (7) and (8) are substituted in equation (6) we find that

$$d^i = -\tfrac{2}{3} e_i (R_{jikh})_0 a \xi^h \lambda^j \lambda^k = \tfrac{2}{3} e_i (R_{ijkh})_0 a \xi^h \lambda^j \lambda^k.$$

Hence to within terms of the fourth degree in a and b we have

(9) $$y^i = \xi^i a + b[\alpha^i + \tfrac{1}{6} e_i (R_{jihl})_0 \xi^j \xi^h \alpha^l a^2] + \tfrac{1}{3} b^2 e_i (R_{ijkh})_0 \alpha^j \alpha^k \xi^h a.$$

If b_0^2 is equal to $\sum_i e_i(y^i - \xi^i a)^2$ to within terms of the sixth degree in a and b, we have

$$b_0^2 = b^2[1 + \tfrac{1}{3}(R_{jihl})_0 \xi^j \alpha^i \xi^h \alpha^l a^2].$$

From (25.9) we have

$$(R_{jihl})_0 \xi^j \alpha^i \xi^h \alpha^l = KB^2,$$

where K is the curvature at O for the orientation determined by the unit vectors ξ^i and α^i, and

(10) $$B^2 = (g_{jh} g_{il} - g_{jl} g_{ih}) \xi^j \alpha^i \xi^h \alpha^l.$$

Hence

$$b_0^2 = b^2(1 + \tfrac{1}{3} KB^2 a^2),$$

from which we have to the same approximation

(11) $$b^2 = b_0^2(1 - \tfrac{1}{3} KB^2 a^2).$$

At the point O draw the geodesic tangent to the vector α^i and denote by Q the point on the geodesic at the distance b from O. Its normal coordinates z^i are given by

$$z^i = \alpha^i b.$$

Denote by a' the length of the geodesic joining Q and P. The small figure $OMPQ$ has been called a parallelogramoid by Levi-Civita, who derived the interesting result now to be obtained.

Since $a_0'^2$ is equal to $\sum_i e_i(y^i - z^i)^2$, to within terms of the sixth order in a and b we have from (9)

$$a_0'^2 = a^2[1 - \tfrac{2}{3}(R_{ijhk})_0 \xi^i \alpha^j \xi^h \alpha^k b^2] = a^2(1 - \tfrac{2}{3} KB^2 b^2).$$

Analogously to (11) we have

$$a'^2 = a_0'^2(1 - \tfrac{1}{3} KB^2 b^2)$$

Substituting in this equation the above expression for $a_0'^2$, we obtain

(12) $$a'^2 = a^2 - a^2 b^2 B^2 K.$$

When the fundamental form of V_n is positive definite B^2 as defined by (10) is $\sin^2 \varphi$ where φ is the angle between the unit vectors ξ^i

and α^i (see Ex. 3, p. 47). In this case we have from (12) the equation of Levi-Civita.*

$$\frac{a^2 - a'^2}{A^2} = K$$

where $A = ab \sin \varphi$ is the area of the parallelogramoid.

Appendix 4

The argument following equation (20.1) is not valid in view of the statement in Appendix 2. By means of equations (7.14) one obtains.

$$\frac{d^2x^i}{ds^2} + \begin{Bmatrix} i \\ jk \end{Bmatrix} \frac{dx^j}{ds}\frac{dx^k}{ds} = \left(\frac{d^2x'^\lambda}{ds^2} + \begin{Bmatrix} \lambda \\ \mu\nu \end{Bmatrix}' \frac{dx'^\mu}{ds}\frac{dx'^\nu}{ds}\right)\frac{\partial x^i}{\partial x'^\lambda},$$

from which follows $\mu^i = \mu'^\lambda \dfrac{\partial x^i}{\partial x'^\lambda}$. Consequently μ^i is a contravariant vector.

Appendix 5

Equating to zero the derivative with respect to s of the expression $g_{ij}\lambda^i \dfrac{dx^j}{ds}$ we have in consequence of (7.4)

$$\left[\lambda^i([ik, j] + [jk, i]) + g_{ij}\frac{\partial \lambda^i}{\partial x^k}\right]\frac{dx^j}{ds}\frac{dx^k}{ds} + g_{il}\lambda^i \frac{d^2x^l}{ds^2} = 0,$$

which by means of (17.8), (7.3), and (11.2) is reducible to

$$g_{ij}\frac{dx^j}{ds}\lambda^i_{,k}\frac{dx^k}{ds} = 0.$$

Since λ^i are components of a unit vector we have $g_{ij}\lambda^i\lambda^j = e$, from which by differentiating with respect to s we have in consequence of (7.4)

$$g_{ij}\lambda^j\lambda^i_{,k}\frac{dx^k}{ds} = 0.$$

From these two equations we have that the quantity $\lambda^i_{,k}\dfrac{dx^k}{ds}$ is either a zero vector or a vector orthogonal to both the geodesic C and to the vector λ^i at each point of C. The latter is impossible. Hence we have (21.1).

* 1917, 1 pp. 198–201; also Cartan, 1946, 4, pp. 245–248.

Appendix 6

Equations (21.1) written in the form

$$\frac{d\lambda^i}{ds} + \lambda^l \begin{Bmatrix} i \\ lk \end{Bmatrix} \frac{dx^k}{ds} = 0$$

admit a solution $\lambda^i(s)$ determined by an arbitrary direction at a point of C. Note: when equation (21.1) is referred to later on this page it is to the equation in the above form.

Appendix 7

Because of (23.7) the conditions of integrability of (23.8) are satisfied identically. Hence these equations are completely integrable, and consequently the solution involves p arbitrary constants. See *Eisenhart*, 1933, 1, pp. 1, 2 or 1940, pp. 114, 115.

Appendix 8

Denoting by $\lambda^i_{\sigma|}$ and $\lambda'^i_{\sigma|}$ the components in x^i and x'^i of any p independent fields of vectors not limited to sets of solutions of (23.1) we have (23.9).

Appendix 9

This note replaces the material on page 70 beginning with "If it etc." on line 17, all of page 71, and page 72 to § 24, and deals with the canonical forms of the metric tensor of a V_n which admits r independent fields of parallel vectors, the components λ^i of each field being a solution of equation (23.1), that is

$$\lambda^i_{,j} \equiv \frac{\partial \lambda^i}{\partial x^j} + \lambda^k \begin{Bmatrix} i \\ kj \end{Bmatrix} = 0. \tag{1}$$

Let $\lambda^i_{a|}$ be the components of r such fields, where $a(= 1, \cdots, r)$ indicates the field and $i(= 1, \cdots, n)$ the components. If we put

$$g_{ij}\lambda^i_{a|}\lambda^j_{b|} = c_{ab} \qquad (a, b = 1, \cdots, r) \tag{2}$$

and differentiate with respect to x^k, and note that the covariant derivatives of g_{ij} are zero, we find that the c's are constants. When the metric differential form is indefinite there is the possibility that some of the parallel fields are null vector fields, that is some of the c_{aa} are zero. From the form of equations (1) it is seen that any linear combination of the λ's with real constant coefficients are the components of a parallel field. By means of such transformations it is possible to transform the matrix $\|c_{ab}\|$ when its rank is $p(\leq r)$ into one for which we have

$$(3)\qquad c_{\alpha\alpha} = e_\alpha, \quad c_{\mu\mu} = 0, \quad c_{ab} = 0 \quad \begin{pmatrix} \alpha = 1, \cdots, p; \mu = p+1, r \\ a, b = 1, \cdots, r; a \neq b \end{pmatrix}.$$

where the e's are $+1$ or -1 as the case may be. Thus in the new vector fields $\lambda^i_{\alpha|}$ are unit vectors, $\lambda^i_{\mu|}$ null vectors. and any two fields are mutually orthogonal.

In the first paragraph on page 70 it was shown that a coordinate system exists for which the components of each field are zero except the component with the same subscript and superscript. In what follows it is understood that this is the coordinate system x^i, and we have at once from (2) and (3) that

$$(4)\qquad g_{\alpha\beta} = g_{\alpha\mu} = g_{\mu\nu} = 0 \qquad (\alpha, \beta = 1, \cdots, p, \alpha \neq \beta; \mu, \nu = p+1, \cdots, r).$$

We consider first the unit vectors $\lambda^i_{\alpha|}(\alpha = 1, \cdots, p)$. We have

$$(5)\qquad \lambda_{\alpha|}{}^\alpha = (e_\alpha g_{\alpha\alpha})^{-1/2}, \qquad \lambda_{\alpha|}{}^h = 0 \quad (h = 1, \cdots, n; h \neq \alpha).$$

When these expressions are substituted in equation (1) we obtain

$$(6)\qquad \frac{1}{2}\frac{\partial}{\partial x^j}\log g_{\alpha\alpha} - \begin{Bmatrix} \alpha \\ \alpha j \end{Bmatrix} = 0 \qquad (\alpha \text{ not summed})^*$$

$$\begin{Bmatrix} h \\ \alpha j \end{Bmatrix} = 0 \quad (h, j = 1, \cdots, n; h \neq \alpha).$$

If we multiply the first of these equations by $g_{\alpha k}$ and subtract from it the second multiplied by g_{hk} and summed for h, we obtain

$$(7)\qquad g_{\alpha k}\frac{\partial}{\partial x^j}\log g_{\alpha\alpha} - \frac{\partial g_{jk}}{\partial x^\alpha} - \frac{\partial g_{\alpha k}}{\partial x^j} + \frac{\partial g_{\alpha j}}{\partial x^k} = 0 \quad \begin{pmatrix} \alpha = 1, \cdots, p; \\ j, k = 1, \cdots, n \end{pmatrix}.$$

For $k = \alpha$ this equation is satisfied identically. For $k = \beta = 1, \cdots, p(\beta \neq \alpha)$ we have in consequence of (4)

$$(8)\qquad \frac{\partial g_{j\beta}}{\partial x^\alpha} - \frac{\partial g_{\alpha j}}{\partial x^\beta} = 0.$$

---

* Throughout this note the summation convention does not apply to indices α and β.

From this equation for $j = \alpha$, or $j = \beta$, we have

$$\text{(9)} \qquad \frac{\partial g_{\alpha\alpha}}{\partial x^\beta} = 0, \qquad \frac{\partial g_{\beta\beta}}{\partial x^\alpha} = 0 \quad (\alpha, \beta = 1, \cdots, p; \alpha \neq \beta)$$

In consequence of this result, (8), and (4), and equation (7) with $j = \beta$ we obtain

$$\text{(10)} \qquad \frac{\partial g_{\alpha\sigma}}{\partial x^\beta} = 0 \quad (\alpha, \beta = 1, \cdots, p(\alpha \neq \beta); \sigma = p + 1, \cdots, n)$$

When now we put $j = \alpha$, $k = \sigma$ in (7), the resulting equation may be written

$$\text{(11)} \qquad \frac{\partial}{\partial x^\alpha}[g_{\alpha\sigma}(e_\alpha g_{\alpha\alpha})^{-1/2}] = e_\alpha \frac{\partial}{\partial x^\sigma}(e_\alpha g_{\alpha\alpha})^{1/2} \quad (\sigma = p + 1, \cdots, n).$$

If we put

$$\text{(12)} \qquad (e_\alpha g_{\alpha\alpha})^{1/2} = \frac{\partial \psi_\alpha}{\partial x^\alpha},$$

where in accordance with (9) ψ_α is a function of $x^\alpha, x^{p+1}, \cdots, x^n$ at most, we obtain from (11)

$$\text{(13)} \qquad g_{\alpha\sigma} = e_\alpha \frac{\partial \psi_\alpha}{\partial x^\alpha}\left(\frac{\partial \psi_\alpha}{\partial x^\sigma} + \varphi_{\alpha\sigma}\right),$$

where in accordance with (10) $\varphi_{\alpha\sigma}$ is a function of $x^{p+1}, \cdots, x^n$ at most.

If in (7) we put $j = \sigma$, $h = \tau$ for $\sigma, \tau = p + 1, \cdots, n$ and substitute from (12) and (13), we obtain

$$\text{(14)} \qquad e_\alpha \frac{\partial g_{\sigma\tau}}{\partial x^\alpha} = \left(\frac{\partial \psi_\alpha}{\partial x^\sigma} + \varphi_{\alpha\sigma}\right)\frac{\partial^2 \psi_\alpha}{\partial x^\alpha\, \partial x^\tau} + \left(\frac{\partial \psi_\alpha}{\partial x^\tau} + \varphi_{\alpha\tau}\right)\frac{\partial^2 \psi_\alpha}{\partial x^\alpha\, \partial x^\sigma} + \frac{\partial \psi_\alpha}{\partial x^\alpha}\left(\frac{\partial \varphi_{\alpha\sigma}}{\partial x^\tau} - \frac{\partial \varphi_{\alpha\tau}}{\partial x^\sigma}\right)$$

Since $g_{\sigma\tau}$ is symmetric in its indices, we must have

$$\frac{\partial \varphi_{\alpha\tau}}{\partial x^\sigma} - \frac{\partial \varphi_{\alpha\sigma}}{\partial x^\tau} = 0,$$

from which it follows that

$$\varphi_{\alpha\sigma} = \frac{\partial \varphi_\alpha}{\partial x^\sigma}$$

where φ_α is a function of $x^{p+1}, \cdots, x^n$ at most. Hence φ_α may be incorporated in ψ_α in (12) and (13) so that we have

$$g_{\alpha\sigma} = e_\alpha \frac{\partial \psi_\alpha}{\partial x^\alpha} \frac{\partial \psi_\alpha}{\partial x^\sigma}. \tag{15}$$

Then from (14) we have

$$g_{\sigma\tau} = \sum_\alpha e_\alpha \frac{\partial \psi_\alpha}{\partial x^\sigma} \frac{\partial \psi_\alpha}{\partial x^\tau} + \varphi_{\sigma\tau}, \tag{16}$$

where $\varphi_{\sigma\tau}$ does not involve $x^1, \cdots, x^p$.

From (12), (15), and (16) it follows that the fundamental quadratic form of V_n is

$$\sum_\alpha e_\alpha (d\psi_\alpha)^2 + \varphi_{\sigma\tau}\, dx^\sigma\, dx^\tau.$$

If now we effect the non-singular transformation

$$x'^\alpha = \psi_\alpha, \qquad x'^\sigma = x^\sigma \quad (\alpha = 1, \cdots, p; \sigma = p + 1, \cdots, n)$$

in the new coordinate system the fundamental form is

$$\sum_\alpha e_\alpha (dx^\alpha)^2 + g_{\sigma\tau}\, dx^\sigma\, dx^\tau, \qquad (\alpha = 1, \cdots, p; \sigma, \tau = p + 1, \cdots, n) \tag{17}$$

where $g_{\sigma\tau}$ are independent of $x^1, \cdots, x^p$. Also in this coordinate system the components (5) of the parallel unit vector fields are given by

$$\lambda^i_{\alpha|} = \delta^i_\alpha \qquad (\alpha = 1, \cdots, p). \tag{18}$$

It remains to be shown that for these values and the form (17) equations (1) are satisfied. In fact, the conditions are

$$\left\{ {i \atop \alpha j} \right\} = g^{ik}[\alpha j, k] = \frac{1}{2} g^{ik} \left(\frac{\partial g_{jk}}{\partial x^\alpha} + \frac{\partial g_{\alpha k}}{\partial x^j} - \frac{\partial g_{\alpha j}}{\partial x^k} \right) = 0$$

Since $\alpha = 1, \cdots, p$, when j and k have values $1, \cdots, p$ the expression in parenthesis is zero; the same is true when either takes these values and the other $p + 1, \cdots, n$; also since $g_{\sigma\tau}$ does not involve $x^1, \cdots, x^p$, the quantity in parenthesis vanishes when both j and k take the values $p + 1, \cdots, n$.

If then the rank of the matrix $\|c_{ab}\|$ is r, we have the theorem:

When a V_n admits r fields of parallel vectors, none of which is a null vector-field orthogonal to all the other fields, each field is expressible linearly and homogeneously with constant coefficients in terms of r mutually *orthogonal unit vector-fields; a coordinate system exists in terms of which the components of these unit vectors are given by* (18) *and the fundamental form of* V_n *is given by* (17) *with* $p = r$.

When the fundamental form of V_n is definite, there being no possibility of null vectors, the above theorem gives the canonical form of the metric tensor of a V_n admitting p independent parallel vector fields. When however the fundamental form is indefinite there is the possibility of p independent unit vector-fields and $r - p$ null fields, each field being orthogonal to the other fields of both sets. In this case in addition to the unit vectors of components (18) we have also the null vectors of components

$$\lambda^{\mu}_{\mu|} = \varphi_{\mu}, \qquad \lambda^{i}_{\mu|} = 0 \qquad (i \neq \mu),$$

for $\mu = p + 1, \cdots, r$ where φ_{μ} is to be determined. For these vectors we obtain from equations (1) in a manner similar to that which led to equation (7) the equations

$$(19) \qquad 2g_{\mu k}\frac{\partial \log \varphi_{\mu}}{\partial x^{j}} + \frac{\partial g_{\mu k}}{\partial x^{j}} + \frac{\partial g_{jk}}{\partial x^{\mu}} - \frac{\partial g_{\mu j}}{\partial x^{k}} = 0 \quad \begin{pmatrix} \mu = p+1, \cdots, r, \\ j, k = 1, \cdots, n \end{pmatrix}.$$

From (17) we have $g_{\alpha\sigma} = 0$ for $\sigma > p$ and that $g_{\sigma\tau}$ for $\sigma, \tau > p$ are independent of $x^1 \cdots, x^p$. In consequence of (4) and the above we have from (19) for $j = \alpha(= 1, \cdots, p)$ and $k = \rho(= r + 1, \cdots, n)$

$$g_{\mu\rho}\frac{\partial \log \varphi_{\mu}}{\partial x^{\alpha}} = 0 \qquad (\alpha = 1, \cdots, p).$$

Since the determinant of g_{ij} is not zero all of the functions $g_{\mu\rho}$ are not equal to zero. Consequently the functions φ_{μ} do not involve $x^1, \cdots, x^p$.

Since $g_{\mu\nu} = 0$ for $\mu, \nu = p + 1, \cdots, r$ equation (19) for $k = \nu$, $j = \rho$ reduces to

$$(20) \qquad \frac{\partial g_{\nu\rho}}{\partial x^{\mu}} - \frac{\partial g_{\mu\rho}}{\partial x^{\nu}} = 0 \quad \begin{pmatrix} \mu, \nu = p+1, \cdots, r \\ \rho = r+1, \cdots, n \end{pmatrix}.$$

In consequence of these equations, equations (19) for $j = \nu$. $k = \rho$ reduce to

$$(21) \qquad \frac{\partial}{\partial x^{\nu}} g_{\mu\rho}\varphi_{\mu} = 0.$$

If from equation (19) for $j = \rho$, $k = \tau$ for $\rho, \tau = r + 1, \cdots, n$ one subtracts the equation for $j = \tau$, $k = \rho$, one obtains

$$\frac{\partial}{\partial x^\rho} \varphi_\mu g_{\mu\tau} = \frac{\partial}{\partial x^\tau} \varphi_\mu g_{\mu\rho}.$$

Accordingly we put

$$\varphi_\mu g_{\mu\rho} = \frac{\partial \theta_\mu}{\partial x^\rho}. \tag{22}$$

In view of equation (21) and the fact that φ_μ and $g_{\mu\rho}$ do not involve $x^1, \cdots, x^p$ we have that θ_μ are functions of $x^{r+1}, \cdots, x^n$ at most. Since the determinant of g_{ij} is different from zero, the functions θ_μ must be independent, that is the rank of the jacobian matrix $\left\| \frac{\partial \theta_\mu}{\partial x^\rho} \right\|$ must be $r - p$. Consequently $r - p \leq n < p - (r - p)$, that is $r - p \leq \frac{1}{2}(n - p)$. Consequently $r - p \leq n - r$.

From (20) and (22) we have

$$\frac{\partial}{\partial x^\nu} \left(\frac{1}{\varphi_\mu} \right) \frac{\partial \theta_\mu}{\partial x^\rho} = \frac{\partial}{\partial x^\mu} \left(\frac{1}{\varphi_\nu} \right) \frac{\partial \theta_\nu}{\partial x^\rho}$$

Since these equations must hold for $\rho = r + 1, \cdots, n$ and the θ's are independent, we have that φ_μ is a function of $x^\mu, x^{r+1}, \cdots, x^n$ at most. If then we effect the non-singular transformation

$$x'^\alpha = x^\alpha, \qquad x'^\mu = \int \frac{dx^\mu}{\varphi_\mu}, \qquad x'^\rho = x^\rho \quad (\rho = r + 1, \cdots, n),$$

the components of the null vectors are

$$\lambda^i_{\mu|} = \delta^i_\mu$$

In this coordinate system $\varphi_\mu = 1$ and $g_{\mu\rho} = \frac{\partial \theta_\mu}{\partial x^\rho}$. Then from (19) we have that $g_{\rho\tau}$ are at most functions of $x^{r+1}, \cdots, x^n$. In this coordinate system we have in the fundamental form the terms $\sum_\mu g_{\mu\rho}\, dx^\mu\, dx^\rho$, which from (22) with $\varphi_\mu = 1$ is $\sum_\mu dx^\mu\, d\theta_\mu$. Since the rank of the matrix $\left\| \frac{\partial \theta_\mu}{\partial x^\rho} \right\|$ is $r - p$, there is no loss in generality in assuming that the jacobian $\left| \frac{\partial \theta_\mu}{\partial x^\sigma} \right|$ for $\sigma = r + 1, \cdots, 2r - p$ is different from zero.

Accordingly if we make the non-singular transformation

$$x'^h = x^h, \qquad x'^{r-p+\mu} = \theta\mu \quad \begin{pmatrix} h = 1, \cdots, r, 2r - p + 1, \cdots, n \\ \mu = p + 1, \cdots, r \end{pmatrix},$$

the fundamental form of V_n is (dropping primes)

$$(23) \quad \sum_\alpha e_\alpha (dx^\alpha)^2 + 2 \sum_\epsilon dx^{p+\epsilon}\, dx^{r+\epsilon} + g_{\rho\tau}\, dx^\rho\, dx^\tau$$

$$\begin{pmatrix} \alpha = 1, \cdots, p; \epsilon = 1, \cdots, r - p \\ \rho, \tau = r + 1, \cdots, n \end{pmatrix}$$

It can be shown as in the preceding case that for a V_n with this fundamental form equations (1) are satisfied by $\lambda^i_{\mu|} = \delta^i_\mu$. Hence we have the theorem:

When a V_n admits r independent fields of parallel vectors such that p of the fields are expressible linearly and with constant coefficients in terms of p mutually orthogonal unit fields, and $r - p$ fields consist of null vectors, each field being orthogonal to the other $r - 1$ fields, there exists a coordinate system in terms of which the fundamental metric form is given by (23), *where $g_{\rho\tau}$ are functions of $x^{r+1}, \cdots, x^n$ at most, and the components of the vector fields are δ^i_β for $\beta = 1, \cdots, r$.*

Appendix 10

The Greek letters take the values $1 \cdots, n$, the Latin $1, \cdots, p(< n)$, the upper index indicating the column, the lower the row.

Appendix 11

In (32.15) and (32.17) replace $\dfrac{1}{\rho_p}$ by $\dfrac{e_p e_{p+1}}{\rho_p}$. Replace right-hand member of (32.16) by $e_p \left(- \dfrac{\lambda_{p-1|}{}^i}{\rho_{p-1}} + \dfrac{\lambda_{p+1|}{}^i}{\rho_p} \right)$. In the equation preceding (32.18) replace $\dfrac{1}{\rho_p}$ by $\dfrac{e_{p+1}}{\rho_p}$ and $\dfrac{\partial x^j}{\partial s}$ by $\dfrac{dx^j}{ds}$.

Appendix 12

Put $c_{ij} Y^i_k Y^j_l = b_{kl}$ where the Y's are solutions of (37.9) determined by initial values such that (37.10) and (37.11) are satisfied. From (37.9) it follows that the first derivatives of b_{kl} are linear and homogeneous in the b's and these derivatives are equal to zero for the initial values. The same holds true for all higher derivatives of the b's, and consequently (37.10) and (37.11) hold for all values of the x's.

Appendix 13

Separable Systems of Stäckel. A necessary and sufficient condition that

$$a_{ij}\frac{dx^i}{ds}\frac{dx^j}{ds} = \text{const.}$$

be a quadratic first integral of the equation of geodesics of a Riemannian V_n is, as derived in § 39,

$$a_{ij,k} + a_{jk,i} + a_{ki,j} = 0, \tag{1}$$

where a comma followed by an index indicates covariant differentiation; there is no less in assuming that a_{ij} is symmetric in the indices.

If ρ_i are the roots of the determinant equation

$$|a_{ij} - \rho g_{ij}| = 0, \tag{2}$$

the equations

$$(a_{ij} - \rho_h g_{ij})\lambda^i_{h|} = 0, \tag{3}$$

as shown in § 33, determine an orthogonal ennuple of contravariant vectors of components $\lambda^i_{h|}$, where h indicates the vector and i the component. Ordinarily the vector-fields so defined are not normal in the sense that a vector-field admits a family of hypersurfaces orthogonal to the vectors.

We assume that a_{ij} is such that these vector-fields are normal and that the hypersurfaces orthogonal to them are taken as parametric; and we write the fundamental form thus

$$ds^2 = e_1 H_1^2(dx^1)^2 + \cdots + e_n H_n^2(dx^n)^2, \tag{4}$$

where the e's are plus or minus one as the case may be. In this case $\lambda^i_{h|} = 0$ for $i \neq h$ and $a_{ij} = 0$ for $i \neq j$. Making use of (15.7) we find that equations (1) for $j = k = i$ and $j = i$, $k = j$ respectively reduce to

$$\begin{aligned} &\frac{\partial \log \sqrt{a_{ii}}}{\partial x^i} = \frac{\partial \log H_i}{\partial x^i}, \\ &\frac{\partial a_{ii}}{\partial x^j} - 2a_{ii}\frac{\partial \log H_i^2}{\partial x^j} + a_{jj}\frac{1}{H_j^2}\frac{\partial H_i^2}{\partial x^j} = 0, \end{aligned} \tag{5}$$

and equations (1) for i, j, k different are satisfied identically. From the first set of (5) we have

$$a_{ii} = \rho_i H_i^2, \tag{6}$$

where ρ_i, thus defined, is independent of x^i. The second set of (5) reduce to

$$\frac{\partial}{\partial x^j} \log \frac{\rho_i - \rho_j}{H_i^2} = 0, \tag{7}$$

from which it follows that $(\rho_i - \rho_j)/H_i^2$ is independent of x^j. Writing these results in the form

$$\frac{\partial \rho_i}{\partial x^j} = (\rho_i - \rho_j)\frac{\partial \log H_i^2}{\partial x^j}, \qquad \frac{\partial \rho_i}{\partial x^i} = 0, \tag{8}$$

and expressing the condition of integrability of this system of equations, we obtain

$$(\rho_i - \rho_j)\left(\frac{\partial^2 \log H_i^2}{\partial x^i\, \partial x^j} + \frac{\partial \log H_i^2}{\partial x^j}\frac{\partial \log H_j^2}{\partial x^i}\right) = 0$$

and

$$(\rho_j - \rho_k)\left(\frac{\partial^2 \log H_i^2}{\partial x^j\, \partial x^k} - \frac{\partial \log H_i^2}{\partial x^j}\frac{\partial \log H_i^2}{\partial x^k} + \frac{\partial \log H_i^2}{\partial x^j}\frac{\partial \log H_j^2}{\partial x^k} + \frac{\partial \log H_i^2}{\partial x^k}\frac{\partial \log H_k^2}{\partial x^j}\right) = 0.$$

In order that (8) may admit a solution with all the ρ's different we must have

$$\frac{\partial^2 \log H_i^2}{\partial x^i\, \partial x^j} + \frac{\partial \log H_i^2}{\partial x^j}\frac{\partial \log H_j^2}{\partial x^i} = 0, \tag{9}$$

$$\frac{\partial^2 \log H_i^2}{\partial x^j\, \partial x^k} - \frac{\partial \log H_i^2}{\partial x^j}\frac{\partial \log H_i^2}{\partial x^k} + \frac{\partial \log H_i^2}{\partial x^j}\frac{\partial \log H_j^2}{\partial x^k} + \frac{\partial \log H_i^2}{\partial x^k}\frac{\partial \log H_k^2}{\partial x^j} = 0. \tag{10}$$

Since these equations are consistent, it follows that when they are satisfied equations (8) are completely integrable. One solution is $\rho_i = \rho_j = a$, a constant. We denote by ρ_i^α (for $\alpha = 2, \cdots, n$) $n - 1$ other solutions such that the determinant of the n solutions is not zero. This may be indicated in the determinant form

$$|\rho_i^\alpha - \rho_j^\alpha| \neq 0, \tag{11}$$

where i is fixed, and $\alpha = 2, \cdots, n; j = 1, \cdots, n; j \neq i$. In this case the equations of the geodesics admit $n - 1$ quadratic first integrals whose coefficients are

$$(12) \qquad a_{ii}^{\alpha} = \rho_i^{\alpha} H_i^2, \qquad a_{ij}^{\alpha} = 0.$$

It is our purpose to show that the preceding conditions determine the Stäckel form of the metric tensor of V_n as given in § 39 but not there derived. To this end we denote by φ_{ij} n^2 functions such that their determinant φ is not zero, and we denote by φ^{ij} the cofactor of φ_{ij} in φ. We put

$$(13) \qquad H_i^2 = \frac{\varphi}{\varphi^{i1}}, \qquad \rho_i^{\alpha} = \frac{\varphi^{i\alpha}}{\varphi^{i1}},$$

and understand that the φ's are such that ρ_i^{α} are independent of x^i.

Also we put

$$(14) \qquad b_{ij}^{\alpha} = \frac{\rho_i^{\alpha} - \rho_j^{\alpha}}{H_i^2} = \frac{\varphi^{j1}\varphi^{i\alpha} - \varphi^{i1}\varphi^{j\alpha}}{\varphi\varphi^{j1}},$$

and have from (7) that b_{ij}^{α} are independent of x^i. We have that

$$\varphi^{j1}\varphi^{i\alpha} - \varphi^{i1}\varphi^{j\alpha} = \varphi M_{j1i\alpha},$$

where $M_{j1i\alpha}$ is the algebraic complement of $\varphi_{j1}\varphi_{i\alpha} - \varphi_{i1}\varphi_{j\alpha}$ in the determinant φ*. Consequently we have

$$(15) \qquad \varphi^{j1} b_{ij}^{\alpha} = M_{j1i\alpha} \quad (i, j = 1, \cdots, n; \alpha = 2, \cdots, n).$$

From the definition of $M_{j1i\alpha}$ we have

$$\varphi^{j\alpha} = \sum_{i(\neq j)} \varphi_{i1} M_{j1i\alpha},$$

and consequently

$$\frac{\varphi^{j\alpha}}{\varphi^{j1}} = \sum_{i(\neq j)} \varphi_{i1} b_{ij}^{\alpha}.$$

Differentiating with respect to x^j, we have

$$0 = \sum_{i(\neq j)} \frac{\partial \varphi_{i1}}{\partial x^j} b_{ij}^{\alpha} \quad (\alpha = 2, \cdots, n; j = 1, \cdots, n).$$

For a given j the determinant of b_{ij}^{α} is not zero, in consequence of (11) and (14). Hence a function φ_{i1} is a function of x^i at most.

* Cf. *Kowalewski*, Einführung in die Determinantentheorie, p. 80.

From (14) we have

$$(16) \qquad \varphi^{j1} b^{\alpha}_{ij} = -\varphi^{i1} b^{\alpha}_{ji}.$$

In consequence of this result and (15) we have

$$(17) \qquad \frac{M_{j1i\beta}}{M_{j1i2}} = \frac{b^{\beta}_{ij}}{b^{2}_{ij}} = \frac{b^{\beta}_{ji}}{b^{2}_{ji}} \equiv \sigma_{ij\beta} = \sigma_{ji\beta} \qquad \begin{pmatrix} \beta = 3, \cdots, n \\ i \neq j \end{pmatrix}$$

Since the second term is independent of x^j and the third of x^i, it follows that $\sigma_{ij\beta}$, as defined in (17), is independent of x^i and x^j. From the identities

$$\sum_{\alpha}^{2, \cdots, n} \varphi_{k\alpha} M_{j1i\alpha} = 0 \qquad (i, j, k \neq),$$

we have with the aid of (17)

$$(18) \qquad \varphi_{k2} + \sum_{\beta}^{3, \cdots, n} \varphi_{k\beta} \sigma_{ij\beta} = 0.$$

Differentiating with respect to x^i, we have

$$(19) \qquad \frac{\partial \varphi_{k2}}{\partial x^i} + \sum_{\beta}^{3, \cdots, n} \frac{\partial \varphi_{k\beta}}{\partial x^i} \sigma_{ij\beta} = 0.$$

For a given i and k, there are $n - 2$ equations (18) satisfied by the $n - 1$ quantities $\varphi_{k2}, \cdots, \varphi_{kn}$; and these same equations are satisfied by the derivatives of these quantities with respect to x^i, hence we have

$$\frac{\partial \varphi_{k\alpha}}{\partial x^i} = \mu_{ik} \varphi_{k\alpha},$$

or

$$\frac{\partial}{\partial x^i} \left(\frac{\varphi_{k\alpha}}{\varphi_{k\gamma}} \right) = 0,$$

for $\gamma \neq \alpha$. Such equations hold for $i = 1, \cdots, n$; $i \neq k$. Hence we have

$$(20) \qquad \varphi_{i\alpha} = e^{\nu_i} \psi_{i\alpha},$$

where $\psi_{i\alpha}$ are functions of x^i at most and ν_i are to be determined.

From (15) we have

$$b^{\alpha}_{j1} M_{11i\alpha} = b^{\alpha}_{i1} M_{11j\alpha} \qquad (i, j = 2, \cdots, n).$$

Substituting from (20) we obtain

$$e^{\nu_j}b^{\alpha}_{j1}N_{1i} = e^{\nu_i}b^{\alpha}_{i1}N_{1j},$$

where N_{1i} is independent of x^1 and x^i. Differentiating with respect to x^1, we have

$$\frac{\partial}{\partial x^1}(\nu_i - \nu_j) = 0 \qquad (i, j = 2, \cdots, n).$$

Again from (15) and (16) we have

$$b^{\alpha}_{j\alpha}M_{\alpha 1\beta\beta} + b^{\beta}_{\beta\alpha}M_{j1\alpha\alpha} = 0 \quad \left(\begin{matrix}\alpha, \beta = 2, \cdots, n; \alpha \neq \beta \\ j = 1, \cdots, n; j \neq \alpha\end{matrix}\right).$$

Substituting from (20) we obtain

$$e^{\nu_j}b^{\alpha}_{j\alpha}N_{\alpha\beta} + e^{\nu_\beta}b^{\beta}_{\beta\alpha}N_{\alpha j} = 0.$$

Differentiating with respect to x^α, we have

$$\frac{\partial}{\partial x^\alpha}(\nu_j - \nu_\beta) = 0.$$

Combining these results, we have

$$\frac{\partial(\nu_i - \nu_j)}{\partial x^k} = 0 \quad (i, j, k = 1, \cdots, n; i, j, k \neq).$$

From the preceding equations we have

$$\text{(21)} \qquad \nu_i - \nu_j = f_{ij}, \qquad \nu_i - \nu_k = f_{ik}, \qquad \nu_j - \nu_k = f_{jk},$$

where f_{ij} is at most a function of x^i and x^j. From these equations we have

$$f_{ij} - f_{ik} + f_{jk} = 0.$$

Differentiating with respect to x^i, we obtain

$$\frac{\partial f_{ij}}{\partial x^i} = \frac{\partial f_{ik}}{\partial x^i}.$$

Since the first term does not involve x^k and the second x^j, we have

$$f_{ij} = \sigma_i - \sigma_j, \qquad f_{ik} = \sigma_i - \sigma_k,$$

where σ_i is as function of x^i alone. Hence equations (21) may be replaced by

$$\nu_i = \nu + \sigma_i,$$

where ν is undetermined, and (20) becomes

$$\varphi_{i\alpha} = e^{\nu}\theta_{i\alpha} \quad (i = 1, \cdots, n; \alpha = 2, \cdots, n),$$

where $\theta_{i\alpha}$ are functions of x^i alone. Since we have shown that φ_{i1} is a function of x^i alone, it follows that when the above expressions are substituted in (13) the factor e^{ν} disappears, and consequently the general solution of the problem is obtained when each function φ_{ij} is a function of x^i alone, which is the Stäckel form. Hence we have:

A necessary and sufficient condition that the metric tensor of V_n can be given the Stäckel form is that the equations of geodesics admit $n - 1$ independent quadratic first integrals, that the roots of the characteristic equations (2) *for each of these integrals be simple, that* (11) *hold, and that the vector-fields determined by* (3) *be normal and be the same vector-fields for all the first integrals.*

Also we have:

A necessary and sufficient condition that the fundamental form (4) *be in the Stäckel form is that equations* (9) *and* (10) *be satisfied.*

From (9) we have

$$\frac{\partial^2}{\partial x^i\,\partial x^j}\log\frac{H_i^2}{H_j^2} = 0. \tag{22}$$

from which it follows that

$$H_i^2 = \psi_{ij}^2\theta_{ij}, \qquad H_j^2 = \psi_{ji}^2\theta_{ij},$$

where ψ_{ij} is independent of x^j and ψ_{ji} of x^i. Substituting these expressions in (9) we find that θ_{ij} is such that

$$H_i^2 = \psi_{ij}^2(\tau_{ij} + \tau_{ji}), \qquad H_j^2 = \psi_{ji}^2(\tau_{ij} + \tau_{ji}), \tag{23}$$

where τ_{ij} is independent of x^j and τ_{ji} of x^i.

Stäckel systems in a flat space

For a flat space we have $R_{jiik} = 0$, where R_{jiik} is given by (37.4). In consequence of (10) this condition is

$$\frac{\partial^2 \log H_i^2}{\partial x^j\,\partial x^k} = 0, \tag{24}$$

from which and (10) we have

$$\frac{\partial \log H_i^2}{\partial x^j}\frac{\partial \log H_i^2}{\partial x^k} - \frac{\partial \log H_i^2}{\partial x^j}\frac{\partial \log H_j^2}{\partial x^k} - \frac{\partial \log H_i^2}{\partial x^k}\frac{\partial \log H_k^2}{\partial x^j} = 0. \tag{25}$$

Substituting the expressions (23) for H_i^2 and H_j^2 in (24) and the equation resulting from (24) when i and j are interchanged we obtain respectively

$$(26)\qquad \frac{\partial\tau_{ji}}{\partial x^j} = (\tau_{ij} + \tau_{ji})\theta_{ji}(x^i_{,}, x^j), \qquad \frac{\partial\tau_{ij}}{\partial x^i} = (\tau_{ij} + \tau_{ji})\theta_{ij}(x^i_{,}, x^j).$$

In order to determine the functions θ_{ji} and θ_{ij}, we differentiate these equations with respect to x^i and x^j respectively, in which case the left-hand numbers are zero, and reduce the results by means of (26); we obtain

$$\frac{\partial\theta_{ji}}{\partial x^i} + \theta_{ji}\theta_{ij} = 0, \qquad \frac{\partial\theta_{ij}}{\partial x^j} + \theta_{ji}\theta_{ij} = 0.$$

Since the two partial derivatives are equal we put

$$\theta_{ji} = \frac{\partial \log \alpha}{\partial x^j}, \qquad \theta_{ij} = \frac{\partial \log \alpha}{\partial x^i},$$

and on substitution find that $\alpha = \alpha_i + \alpha_j$, where α_i and α_j are at most functions of x^i and x^j respectively. Then from (26) we have

$$\tau_{ij} + \tau_{ji} = (\alpha_i + \alpha_j)\omega_{ij},$$

where ω_{ij} is independent of x^i and x^j. In consequence of this result we have from (23)

$$(27)\quad H_i^2 = X_i \prod_{j(\neq i)} (\sigma_{ij} + \sigma_{ji}), \qquad H_j^2 = X_j \prod_{i(\neq j)} (\sigma_{ji} + \sigma_{ij}),$$

$$H_k^2 = X_k \prod_{i(\neq k)} (\sigma_{ki} + \sigma_{ik})$$

where X_i, X_j, X_k are functions of x^i, x^j, x^k respectively, and σ_{ij} is a function of x^i at most, and similarly each σ is a function of x^h at most where h is the first subscript of σ. These expressions satisfy (24).

The functions σ_{ij} must be such that equations (25) be satisfied, and also $R_{jiij} = 0$, as given by (37.4). For euclidean 3-space the results are the following other than the Cartesian case.*

I. $$H_1 = H_2 = 1, \qquad H_3 = x^1,$$

the cartesian coordinates x, y, z being related to the cylindrical polar

* For proof see *Eisenhart*, 1934, 3.

coordinates x^i by the equations

$$x = x^1 \cos x^3, \qquad y = x^1 \sin x^3, \qquad z = x^2.$$

II. $$H_1 = 1, \qquad H_2 = x^1, \qquad H_3 = x^1 \sin x^2,$$

where x^i are polar coordinates, and

$$x = x^1 \sin x^2 \cos x^3, \qquad y = x^1 \sin x^2 \sin x^3, \qquad z = x^1 \cos x^2.$$

III. $$H_1^2 = 1, \qquad H_2^2 = H_3^2 = a^2(\sinh^2 x^2 + \sin^2 x^3),$$

where a is an arbitrary constant, and x^i are elliptic cylinder coordinates, and

$$x = x^1, \qquad y = a \cosh x^2 \cos x^3, \qquad z = a \sinh x^2 \sin x^3.$$

IV. $$H_1^2 = 1, \qquad H_2^2 = H_3^2 = (x^2)^2 + (x^3)^2,$$

where x^i are parabolic cylinder coordinates, and

$$x = x^1, \qquad y = \tfrac{1}{2}((x^2)^2 - (x^3)^2), \qquad z = x^2x^3.$$

V. $$H_1^2 = 1, \qquad H_2^2 = H_3^2 = (x^1)^2[k^2 \operatorname{cn}^2 (x^2,k) + k'^2 \operatorname{cn}^2 (x^3,k')],$$

where x^i are sphere-conical coordinates,

$$x = x^1 \operatorname{dn} (x^2,k) \operatorname{sn} (x^3,k'), \qquad y = x^1 \operatorname{sn} (x^2,k) \operatorname{dn} (x^3,k'),$$
$$z = x^1 \operatorname{cn} (x^2,k) \operatorname{cn} (x^3,k')$$

and the constants k and k' entering in the elliptic functions are in the relation $k^2 + k'^2 = 1$.

VI. $$H_1^2 = H_3^2 = (x^1)^2 + (x^3)^2, \qquad H_2^2 = (x^2x^3)^2,$$

where x^i are parabolic coordinates, and

$$x = x^1x^3 \cos x^2, \qquad y = x^1x^3 \sin x^2, \qquad z = \tfrac{1}{2}((x^1)^2 - (x^3)^2).$$

VII. $$H_1^2 = H_3^2 = a^2(\sinh^2 x^1 + \sin^2 x^3),$$
$$H_2^2 = a^2 \sinh^2 x^1 \sin^2 x^3,$$

where a is an arbitrary constant, x^i are prolate spherical coordinates, and

$$x = a \sinh x^1 \sin x^3 \cos x^2, \qquad y = a \sinh x^1 \sin x^3 \sin x^2,$$
$$z = a \cosh x^1 \cos x^3.$$

VIII. $H_1^2 = H_3^2 = a^2(\cosh^2 x^1 - \sin^2 x^3),$

$$H_2^2 = a^2 \cosh^2 x^1 \sin^2 x^3,$$

where x^i are oblate spheroidal coordinates, and

$$x = a \cosh x^1 \sin x^3 \cos x^2, \qquad y = a \cosh x^1 \sin x^3 \sin x^2,$$
$$z = a \sinh x^1 \cos x^3.$$

IX. $$H_i^2 = \frac{(x^i - x^j)(x^i - x^k)}{f(x^i)},$$

$$f(x^i) = 4(\alpha - x^i)(\beta - x^i)(\gamma - x^i),$$

where $\alpha > x^1 > \beta > x^2 > \gamma > x^3$; the x^i are confocal elliptic coordinates, and

$$x^2 = \frac{\prod_i (\alpha - x^i)}{(\alpha - \beta)(\alpha - \gamma)}, \qquad y^2 = \frac{\prod_i (\beta - x^i)}{(\beta - \alpha)(\beta - \gamma)},$$
$$z^2 = \frac{\prod_i (\gamma - x^i)}{(\gamma - \alpha)(\gamma - \beta)}.$$

X. $$H_i^2 = \frac{(x^i - x^j)(x^i - x^k)}{f(x^i)}, \qquad f(x^i) = 4(\alpha - x^i)(\beta - x^i),$$

where $x^1 > \alpha > x^2 > \beta > x^3$; the x^i are confocal parabolic coordinates, and

$$x = \frac{x^1 + x^2 + x^3 - \alpha - \beta}{2}, \qquad y^2 = \frac{\prod_i (\beta - x^i)}{\alpha - \beta}, \qquad z^2 = \frac{\prod_i (\alpha - x^i)}{\beta - \alpha}$$

In each case the coordinate surfaces $x^i = \text{const.}$ are quadric surfaces including the cases when one or more families of surfaces consist of planes. Hence:

A necessary and sufficient condition that a triply orthogonal system of surfaces in euclidean 3-space be a Stäckel coordinate system is that they be any system of confocal quadrics including the cases when one or more families of the system consists of planes.

Appendix 14

In four places replace $\xi^\alpha_{1|}$ by $\xi_{n+1|}{}^\alpha$.

Appendix 15

Replace the quantity in parenthesis in (43.8) by $\xi^\beta_{;\mu}\, y^\mu_{,j}$, where $\xi^\beta{}_{;\mu}$ denotes the covariant derivative of ξ^β with respect to the metric tensor $a_{\alpha\beta}$ of V_{n+1}. Do the same for the next two equations on this page, and replace (43.10) by

$$\xi^{\beta}_{;\alpha}\, y^{\alpha}_{,j} = -\Omega_{lj}\, g^{lm}\, y^{\beta}_{,m}. \tag{43.10}$$

Appendix 16

The argument is open to the same objection as applies to equation (17.11) in Appendix 2. Differentiating the preceding equation one has

$$\frac{d^2y^\alpha}{ds^2} = \frac{\partial y^\alpha_{,i}}{\partial x^j}\frac{dx^i}{ds}\frac{dx^j}{ds} + y^\alpha_{,i}\frac{d^2x^i}{ds^2}.$$

By means of

$$\frac{\partial y^\alpha_{,i}}{\partial x^j} = y^\alpha_{,ij} + y^\alpha_{,k}\begin{Bmatrix} k \\ ij \end{Bmatrix}$$

and (43.4) one obtains equation (44.1).

Appendix 17

Replace (47.6) and (47.7) by

$$a_{\alpha\beta}y^\alpha_{,i}y^\gamma_{,j}\xi^\beta_{\sigma|;\gamma} = -\Omega_{\sigma|ij}, \tag{47.6}$$

$$a_{\alpha\beta}\xi^\alpha_{\tau|}y^\gamma_{,j}\xi^\beta_{\sigma|;\gamma} = \mu_{\tau\sigma|j}, \tag{47.7}$$

where $\xi^\beta_{\sigma|;\gamma}$ is the covariant derivative of $\xi^\beta_{\sigma|}$ with respect to the metric tensor $a_{\alpha\beta}$ of V_m.

Replace what follows equation (47.8) down to and including (47.9) by the following: Since $\xi^\beta_{\sigma|;\gamma}y^\gamma_{,j}$ are the components of a vector in V_m, we put

$$\xi^\beta_{\sigma|;\gamma}y^\gamma_{,j} = A^k_{\sigma j}y^\beta_{,k} + B^\rho_{\sigma j}\xi^\beta_{\rho|}.$$

Multiplying by $a_{\alpha\beta}y^\beta_{,i}$, summing for β, and making use of (47.2) and (47.6), as written above, we obtain $A^k_{\sigma j}g_{ik} = -\Omega_{\sigma|ij}$. Again multiplying this equation by $a_{\alpha\beta}\xi^\alpha_{\tau|}$, summing for β, and making use of (47.1), (47.2), and the above (47.7), we obtain $e_\tau B^\tau_{\sigma j} = \mu_{\tau\sigma|j}$. Hence in place of equation (47.9) we have

$$\xi^\beta_{\sigma|;\gamma}y^\gamma_{,j} = -\Omega_{\sigma|ij}g^{ik}y^\beta_{,k} + \sum_\tau e_\tau\mu_{\tau\sigma|j}\xi^\beta_{\tau|}. \tag{47.9}$$

Appendix 18

If the associate direction of the normal vector $\xi_{\sigma|}{}^\beta$ in the direction $\xi^\beta_{h|}$ on V_n is to coincide with the direction, the right-hand member of equation (50.1) must equal $\rho\xi_{h|}{}^\beta$, that is $\rho\lambda_{h|}{}^j y^\beta_{,j}$. This condition is

$$[-(\Omega_{\sigma|lj}g^{lk} + \rho\delta_j^k)y^\beta_{,k} + \sum_\tau e_\tau \mu_{\tau\sigma|j}\xi_{\tau|}{}^\beta]\lambda^j_{h|} = 0.$$

From (47.7) in Appendix 17 we have

$$a_{\alpha\beta}\xi^\alpha_{\tau|}\xi^\beta_{\sigma|;\gamma}\xi_{h|}{}^\gamma = \mu_{\tau\sigma|j}\lambda^j_{h|}.$$

If then $\xi_{\sigma|}{}^\beta{}_{;\gamma}\xi_{h|}{}^\gamma = \rho\xi_{h|}{}^\beta$, we have $\mu_{\tau\sigma|j}\lambda^j_{h|} = 0$, and the above condition reduces to

$$(\Omega_{\sigma|lj}g^{lk} + \rho\delta_j{}^k)\lambda^j_{h|} = 0,$$

since the vectors $y^\beta{}_{,k}$ are independent. In turn this equation is equal to

$$(\Omega_{\sigma|ij} + \rho g_{ij})\lambda^j_{h|} = 0.$$

Hence in place of the theorem as stated we have:

If the associate direction of a normal to V_n in a V_m for a curve in V_n is tangent to the curve, the latter is a line of curvature for the given normal.

Appendix 19

If (54.1) is satisfied it follows from (43.10) in Appendix 15 that

$$\xi^\alpha{}_{;\beta}y^\beta_{,i} = 0. \tag{54.3}$$

Add to the theorem the clause: *with respect to a displacement in the hypersurface.*

Appendix 20

The results of § 56 for the case where the fundamental form of V_n is positive definite have been obtained by Vranceano using a different method.* In this note his method is applied to a V_n for which the fundamental form is positive definite or indefinite with the understanding that the matrix of g_{ij} is of rank n.

Using the notation of §§ 55, 56 we put

$$z^\alpha = f^\alpha(x^1, \cdots, x^n) + \sum_\sigma x^\sigma \eta^\alpha_{\sigma|} \qquad \begin{pmatrix} \alpha = 1, \cdots, n+p; \\ \sigma = n+1, \cdots, n+p \end{pmatrix}, \tag{1}$$

where $\eta^\alpha_{\sigma|}$ are the components of p mutually orthogonal unit vectors normal to V_n, so that we have

$$\sum_\alpha c_\alpha \frac{\partial z^\alpha}{\partial x^i}\eta^\alpha_{\sigma|} = 0, \qquad \sum_\alpha c_\alpha(\eta^\alpha_{\sigma|})^2 = e_\sigma, \qquad \sum_\alpha c_\alpha \eta^\alpha_{\sigma|}\eta^\alpha_{\tau|} = 0 \quad (\sigma \neq \tau). \tag{2}$$

If the quantities x^σ are considered as p new variables, equations (1) may be considered to be a transformation of coordinates in S_{n+p}, which

* 1930, 8.

is non-singular in the neighborhood of V_n, since along $V_n(x^\sigma = 0)$ the jacobian of the transformation is the determinant of the quantities $\frac{\partial f^\alpha}{\partial x^i}$ and $\eta^\alpha_{\sigma|}$; this cannot be zero, since there is no linear relation between these quantities.

From (1), (55.1), (55.3), and (2) we have

$$ds^2 = \sum_\alpha c_\alpha (dz^\alpha)^2 = g_{ij}\, dx^i\, dx^j + 2x^\sigma \sum_\alpha c_\alpha\, df^\alpha\, d\eta^\alpha_{\sigma|}$$
$$+ 2\, dx^\sigma x^\tau \sum_\alpha c_\alpha \eta^\alpha_{\sigma|}\, d\eta^\alpha_{\tau|} + x^\sigma x^\tau \sum_\alpha c_\alpha\, d\eta^\alpha_{\sigma|}\, d\eta^\alpha_{\tau|} + \sum_\sigma c_\sigma (dx^\sigma)^2.$$

If we write

$$ds^2 = a_{\alpha\beta}\, dx^\alpha\, dx^\beta,$$

and make use of (56.3) we obtain the following expressions for the a's:*

$$a_{ij} = g_{ij} - 2x^\sigma b_{\sigma|ij} + x^\sigma x^\tau c_{\sigma\tau ij},$$

(3) $$2c_{\sigma\tau ij} = g^{kl}(b_{\sigma|ki} b_{\tau|lj} + b_{\sigma|lj} b_{\tau|ki}) + \sum_\rho e_\rho(\gamma_{\rho\sigma|i}\gamma_{\rho\tau|j} + \gamma_{\rho\sigma|j}\gamma_{\rho\tau|i})$$

$$a_{i\sigma} = a_{\sigma i} = x^\tau \gamma_{\sigma\tau|i}, \qquad a_{\sigma\tau} = 0\ (\sigma \neq \tau), \qquad a_{\sigma\sigma} = e_\sigma$$
$$(\rho, \sigma, \tau = n + 1, \cdots, n + p)$$

where the index i takes the values $1, \cdots, n$.

Indicating by $[\alpha\beta, \gamma]_a$ and $[ij, k]$ the Christoffel symbols of the first kind formed with respect to the a's and g's respectively, we obtain the following results, noting (56.4):

$$[ij, k]_a = [ij, k] - 2x^\sigma b_{\sigma|ijk} + x^\sigma x^\tau c_{\sigma\tau ijk},$$

where

$$b_{\sigma|ijk} = \frac{1}{2}\left(\frac{\partial b_{\sigma|ik}}{\partial x^j} + \frac{\partial b_{\sigma|jk}}{\partial x^i} - \frac{\partial b_{\sigma|ij}}{\partial x^k}\right),$$

$$c_{\sigma\rho ijk} = \frac{1}{2}\left(\frac{\partial c_{\sigma\tau ik}}{\partial x^j} + \frac{\partial c_{\sigma\tau jk}}{\partial x^i} - \frac{\partial c_{\sigma\tau ij}}{\partial x^k}\right),$$

$$[ij, \rho]_a = \frac{1}{2}\left[x^\tau\left(\frac{\partial \gamma_{\rho\tau|i}}{\partial x^j} + \frac{\partial \gamma_{\rho\tau|j}}{\partial x^i}\right) + 2b_{\rho|ij} - 2x^\tau c_{\rho\tau ij}\right],$$

$$[i\rho, j]_a = \frac{1}{2}\left[x^\tau\left(\frac{\partial \gamma_{\rho\tau|j}}{\partial x^i} - \frac{\partial \gamma_{\rho\tau|i}}{\partial x^j}\right) - 2b_{\rho|ij} + 2x^\tau c_{\rho\tau ij}\right],$$

$$[\sigma\tau, i] = 0, \qquad [\sigma i, \tau] = \gamma_{\tau\sigma|i}, \qquad [\sigma\tau, \rho] = 0$$
$$\begin{pmatrix} i, j, k = 1, \cdots, n; \\ \rho, \sigma, \tau = n + 1, \cdots, n + p \end{pmatrix}.$$

* Here, and in what follows, the greek letter nu as used in §56 is replaced by gamma in such terms as $\nu_{\rho\sigma|i}$.

Denoting by $A_{\alpha\beta\gamma\delta}$ the Riemann curvature tensor for the a's we have from (8.8)

$$A_{\alpha\beta\gamma\delta} = \frac{\partial}{\partial x^\gamma}[\beta\delta, \alpha]_a - \frac{\partial}{\partial x^\delta}[\beta\gamma, \alpha]_a + a^{\sigma\tau}([\beta\gamma, \sigma]_a[\alpha\delta, \tau]_a - [\beta\delta, \sigma]_a[\alpha\gamma, \tau]_a).$$

At points of V_n

$$a^{ij} = g^{ij}, \qquad a^{i\sigma} = 0, \qquad a^{\sigma\tau} = 0, \qquad a^{\sigma\sigma} = e_\sigma,$$

so that we have at points of V_n

$$(4) \quad A_{\alpha\beta\gamma\delta} = \frac{\partial}{\partial x^\gamma}[\beta\delta, \alpha]_a - \frac{\partial}{\partial x^\delta}[\beta\gamma, \alpha]_a + g^{hl}([\beta\gamma, h]_a[\alpha\delta, l]_a - [\beta\delta, h]_a[\alpha\gamma, l]_a) + \sum_\sigma e_\sigma([\beta\gamma, \sigma]_a[\alpha\delta, \sigma]_a - [\beta\delta, \sigma]_a[\alpha\gamma, \sigma]_a) = 0 \quad (h, l = 1, \cdots n),$$

since $A_{\alpha\beta\gamma\delta} = 0$ throughout S_{n+p}.

For $\alpha = i, \beta = j, \gamma = k, \delta = l$, this equation becomes at points of V_n in consequence of the above expressions for the Christoffel symbols

$$(5) \quad R_{ijkl} = \sum_\sigma e_\sigma(b_{\sigma|ik}b_{\sigma|jl} - b_{\sigma|il}b_{\sigma|jk}),$$

which are equations (56.5). For $\alpha = k, \beta = j, \gamma = i, \delta = \sigma$, we obtain

$$(6) \quad b_{\sigma|ij,k} - b_{\sigma|ik,j} = \sum_\tau e_\tau(\gamma_{\tau\sigma|k}b_{\tau|ij} - \gamma_{\tau\sigma|j}b_{\tau|ik}),$$

which are equations (56.6). For $\alpha = \tau, \beta = \sigma, \gamma = j, \delta = k$, we obtain

$$(7) \quad \gamma_{\tau\sigma|j,k} - \gamma_{\tau\sigma|k,j} + \sum_\rho e_\rho(\gamma_{\rho\tau|j}\gamma_{\rho\sigma|k} - \gamma_{\rho\tau|k}\gamma_{\rho\sigma|j}) + g^{hl}(b_{\tau|lj}b_{\sigma|hk} - b_{\tau|lk}b_{\sigma|hj}) = 0,$$

which are equations (56.7). For $\alpha = \tau, \beta = i, \gamma = \sigma, \delta = j$ we obtain

$$\tfrac{1}{2}(\gamma_{\tau\sigma|j,i} - \gamma_{\tau\sigma|i,j}) - c_{\tau\sigma ij} + g^{hl}b_{\sigma|hi}b_{\tau|jl} + \sum_\rho e_\rho\gamma_{\rho\sigma|i}\gamma_{\rho\tau|j} = 0.$$

When the expression for $c_{\tau\sigma ij}$ from (3) is substituted in this equation, one obtains equation (7) with k replaced by i. When three or four of the indices take values greater than n equations (4) are satisfied identically.

Appendix 21

This note deals with a non-flat V_n whose fundamental quadratic form $g_{ij}\,dx^i\,dx^j$ may be positive definite or indefinite, provided that the matrix of g_{ij} is of rank n. We consider the V_n as immersed in a flat space S_m of coordinates z^α, as in § 55, the fundamental form of S_m being $\sum_\alpha c_\alpha(dz^\alpha)^2$, where the c's are plus or minus one according to the character of V_n and S_m, the equations of V_n in S_m being

$$z^\alpha = f^\alpha(x^1, \cdots, x^n).$$

From equations (55.3), namely

$$(1) \qquad \sum_\alpha c_\alpha z^\alpha_{,i} z^\alpha_{,j} = g_{ij}, \qquad z^\alpha_{,i} \equiv \frac{\partial z^\alpha}{\partial x^i},$$

we obtain by covariant differentiation based upon g_{ij}

$$\sum_\alpha c_\alpha (z^\alpha_{,ik} z^\alpha_{,j} + z^\alpha_{,i} z^\alpha_{,jk}) = 0$$

where $z^\alpha_{,ik}$ is the covariant derivative of $z^\alpha_{,i}$ that is

$$z^\alpha_{,ik} \equiv \frac{\partial^2 z^\alpha}{\partial x^i \, \partial x^k} - \frac{\partial z^\alpha}{\partial x^l} \begin{Bmatrix} l \\ ik \end{Bmatrix}.$$

If we subtract this equation from the sum of the two equations obtained from it on interchanging i and k and j and k respectively, we obtain

$$(2) \qquad \sum_\alpha c_\alpha z^\alpha_{,k} z^\alpha_{,ij} = 0.$$

From the Ricci identities [cf. (11.14)]

$$(3) \qquad z^\alpha_{,ijk} - z^\alpha_{,ikj} = z^\alpha_{,h} R^h_{ijk} = g^{hl} z^\alpha_{,l} R_{hijk}$$

it follows that the quantities $z^\alpha_{,ij}$ cannot be zero for a non-flat V_n.

From equation (2) and the theorem on page 145 it follows that the quantities $z^\alpha_{,ij}$ are expressible linearly in terms of $m - n$ mutually orthogonal unit vectors in S_m normal to V_n. That process is followed in § 56. In this note we consider the case when the quantities $z^\alpha_{,ij}$ are linearly expressible in terms of $q_1 < m - n$ unit vectors $\eta^\alpha_{\sigma|}$ which in accordance with § 13 may be taken to be mutually orthogonal in all generality, q_1 being the least number of such vectors in terms of which $z^\alpha_{,ij}$ is expressible. We have

$$(4) \qquad \sum_\alpha c_\alpha(\eta^\alpha_{\sigma|})^2 = e_\sigma, \qquad \sum_\alpha c_\alpha \eta^\alpha_{\sigma|}\eta^\alpha_{\tau|} = 0, \qquad \sum_\alpha c_\alpha \eta^\alpha_{\sigma|} z^\alpha_{,i} = 0$$

$$(\sigma, \tau = 1, \cdots, q_1; \sigma \neq \tau),$$

where the e's are plus or minus one as the case may be. Accordingly we have

$$z^\alpha_{,ij} = \sum_\sigma e_\sigma b_{\sigma|ij} \eta_{\sigma|}{}^\alpha, \tag{5}$$

where the quantities $b_{\sigma|ij}$ are components of q_1 tensors symmetric in their indices, as follows from (5) and the fact that $z^\alpha_{,ij}$ is symmetric in i and j. The vectors $\eta^\alpha_{\sigma|}$ and all vectors linearly expressible in terms of these are said to constitute the *first normal complex* N_1 to V_n in S_m.

If the third set of equations (4) are differentiated covariantly the result is reducible by (4) and (5) to

$$b_{\sigma|ik} + \sum_\alpha c_\alpha z^\alpha_{,i} \eta^\alpha_{\sigma|,k} = 0 \qquad (\sigma = 1, \cdots, q_1). \tag{6}$$

From equation (1) we have

$$\sum_\alpha c_\alpha z^\alpha_{,i} z^\alpha_{,j} g^{hj} = \delta^h_i,$$

by means of which equation (6) is equivalent to

$$\sum_\alpha c_\alpha (\eta^\alpha_{\sigma|,k} + b_{\sigma|hk} g^{hj} z^\alpha_{,j}) z^\alpha_{,i} = 0.$$

From the form of these equations it follows that the expression in parenthesis is linearly expressible in vectors normal to V_n, and consequently in terms of $\eta_{\sigma|}{}^\alpha$, and at least q_2 other normal unit vectors $\eta_{\sigma_2|}{}^\alpha$, all the vectors of the two sets being mutually orthogonal. Thus we have

$$\eta^\alpha_{\sigma|,k} = -b_{\sigma|hk} g^{hj} z^\alpha_{,j} + \sum_\tau e_\tau \gamma_{\tau\sigma|k} \eta^\alpha_{\tau|} + \sum_{\sigma_2} e_{\sigma_2} \gamma_{\sigma_2\sigma|k} \eta_{\sigma_2|}{}^\alpha \tag{7}$$

$$\begin{pmatrix} \sigma, \tau = 1, \cdots, q_1 \\ \sigma_2 = q_1 + 1, \cdots, q_1 + q_2 \end{pmatrix}.$$

If all the quantities $\gamma_{\sigma_2\sigma|k}$ are zero, that is if the quantities are expressible in terms of the vectors $\eta_{\sigma|}{}^\alpha$, equations (6) and (7) are in fact equations (56.2) and (56.3) respectively,* and consequently V_n is of class q_1. Otherwise the unit vectors $\eta_{\sigma_2|}{}^\alpha$ and all the vectors linearly expressible in terms of them are said to constitute the *second normal complex* N_2 of V_n in S_m. If N_2 does not exist, we say it is vacuous.

---

* Here, and in what follows, the greek letter nu as used in § 56 is replaced by gamma in such terms as $\gamma_{\tau\sigma|k}$.

From the first and second sets of equations (4) and (7) it follows that

$$(8) \qquad \gamma_{\sigma\tau|k} + \gamma_{\tau\sigma|k} = 0, \qquad \gamma_{\sigma\sigma|k} = 0.$$

From (5) and (7) we have

$$(9) \quad z^{\alpha}_{,ijk} = -\sum_{\sigma} e_{\sigma} b_{\sigma|ij} b_{\sigma|hk} g^{hl} z^{\alpha}_{,l} + \sum_{\tau} (b_{\tau|ij,k} + \sum_{\sigma} e_{\sigma} b_{\sigma|ij} \gamma_{\tau\sigma|k}) \eta_{\tau|}{}^{\alpha} + \sum_{\sigma} e_{\sigma} b_{\sigma|ij} \sum_{\sigma_2} \gamma_{\sigma_2\sigma|k} \eta_{\sigma_2|}{}^{\alpha}.$$

When this expression and the one obtained from it on interchanging j and k are substituted in (3), we obtain an expression linear in $z^{\alpha}_{,l}$, $\eta_{\sigma|}{}^{\alpha}$, and $\eta^{\alpha}_{\sigma_2|}$ equal to zero. Since there can be no linear relation between these quantities, all the coefficients must be zero. Accordingly we have the following sets of equations in which a term with carets (^) over two indices stands for this term minus the similar term with these indices interchanged, as in (10):

$$(10) \qquad R_{hijk} = \sum_{\sigma} e_{\sigma}(b_{\sigma|hj} b_{\sigma|ik} - b_{\sigma|hk} b_{\sigma|ij}) = \sum_{\sigma} e_{\sigma} b_{\sigma|h\hat{j}} b_{\sigma|i\hat{k}},$$

$$(11) \qquad b_{\sigma|i\hat{j},\hat{k}} = \sum_{\tau} e_{\tau} b_{\sigma|i\hat{j}} \gamma_{\tau\sigma|\hat{k}},$$

$$(12) \qquad \sum_{\sigma} e_{\sigma} b_{\sigma|i\hat{j}} \gamma_{\sigma_2\sigma|\hat{k}} = 0.$$

If N_2 is not vacuous, we have for $\sigma_2, \tau_2 = q_1 + 1, \cdots q_1 + q_2$

$$(13) \quad \sum_{\alpha} c_{\alpha}(\eta_{\sigma_2|}{}^{\alpha})^2 = e_{\sigma_2}, \qquad \sum_{\alpha} c_{\alpha} \eta_{\sigma_2|}{}^{\alpha} \eta_{\tau_2|}{}^{\alpha} = 0 \; (\tau_2 \neq \sigma_2),$$
$$\sum_{\alpha} c_{\alpha} \eta_{\sigma_2|}{}^{\alpha} \eta_{\sigma|}{}^{\alpha} = 0, \qquad \sum_{\alpha} c_{\alpha} \eta_{\sigma_2|}{}^{\alpha} z^{\alpha}_{,i} = 0.$$

Differentiating the last set of these equations covariantly and making use of (5) and the third set of equations (13), we obtain

$$\sum_{\alpha} c_{\alpha} \eta^{\alpha}_{\sigma_2|,k} z^{\alpha}_{,i} = 0.$$

Hence we have

$$(14) \quad \eta^{\alpha}_{\sigma_2|,k} = \sum_{\sigma} e_{\sigma} \gamma_{\sigma\sigma_2|k} \eta_{\sigma|}{}^{\alpha} + \sum_{\tau_2} e_{\tau_2} \gamma_{\tau_2\sigma_2|k} \eta_{\tau_2|}{}^{\alpha} + \sum_{\sigma_3} e_{\sigma_3} \gamma_{\sigma_3\sigma_2|k} \eta^{\alpha}_{\sigma_3|},$$

where $\eta_{\sigma_3|}{}^{\alpha}$, if these exist, are q_3 mutually orthogonal unit vectors orthogonal to the vectors $\eta_{\sigma|}{}^{\alpha}$ and $\eta_{\sigma_2|}{}^{\alpha}$. These vectors and those linearly expressible in terms of them are said to constitute the *third normal complex* N_3.

Expressing the condition of integrability of equations (7), that is $\eta^{\alpha}_{\sigma|,kl} = \eta^{\alpha}_{\sigma|,lk}$, and making use of (5), (7), (14), and (11), we obtain the following equations:

$$(15)\qquad \gamma_{\tau\sigma|\hat{k}.\hat{l}} + \sum_{\rho} e_{\rho}\gamma_{\rho\tau|\hat{k}}\gamma_{\rho\sigma|\hat{l}} + g^{hi}b_{\tau|j\hat{k}}b_{\sigma|h\hat{l}} + \sum_{\sigma_2} e_{\sigma_2}\gamma_{\sigma_2\sigma|\hat{k}}\gamma_{\tau\sigma_2|\hat{l}} = 0,$$

$$(16)\qquad \gamma_{\sigma_2\sigma|\hat{k}.\hat{l}} + \sum_{\tau} e_{\tau}\gamma_{\tau\sigma|\hat{k}}\gamma_{\sigma_2\tau|\hat{l}} + \sum_{\tau_2} e_{\tau_2}\gamma_{\tau_2\sigma|\hat{k}}\gamma_{\sigma_2\tau_2|\hat{l}} = 0,$$

$$(17)\qquad \sum_{\sigma_2} e_{\sigma_2}\gamma_{\sigma_2\sigma|\hat{k}}\gamma_{\sigma_3\sigma_2|\hat{l}} = 0.$$

When N_2 is vacuous equations (10), (11), and (15) are equations (56.5), (56.6), and (56.7) respectively, and consequently V_n is of class q_1.

We consider whether the first normal complex can consist of a single vector. From (12) we have

$$b_{1|ij} = \rho_i\gamma_{\sigma_2 1|j},$$

from which and (10) one obtains $R_{hijk} = 0$. Hence we have:

*The first normal complex N_1 of a non-flat V_n immersed in an $S_m(m > n + 1)$ consists of more than one vector field.**

As a corollary we have

For a V_n of class 2 immersed in a flat-space of order $n + 2$ all the vectors normal to V_n are in the first normal complex N_1.

We consider next the case where the second normal complex N_2 consists of a single vector $\eta_{q_2|}{}^{\alpha}$ $(q_2 = q_1 + 1)$ and N_3 is not vacuous. From (17) it follows that

$$(18)\qquad \gamma_{q_2\sigma|k} = \theta_{\sigma}\mu_k, \qquad \gamma_{\sigma_3 q_2|k} = \theta_{\sigma_3}\mu_k,$$

where the θ's are scalars and μ_k a covariant vector. In this case the last term in (14) becomes $\mu_k \sum_{\sigma_3} e_{\sigma_3}\theta_{\sigma_3}\eta_{\sigma_3|}{}^{\alpha}$ which is a single vector field, which we denote by $\eta_{q_3|}{}^{\alpha}(q_3 = q_2 + 1)$. Hence we have

If N_2 consists of a single vector-field and N_3 is not vacuous, N_3 consists of a single vector-field.

From (14) and the first of (13) it follows that in this case $\gamma_{q_2q_2|k} = 0$. Consequently equation (14) is

$$(19)\qquad \eta^{\alpha}_{q_2|.k} = \sum_{\sigma} e_{\sigma}\gamma_{\sigma q_2|k}\eta_{\sigma|}{}^{\alpha} + e_{q_3}\mu_k\eta_{q_3}{}^{\alpha}.$$

By the process which led to (14) we have

$$(20)\qquad \eta^{\alpha}_{q_3|.k} = \sum_{\sigma} e_{\sigma}\gamma_{\sigma q_3|k}\eta_{\sigma|}{}^{\alpha} + e_{q_2}\gamma_{q_2q_3|k}\eta_{q_2|}{}^{\alpha} + \sum_{\sigma_4} e_{\sigma_4}\gamma_{\sigma_4 q_3|k}\eta_{\sigma_4|}{}^{\alpha},$$

* This theorem for a V_n with positive definite fundamental form was established by *Burstin*, 1929, 6, p. 113.

where the vectors $\eta_{\sigma_4|}{}^\alpha$ if they exist constitute the fourth normal complex. Differentiating $\sum_\alpha c_\alpha \eta_{\sigma|}{}^\alpha \eta_{q_2|}{}^\alpha = 0$ and making use of (7), (19), (4), (13) and $\sum_\alpha c_\alpha \eta_{\sigma|}{}^\alpha \eta_{q_3|}{}^\alpha = 0$ we obtain

$$\gamma_{q_2\sigma|k} + \gamma_{\sigma q_2|k} = 0. \tag{21}$$

In consequence of this result, (12), (16) and (18) we find that the condition of integrability of equations (19) reduce to the following:

$$\mu_{\hat{k}}\gamma_{\sigma q_3|\hat{l}} = 0, \qquad \mu_{\hat{k}}\gamma_{q_2 q_3|\hat{l}} = 0, \qquad \mu_{\hat{k}.\hat{l}} = 0, \qquad \mu_{\hat{k}}\gamma_{\sigma_4 q_3|\hat{l}} = 0.$$

The last of these equations being similar to equation (17), by the same process based upon the latter equation for an N_3 it can be shown that if N_4 is not vacuous it consists of a single vector. In like manner it is shown from the condition of integrability of equations (20) that if N_5 is not vacuous it consists of a single vector. The preceding results would have followed if N_2 consisted of more than one vector-field and N_3 of only one. Hence we have:

*If any normal complex other than the first consists of only one vector-field, the same is true of all subsequent complexes until a vacuous one is reached.**

When N_2 consists of a single vector $\eta_{q_2|}{}^\alpha$, it follows from equations (18) and (21) that the last term in equation (15) is equal to zero, and this equation, (10), and (11) are respectively equations (56.7), (56.5), and (56.6). Hence we have:

When the first normal complex of a V_n consists of q_1 mutually orthogonal unit vectors and the second complex consists of a single unit vector, V_n is class q_1.

In the foregoing discussion each normal complex was spanned by mutually orthogonal unit vectors. When the fundamental form of V_n is positive definite, this is necessarily the case. When however the fundamental form of V_n is indefinite there is the possibility of null vectors being included among the vectors determining a complex. This alters the treatment of the problem in various ways (see 1937, 2).

Appendix 22

The following theorem is due to T. Y. Thomas:†

For a V_n ($n > 3$) of class one for which the determinant of the coefficients

---

* Cf. *Burstin*, 1929, 6; also *Mayer*, 1928, 2, and *Tucker*, 1931, 3.

† 1936, 1, p. 190.

b_{ij} of the second fundamental form is not equal to zero, the Codazzi equations (59.4) *are a consequence of the Gauss equations* (59.3).

Differentiating covariantly equation (59.3), that is

$$R_{hijk} = e(b_{hj}b_{ik} - b_{hk}b_{ij}),$$

one obtains

$$R_{hijk,l} = e(b_{hj,l}b_{ik} + b_{hj}b_{ik,l} - b_{hk,l}b_{ij} - b_{hk}b_{ij,l}).$$

Permuting the indices j, k, l cyclically, we have the two equations

$$R_{hikl,j} = e(b_{hk,j}b_{il} + b_{hk}b_{il,j} - b_{hl,j}b_{ik} - b_{hl}b_{ik,j}),$$
$$R_{hilj,k} = e(b_{hl,k}b_{ij} + b_{hl}b_{ij,k} - b_{hj,k}b_{il} - b_{hj}b_{il,k}).$$

When these three equations are added the left-hand member is zero, because of the Bianchi identities (26.3), and we have

(1) $$b_{ik}B_{hjl} + b_{hj}B_{ikl} + b_{il}B_{hkj} + b_{hk}B_{ilj} + b_{ij}B_{hlk} + b_{hl}B_{ijk} = 0,$$

where

$$B_{hjl} = b_{hj,l} - b_{hl,j}.$$

If the matrix $\|b_{hj}\|$ is of rank n, quantities b^{hj} are uniquely determined such that

$$b^{hj}b_{hk} = \delta^j_k, \qquad b^{hj}b_{kj} = \delta^h_k.$$

If we multiply equation (1) by b^{hj} and sum for h and j we obtain, noting that the B's are skew-symmetric in the last two indices,

$$(n - 3)B_{ikl} + b_{ik}b^{hj}B_{hjl} - b_{il}b^{hj}B_{hjk} = 0$$

Multiplying this equation by b^{ik} and summing for i and k we obtain

$$2(n - 2)b^{hj}B_{hjl} = 0.$$

When $n > 3$ we have from these two sets of equations that $B_{ikl} = 0$, as was to be proved.

Appendix 23

By definition a V_n is an Einstein space, if

(1) $$R_{ij} = \rho g_{ij}.$$

Every V_2 is an Einstein space (Ex. 2, p. 47). For $n = 3$ an Einstein space is of constant curvature (Ex. 2, p. 92). It will now be determined under what conditions for $n \geqq 4$ an Einstein space of class one is of constant Riemannian curvature.

The Riemannian curvature r_{hk} of a V_n of class one in S_{n+1} for the orientation determined by the unit vectors $\lambda_{h|}{}^i$ and $\lambda_{k|}{}^i$ tangent to the lines of curvature of V_n in terms of the principal normal curvatures c_h and c_k for these lines of curvature is given by (cf. 59.9)

$$(2) \qquad r_{hk} = ec_h c_k,$$

where e is $+1$ or -1 according to the character of the normal to V_n in S_{n+1}. From (34.2) we have

$$(3) \qquad \sum_{k(\neq h)} r_{hk} = -e_h R_{ij} \lambda_{h|}{}^i \lambda_{h|}{}^j.$$

From (1), (2), and (3) one has

$$(4) \qquad \sum_{k(\neq h)} c_h c_k = -e\rho.$$

Let $S = \sum_k c_k$, then from equation (4) one obtains

$$c_h^2 - Sc_h - e\rho = 0;$$

that is each c_h is a solution of the quadric

$$(5) \qquad x^2 - Sx - e\rho = 0.$$

If all the c's are equal, V_n is of constant curvature as follows from the theorem of Schur (§ 26). Assume that they are not all equal, say $c_1 \neq c_2$. The others are equal to c_1 or c_2, being roots of equation (5). Let c_1 occur p times and c_2 q times $(p + q = n)$. Then from (5) we have

$$(6) \qquad c_1 + c_2 = S, \qquad c_1 c_2 = -e\rho,$$

and from the first of these and $S = pc_1 + qc_2$ we have

$$(p - 1)c_1 + (q - 1)c_2 = 0.$$

Since neither c_1 nor c_2 can be zero from the second of equations (6), it follows that p and q are each greater than 1, in which case c_1 and c_2 differ in sign. This is not possible if $e\rho$ is negative. Hence, the assumption that there be different c's is not valid and we have:

*An Einstein space of class one for $n \geqq 4$ is a space of constant Riemannian curvatures if $e\rho$ is negative.**

Appendix 24

In equations (64.15) and what follows $z^\alpha (\alpha = 1, \cdots, n + 2)$ are generalized cartesian coordinates of the flat space S_{n+2} of which V_{n+1}

---

* See 1937, 3 when the fundamental quadratic form of V_n is positive definite, noting that $B_{ij} = -R_{ij}$.

is a hypersurface. Let ξ^α be a normal to V_n in S_{n+2} not in V_{n+1}. From (64.10) we have $c_\alpha \xi^\alpha \eta^\alpha_{0,i} = 0$, and consequently

$$c_\alpha \xi^\alpha \left(z^\alpha_{0,i} + R\eta^\alpha_{0,i} \tan \frac{w}{R} \right) \frac{dx^i}{ds} = 0.$$

Since the determinant of the quantities $z^\alpha_{0,i}$, $\eta_0{}^\alpha$, ξ^α is different from zero, equations (64.17) follow, and consequently the theorem is proved for the case (64.15).

Appendix 25

If φ^1 is a solution of $\xi^i \dfrac{\partial \varphi}{\partial x^i} = 1$, and $\varphi^2, \cdots, \varphi^n$ are independent solutions of $\xi^i \dfrac{\partial \varphi}{\partial x^i} = 0$, then in the coordinates $x'^i = \varphi^i$ we have $\xi'^1 = 1$, $\xi'^2 = \cdots = \xi'^n = 0$. Hence the theorem:

Appendix 26

Accordingly V_{τ_0} is included in the variety whose equations are obtained by equating to zero the minors of M of order $\tau_0 + 1$.

Appendix 27

By means of the Bianchi identities (26.2) and the identities (8.10) equations (69.3) can be written

$$\begin{aligned}(69.3) \quad -\xi^m R_{ijkl,m} + \xi_{m,l} R^m{}_{kij} - \xi_{m,k} R^m{}_{lij} + \xi_{i,m} R^m{}_{jkl} + \xi_{m,j} R^m{}_{ikl} \\ + \tfrac{1}{2}(g_{il}\psi_{,jk} - g_{ik}\psi_{,jl} + g_{jk}\psi_{,il} - g_{jl}\psi_{,ik}) = 0.\end{aligned}$$

Multiplying these equations by g^{il} and summing for i and l, we get

$$\xi^m R_{jk,m} + \xi_{m,k} R^m{}_j + \xi_{m,j} R^m{}_k = \tfrac{1}{2}(n-2)\psi_{,jk} + \tfrac{1}{2} g_{jk}\, \Delta_2 \psi.$$

Multiplying by g^{jk} and summing for j and k, we have, using (69.1),

$$(n-1)\, \Delta_2 \psi = \xi^m R_{,m} + 2\xi_{m,p} R^{mp} = \xi^m R_{,m} + \psi R.$$

Substituting this expression for $\Delta_2\psi$ in the preceding equation, we obtain

$$\frac{1}{2}(n-2)\psi_{,jk} = \xi^m \left(R_{jk,m} - \frac{1}{2(n-1)} g_{jk} R_{,m} \right) + \xi_{m,k} R^m{}_j + \xi_{m,j} R^m{}_k - \frac{1}{2(n-1)} g_{jk} \psi R.$$

When this and similar expressions are substituted in (69.3), the result is reducible in consequence of (69.1) to

(a) $$-\xi^m C_{ijkl.m} + \xi_{m,p}(\delta_l{}^p C^m{}_{kij} - \delta_k{}^p C^m{}_{lij} + \delta_i{}^m C^p{}_{jkl} + \delta_j{}^p C^m{}_{ikl}) = 0,$$

where C_{ijkl} is the Weyl conformal curvature tensor (28.12).

For a V_3 the conformal curvature tensor is a zero tensor (p. 91). Hence equations (a) are satisfied identically for a V_3.

In order that equations (a) be satisfied identically for a $V_n (n > 3)$, it is necessary and sufficient that $C_{ijkl.m}$ and the anti-symmetric part of the remainder of equations (a) be zero, that is

(b) $$\delta_l{}^p C^m{}_{kij} - \delta_l^m C^p_{kij} - \delta_k^p C^m{}_{lij} + \delta_k{}^m C^p{}_{lij} + \delta_i{}^m C^p{}_{jkl} - \delta_i^p C^m_{jkl} + \delta_j^p C^m_{ikl} - \delta_j^m C^p{}_{ikl} = 0.$$

From (28.12) we have

$$C^l{}_{ijl} = 0, \qquad C^l{}_{ljk} = 0, \qquad C^h{}_{ijk} + C^h{}_{jki} + C^h{}_{kij} = 0.$$

If p in equations (b) is replaced by l and summed for l, we get

$$(n - 1)C^m{}_{kij} = 0.$$

Hence a necessary and sufficient condition that equations (a) for $V_n (n > 3)$ be satisfied identically is that V_n be conformally flat.

Eliminating the function ψ from the equations (69.1), that is

(c) $$\xi_{i,j} + \xi_{j,i} = g_{ij}\psi,$$

we obtain $\dfrac{n(n+1)}{2} - 1$ equations connecting the ξ's and their first derivatives. When these equations are differentiated we obtain $\dfrac{n^2(n+1)}{2} - n$ equations. Thus there are $\dfrac{n+1}{2}(n^2 + n - 2)$ equations in the ξ's and their first and second derivatives, in number $\dfrac{n+1}{2}(n^2 + 2n)$. Hence according as equations (a) are satisfied identically or not there are $\dfrac{(n+1)(n+2)}{2}$ or fewer linearly independent solutions of (c). In particular we have:*

For any Riemannian space V_3 there are ten linearly independent infinitesimal conformal transformations; for any conformally flat space $V_n (n > 3)$ there are $(n + 1)(n + 2)/2$ linearly independent infinitesimal conformal transformations.

---

* Cf. *Taub*, 1949, 1.

Appendix 28

The proof of the theorem, which is essentially that given by Fubini, establishes a coordinate system of the kind described for any minimum invariant variety V_{n-k} but not necessarily for V_n. For a proof see *Eisenhart* 1932, 2 or 1933, 1, § 55.

Appendix 29

In § 77 it is proved that any simply transitive group of order n is a group of motions of an infinity of Riemannian spaces. Denoting by $\xi_{j|}{}^i$ the components of the group and using the functions Γ^i_{jk} defined in terms of them by (77.4) we consider the equations

$$\frac{\partial \lambda^i}{\partial x^k} + \lambda^j \Gamma^i_{jk} = 0. \tag{1}$$

The condition of integrability of these equations is found to be

$$\lambda^j B^i{}_{jkl} = 0,$$

where

$$B^i{}_{jkl} = \frac{\partial \Gamma^i_{jk}}{\partial x^l} - \frac{\partial \Gamma^i_{jl}}{\partial x^k} + \Gamma^h_{jk}\Gamma^i_{hl} - \Gamma^h_{jl}\Gamma^i_{hk}.$$

In § 77 it is shown that $B^i{}_{jkl} = 0$. Hence equations (1) are completely integrable, that is there exist n independent solutions $\lambda_{\alpha|}{}^i$, where $\alpha = 1, \cdots, n$ indicates the vector and $i = 1, \cdots, n$ the component. The quantities $\lambda_{\alpha|}{}^i$ are the vectors of another simply transitive group, the reciprocal group of the one with components $\xi_{j|}{}^i$.*

If a V_n admits a motion defined by

$$\bar{x}^i = f^i(x, t),$$

then as explained in § 27, we must have

$$g_{ij}(x) = g_{kl}(\bar{x}) \frac{\partial \bar{x}^k}{\partial x^i} \frac{\partial \bar{x}^l}{\partial x^j}. \tag{2}$$

For an infinitesimal motion

$$\bar{x}^i = x^i + \xi^i \delta t$$

equations (2) reduce to the Killing equations (70.1), namely

$$\xi^h \frac{\partial g_{ij}}{\partial x^h} + g_{ih} \frac{\partial \xi^h}{\partial x^j} + g_{jh} \frac{\partial \xi^h}{\partial x^i} = 0.$$

* Cf. 1933, 1, p. 113.

Equation (2) means that the g's are carried into themselves by a motion. The same applies to any tensor whose components are functions of the g's and their derivatives, and in particular to the Ricci tensor R_{ij}. Hence we have

$$\xi^h \frac{\partial R_{ij}}{\partial x^h} + R_{ih} \frac{\partial \xi^h}{\partial x^j} + R_{jh} \frac{\partial \xi^h}{\partial x^i} = 0. \tag{3}$$

In § 77 by making use of the Killing equations we derived equations (77.3), namely

$$\frac{\partial g_{ij}}{\partial x^k} - g_{ih}\Gamma^h_{jk} - g_{jh}\Gamma^h_{ik} = 0. \tag{4}$$

Proceeding in like manner with equations (3) for all the vectors $\xi_{\lambda|}{}^i$ we obtain

$$\frac{\partial R_{ij}}{\partial x^k} - R_{ih}\Gamma^h_{jk} - R_{jh}\Gamma^h_{ik} = 0. \tag{5}$$

Making use of (4), (5) and

$$\frac{\partial \lambda_{\alpha|}{}^i}{\partial x^k} + \lambda_{\alpha|}{}^j\Gamma^i_{jk} = 0, \tag{6}$$

we obtain

$$\frac{\partial}{\partial x^k}(g_{ij}\lambda_{\alpha|}{}^i\lambda_{\alpha|}{}^j) = 0, \qquad \frac{\partial}{\partial x^k}(g_{ij}\lambda_{\alpha|}{}^i\lambda_{\beta|}{}^j) = 0, \tag{7}$$

$$\frac{\partial}{\partial x^k}(R_{ij}\lambda_{\alpha|}{}^i\lambda_{\beta|}{}^j) = 0 \quad (\alpha \neq \beta).$$

Since equations (1) are completely integrable, each set of solutions is determined by initial values. If the initial values of the n^2 quantities $\lambda_{\alpha|}{}^i$ are chosen to satisfy the n^2 conditions

$$g_{ij}\lambda_{\alpha|}{}^i\lambda_{\alpha|}{}^j = e_\alpha, \qquad g_{ij}\lambda_{\alpha|}{}^i\lambda_{\beta|}{}^j = 0, \qquad R_{ij}\lambda_{\alpha|}{}^i\lambda_{\beta|}{}^j = 0 \qquad (\alpha \neq \beta), \tag{8}$$

where e_α are $+1$ or -1, the resulting solutions satisfy these conditions for all values of the x's. From these equations, (34.4), and (33.10) it follows that $\lambda_{|i}{}^\alpha$ are the components of the Ricci principal directions. Hence:*

When a V_n admits a simply transitive group of motions, the components of the reciprocal transitive group can be chosen so that they are the Ricci principal directions of V_n.

---

* Cf. 1935, 2, p. 827.

Note that equations (6) are analogous to equations (23.1), differing in that $\left\{{i \atop jk}\right\}$ is replaced by Γ^i_{jk}. Hence if parallelism is defined by means of the quantities Γ^i_{jk}, we have that the n vector-fields $\lambda_{\alpha|}{}^i$, and any field $\lambda^i = \sum_\alpha c_\alpha \lambda_{\alpha|}{}^i$, where the c's are constants, are absolutely parallel. In this sense for a V_n which admits a simply transitive group of motions the reciprocal group provides absolute parallelism. Furthermore in the case of this parallelism, since any two of these fields λ^i and μ^i are such that $g_{ij}\lambda^i\mu^j$ is a constant, as follows from (4) and (6), the angle between the vectors at a point is the same for the parallel vectors at all points in V_n.

Absolute parallelism is not a concept limited to spaces admitting a simply transitive group of motions. In fact, if one takes any orthogonal ennuple of unit contravariant vectors $\lambda_{\alpha|}{}^i$ in any V_n and the covariant components of these vectors, that is $\lambda_{\alpha|i} = g_{ij}\lambda_{\alpha|}{}^j$ and defines quantities Γ^i_{jk} by the equations

$$\Gamma^i_{jk} = -\sum_\alpha e_\alpha \lambda_{\alpha|j} \frac{\partial \lambda_{\alpha|}{}^i}{\partial x^k},$$

then equations (6) follow.*

Because a Riemannian space admits many orthogonal ennuples (cf. § 13), such a definition of absolute parallelism lacks definiteness. For a space for which $R_{ij} \neq \rho g_{ij}$, the Ricci principal directions provide a basis. A natural inquiry is whether an orthogonal ennuple can be chosen so that the geodesics of V_n are the integral curves of the equations

$$\frac{d^2x^i}{ds^2} + \Gamma^i_{jk}\frac{dx^j}{ds}\frac{dx^k}{ds} = 0.\dagger$$

* Cf. *Cartan*, 1930, 7; also *Eisenhart*, 1933, 2.

† Cf. *Cartan* and *Schouten*, 1926, 3; *Eisenhart*, 1927, 3, p. 33.

Bibliography

This bibliography contains only the books and memoirs which are referred to in the text. For a more extensive bibliography, the reader is referred to *Struik*, 1922, 8, pp. 168–185 and to *Schouten*, 1924, 1, pp. 290–300; also to *Sommerville*, 1911, 2.

1854. 1. *Riemann, B.*: Über die Hypothesen, welche der Geometrie zu Grunde liegen. Gesammelte Werke, 1876, pp. 254–269. Also an edition edited by *H. Weyl*, Springer, Berlin, 1919.

1857. 1. *Lamé, G.*: Leçons sur les fonctions inverses des transcendantes et les surfaces isothermes. Mallet-Bachelier, Paris.

1868. 1. *Beltrami, E.*: Teoriá fondamentale degli spazii di curvatura costante. Annali di matematica, ser. 2, vol. 2, pp. 232–255; also Opere matematiche, vol. 1, pp. 406–429. Hoepli, Milano, 1902.

1869. 1. *Christoffel, E. B.*: Über die Transformation der homogenen Differentialausdrücke zweiten Grades. Journal für die reine und angew. Math. (Crelle), vol. 70, pp. 46–70.

1874. 1. *Lipschitz, R.*: Ausdehnung der Theorie der Minimalflächen. Journal für die reine und angew. Math. (Crelle), vol. 78, pp. 1–45.

1880. 1. *Voß, A.*: Zur Theorie der Transformation quadratischer Differentialausdrücke und der Krümmung höherer Mannigfaltigkeiten. Mathematische Annalen, vol. 16, pp. 129–179.

1885. 1. *Killing, W.*: Die nicht-euklidischen Raumformen in analytischer Behandlung. Teubner, Leipzig.

1886. 1. *Schur, F.*: Über den Zusammenhang der Räume konstanten Krümmungsmaßes mit den projektiven Räumen. Mathematische Annalen, vol. 27, pp. 537–567.

1891. 1. *Goursat, E.*: Leçons sur l'intégration des équations aux dérivées partielles du premier ordre. Hermann, Paris.

1892. 1. *Killing, W.*: Über die Grundlagen der Geometrie. Journal für die reine und angew. Math. (Crelle), vol. 109, pp. 121–186.

1893. 1. *Stäckel, P.*: Sur une classe de problèmes de dynamique. Comptes Rendus, vol. 116, pp. 485–487.

2. *Stäckel, P.*: Sur des problèmes de dynamique, qui se réduisent à des quadratures. Comptes Rendus, vol. 116, pp. 1284–1286.

3. *Lie, S.*: Vorlesungen über kontinuierliche Gruppen. Teubner, Leipzig.

1895. 1. *Ricci, G.*: Dei sistemi di congruenze ortogonali in una varietà qualunque. Memorie dei Lincei, ser. 5, vol. 2, pp. 276–322.

1896. 1. *Di Pirro, G.*: Sugli integrali primi quadratici delle equazioni della meccania. Annali di Matematica, ser. 2, vol. 24, pp. 313–334.

2. *Levi-Civita, T.*: Sulle trasformazioni delle equazioni dinamiche. Annali di Matematica, ser. 2, vol. 24, pp. 255–300.

1898. 1. *Darboux, G.*: Leçons sur les systèmes orthogonaux et les coordonnées curvilignes. Gauthier-Villars, Paris.

2. *Ricci, G.*: Lezioni sulla teoria delle superficie. Drucker, Verona-Padova.

1899. 1. *Ricci, G.*: Sui gruppi continui di movimenti in una varietà qualunque a tre dimensioni. Memorie della Societa Italiana della Scienze, ser. 3, vol. 12, pp. 69–92.

1901. 1. *Ricci, G*, and *Levi-Civita, T.*: Méthodes de calcul différentiel absolu et leurs applications. Mathematische Annalen, vol. 54, pp. 125–201, 608.

2. *Hadamard, A.*: Sur les éléments linéares à plusieurs dimensions. Bulletin des Sciences Mathématiques, ser. 2, vol. 25, pp. 37–40.

1902. 1. *Bianchi, L.*: Lezioni di geometria differenziale, 2nd edition, vol. 1. Spoerri, Pisa.

2. *Ricci, G.*: Formole fondamentali nella teoria generale di varietà e della lors curvatura. Rendiconti dei Lincei, ser. 5, vol. 11^1, pp. 355–362.

1903. 1. *Bianchi, L.*: Lezioni di geometria differenziale, 2nd edition, vol. 2. Spoerri, Pisa.

2. *Ricci, G.*: Sulle superficie geodetiche in una varietà qualunque et in particolare nella varietà a tre dimensioni. Rendiconti dei Lincei, ser. 5, vol. 12^1, pp. 409–420.

3. *Fubini, G.*: Sulla teoria degli spazii che ammettono un gruppo conforme. Atti, Torino, vol. 38, pp. 404–418.

4. *Fubini, G.*: Sugli spazii che ammettono un gruppo continuo di movimenti. Annali di Matematica, ser. 3, vol. 8, pp. 39–81.

1904. 1. *Goursat, E.*: A course in mathematical analysis, translated by E. R. Hedrick, vol. 1. Ginn and Company, Boston.

2. *Ricci, G.*: Direzioni e invarianti principali in una varietà qualunque. Atti del Reale Inst. Veneto, vol. 63, pp. 1233–1239.

3. *Bolza, O.*: Lectures on the calculus of variations. University of Chicago Press.

4. *Fubini, G.*: Sugli spazii a quattro dimensioni che ammettono un gruppo continuo di movimenti. Annali di Matematica, ser. 3, vol. 9, pp. 33–90.

1905. 1. *Ricci, G.*: Sui gruppi continui di movimenti rigidi negli iperspazii. Rendiconti dei Lincei, ser. 5, vol. 14^2, pp. 487–491.

2. *Kommerell, K.*: Riemannsche Flächen im ebenen Raum von vier Dimensionen. Mathematische Annalen, vol. 60, pp. 548–596.

1906. 1. *Bromwich, T. J. I'A.*: Quadratic forms and their classification by means of invariant factors. Cambridge Tract, No. 3. Cambridge University Press.

1907. 1. *Bôcher, M.*: Introduction to higher algebra. Macmillan, New York.

1908. 1. *Schmidt, E.*: Über die Auflösung linearer Gleichungen mit unendlich vielen Unbekannten. Rendiconti di Palermo, vol. 25, pp. 53–77.

1909. 1. *Eisenhart, L. P.*: A treatise on the differential geometry of curves and surfaces. Ginn and Company, Boston.

2. *Kowalewski, G.*: Einführung in die Determinantentheorie. Veit und Comp., Leipzig.

3. *Sbrana, U.*: Sulle varietà ad $n-1$ dimensioni deformabili nello spazio euclideo ad n dimensioni. Rendiconti di Palermo, vol. 27, pp. 1–45.

1911. 1. *Wilson, E. B.*: Advanced calculus. Ginn and Company, Boston.
2. *Sommerville, D. M. Y.*: Bibliography of non-euclidean geometry, including the theory of parallels, the foundations of geometry and space of *n* dimensions. Harrison and Sons, London.

1912. 1. *Eisenhart, L. P.*: Minimal surfaces in euclidean four-space. American Journal of Mathematics, vol. 34, pp. 215–236.

1916. 1. *Einstein, A.*: Die Grundlage der allgemeinen Relativitätstheorie. Annalen der Physik, vol. 49, pp. 769–822.
2. *Herglotz, G.*: Zur Einsteinschen Gravitationstheorie. Sitzungsber. sächs. Gesellsch. Wiss. Leipzig, vol. 68, pp. 199-203.
3. *Schwarzschild, K.*: Über das Gravitationsfeld eines Massenpunktes nach der Einsteinschen Theorie. Sitzungsber. der Akad. Wiss. Berlin, pp. 189–196.

1917. 1. *Levi-Civita, T.*: Nozione di parallelismo in una varietà qualunque e consequente specificazione geometrica della curvatura Riemanniana. Rendiconti di Palermo, vol. 42, pp. 173-205.
2. *Severi, F.*: Sulla curvatura delle superficie e varietà. Rendiconti di Palermo, vol. 42, pp. 227–259.
3. *Levi-Civita, T.*: Sulla espressione analitica spettante al tensore gravitazionale nella teoria di Einstein. Rendiconti dei Lincei, ser. 5, vol. 26^1, pp. 381-391.

1918. 1. *Schouten, J. A.*: Die direkte Analysis zur neueren Relativitätstheorie. Verhandelingen Kon. Akad. Amsterdam, vol. 12, no. 6.
2. *Weyl, H.*: Reine Infinitesimalgeometrie. Mathematische Zeitschrift, vol. 2, pp. 384–411.
3. *Kottler, F.*: Über die physikalischen Grundlagen der Einsteinschen Gravitationstheorie. Annalen der Physik, ser. 4, vol. 56, pp. 401–462.
4. *Bianchi, L.*: Lezioni sulla teoria dei gruppi continui finiti di trasformazioni. Spoerri, Pisa.

1919. 1. *Pérès J.*: Le parallelisme de M. Levi.Civita et la courbure Riemanniene. Rendiconti dei Lincei, ser. 5, vol. 28^1, pp. 425–428.
2. *Bompiani, E.*: Surfaces de translation et surfaces minima dans les espaces courbes. Comptes Rendus, vol. 169, pp. 840–843.

1920. 1. *Blaschke, W.*: Frenets Formeln für den Raum von Riemann. Mathematische Zeitschrift, vol. 6, pp. 94–99.

1921. 1. *Weyl, H.*: Space, time, matter. Translated by H. L. Brose. Methuen, London.
2. *Schouten, J. A.*: Über die konforme Abbildung *n*-dimensionaler Mannigfaltigkeiten mit quadratischer Maßbestimmung auf eine Mannigfaltigkeit mit euklidischer Maßbestimmung. Mathematische Zeitschrift, vol. 11, pp. 58-88.
3. *Schouten, J. A.* and *Struik, D. J.*: On some properties of general manifolds relating to Einstein's theory of gravitation. American Journal of Mathematics, vol. 43, pp. 213-216.
4. *Weyl, H.*: Zur Infinitesimalgeometrie: Einordnung der projektiven und der konformen Auffassung. Göttinger Nachrichten, pp. 99–112.

5. *Bompiani, E.*: Studi sugli spazi curvi. La seconda forma fondamentale di una V_m in V_n. Atti del Reale Inst. Veneto, vol. 80, pp. 1113–1145.

6. *Kasner, E.*: Finite representation of the solar gravitational field in flat space of six dimensions. American Journal of Mathematics, vol. 43, pp. 130–133.

7. *Kasner, E.*: The impossibility of Einstein fields immersed in flat space of five dimensions. American Journal of Mathematics, vol. 43, pp. 126–129.

1922. 1. *Eisenhart, L. P.*: Condition that a tensor be the curl of a vector. Bulletin of the Amer. Math. Soc., vol. 28, pp. 425–427.

2. *Dienes, P.*: Sur la connexion du champ tensoriel. Comptes Rendus, vol. 174, pp. 1167–1169.

3. *Eisenhart, L. P.*: Fields of parallel vectors in the geometry of paths. Proceedings of the Nat. Acad. of Sciences, vol. 8, pp. 207–212.

4. *Bianchi, L.*: Sur parallelismo vincolato di Levi-Civita nella metrica degli spazi curvi. Rendiconti Accad. di Napoli, ser. 3, vol. 28, pp. 150–171.

5. *Fermi, E.*: Sopra i fenomeni che avvengono in vicinanza di una linea oraria. Rendiconti dei Lincei, vol. 31^1, pp. 21–23, 51–52.

6. *Eisenhart, L. P.*: Spaces with corresponding paths. Proceedings of the Nat. Acad. of Sciences, vol. 8, pp. 233-238.

7. *Veblen, O.*: Projective and affine geometry of paths. Proceedings of Nat. Acad. of Sciences, vol. 8, pp. 347–350.

8. *Struik, D. J.*: Grundzüge der mehrdimensionalen Differentialgeometrie in direkter Darstellung. Springer, Berlin.

1923. 1. *Eddington, A. S.*: The mathematical theory of relativity. Cambridge University Press.

2. *Birkhoff, G. D.*: Relativity and modern physics. Harvard University Press, Cambridge, Mass.

3. *Synge, J. L.*: Parallel propagation of a vector around an infinitesimal circuit in an affine connected manifold. Annals of Mathematics, ser. 2, vol. 25, pp. 181–184.

4. *Veblen, O.* and *Thomas, T. Y.*: The geometry of paths. Transactions of the Amer. Math. Soc., vol. 25, pp. 551–608.

5. *Eisenhart, L. P.*: Symmetric tensors of the second order whose first covariant derivatives are zero. Transactions of the Amer. Math. Soc., vol. 25, pp. 297–306.

6. *Eisenhart, L. P.*: Orthogonal systems of hypersurfaces in a general Riemann space. Transactions of the Amer. Math. Soc., vol. 25, pp. 259–280.

7. *Brinkmann, H. W.*: On Riemann spaces conformal to euclidean space. Proceedings of the Nat. Acad. of Sciences, vol. 9, pp. 1–3.

1924. 1. *Schouten, J. A.*: Der Ricci-Kalkül. Springer, Berlin.

2. *Brinkmann, H. W.*: Riemann spaces conformal to Einstein spaces. Mathematische Annalen, vol. 91, pp. 269–278.

3. *Bianchi, L.*: Lezioni di geometria differenziale, 3rd edition, vol. 2, part. 2. Zanichelli, Bologna.

4. *Eisenhart, L. P.*: Space-time continua of perfect fluids in general relativity. Transactions of the Amer. Math. Soc., vol. 26, pp. 205–220.

5. *Bompiani, E.*: Spazi Riemanniani luoghi di varietà totalmente geodetiche. Rendiconti di Palermo, vol. 48, pp. 121–134.
6. *Ricci, G.*: Contributo alla teoria delle varietà Riemanniane. Proceedings of the International Mathematical Congress. Toronto.

1925. 1. *Levy, H.*: Tensors determined by a hypersurface in a Riemann space. Transactions of Amer. Math. Soc., vol. 28.
2. *Bliss, G. A.*: Calculus of variations. Carus mathematical monograph, No. 1. Open Court Pub. Co., Chicago.
3. *Eisenhart, L. P.*: Fields of parallel vectors in a Riemannian geometry. Transactions of the Amer. Math. Soc., vol. 27, pp. 563–573.
4. *Levi-Civita, T.*: Lezioni di calcolo differenziale assoluto. Stock, Roma.
5. *Thomas, J. M.*: Conformal correspondence of Riemann spaces. Proceedings of the Nat. Acad. of Sciences, vol. 11, pp. 257–259.
6. *Brinkmann, H. W.*: Einstein spaces which are mapped conformally on each other. Mathematische Annalen, vol. 94, pp. 119–145.
7. *Murnaghan, F. D.*: The generalized Kronecker symbol and its application to the theory of determinants. Amer. Math. Monthly, vol. 32, pp. 233–241.
8. *Levy, H.*: Normal congruences of curves in Riemann space. Bulletin of the Amer. Math. Soc., vol. 31, pp. 39–42.
9. *Thomas, T. Y.*: On the projective and equi-projective geometries of paths. Proceedings of the Nat. Acad. of Sciences, vol. 11, pp. 199-203.
10. *Thomas, J. M.*: Note on the projective geometry of paths. Proceedings of the Nat. Acad. of Sciences, vol. 11, pp. 207–209.
11. *Veblen, O.* and *Thomas, J. M.*: Projective normal coordinates for the geometry of paths. Proceedings of the Nat. Acad. of Sciences, vol. 11, pp. 204–207.
12. *Eisenhart, L. P.*: Linear connections of a space which are determined· by simply transitive continuous groups. Proceedings of the Nat. Acad of Sciences, vol. 11, pp. 246–250.

1926. 1. *Levy, H.*: Symmetric tensors of the second order whose covariant derivatives vanish. Annals of Mathematics, vol. 27.

Additional Bibliography

1891. 2. *Stäckel, P.*: Habilitationsschrift.

1902. 3. *Charlier, C. L.*: Die Mechanik des Himmels. Vol. 1 Veit, Leipzig.

1922. 9. *Ricci, G.*: Reducibilità delle quadriche differenziali e ds^2 della statica einsteiniani. Rendiconti dei Lincei, ser. 5, vol. 31, pp. 63–71.
10. *Ricci, G.*: Della integrazione dei sistemi di equazioni ai differenziali totali. Atti del Reale Inst. Veneto, vol. 81, pt. 2, pp. 179–183.

1926. 2. *Levi-Civita, T.*: Sur l'écart géodésique. Math. Annalen, vol. 97, pp. 291–320.

3. *Cartan, E. and Schouten, J. A.*: On Riemannian geometries admitting absolute parallelism. Proc. Kon. Akad. Wet. Amsterdam, vol. 29, pp. 933–946.
4. *Levy, H.*: Forma canonica dei ds² per i quali si annullano i simboli di Riemann a cinque indici. Rendiconti dei Lincei, ser. 6, vol. 3, pp. 65–69; 124–129.
5. *Cartan, E.*: Sur les espaces de Riemann dans les quels le transport par parallelisma conserve la courbure. Rendiconti dei Lincei, ser. 6, vol. 3, pp. 544–547.
6. *Cartan, E.*: Sur une classe remarquable d'espaces de Riemann. Bull. Math. Soc. de France, vol. 54, pp. 214–264; vol. 55, pp. 114–134.
7. *Janet, M.*: Sur la possibilité de plonger un espace riemannian donné dans espace euclidian. Ann. de la Soc. Polonaise de Math. vol. 5, pp. 38–43.
8. *Cartan, E.*: Sur l'écart géodésique et quelques notions connexes. Rendiconti dei Lincei, ser. 6, vol. 5, pp. 609–613.

1927. 1. *Levi-Civita, T.*: The absolute differential calculus. Blackie, London.
2. *Weatherburn, C. E.*: Differential geometry of three dimensions, vol 1. University Press, Cambridge.
3. *Eisenhart, L. P.*: Non-Riemannian geometry. Amer. Math. Soc. Colloquium Publications, New York.
4. *Veblen, O.*: Invariants of quadratic differential forms. Cambridge Tract 24, University Press, Cambridge.
5. *Robertson, H. P.*: Bemerkung über separierbare systeme in der Wellenmechanik. Math. Annalen, vol. 98, pp. 749–752.
6. *Cartan, E.*: La géométrie des groupes de transformations. Jour. Math. puree et appl., vol. 6, pp. 1–119.
7. *Cartan, E.*: Sur la possibilité de plonger un espace riemannian donné dans un espace euclidian. Ann. de la Soc. Polonaise de Math, vol. 6, pp. 1–7.

1928. 1. *Cartan, E.*: Lecons sur la géométrie des espaces de Riemann. Gauthier-Villars, Paris.
2. *Mayer, W.*: Ueber das vollständige Formensystem der *Fe* in R_n. Monatshefte für Math. und Physik, vol. 35, pp. 87–110.
3. *McConnell, A. J.*: Schmidt's orthogonal ennuple and the Frenet formulas for a curve. Bull. Amer. Math. Soc., vol. 34, pp. 713–714.

1929. 1. *Weyl, H.*: On the foundations of general infinitesimal geometry. Bulletin of the Amer. Math. Soc., vol. 35, pp. 716–725.
2. *Eisenhart, L. P.*: Dynamical trajectories and geodesics. Annals of Mathematics, vol. 30, pp. 591–606.
3. *Schouten, J. A.*: Zur Geometrie der kontinuierlichen Transformationsgruppen. Math. Annalen, vol. 102, pp. 244–272.
4. *Sommerville, D. M.*: An introduction to the geometry of *N* dimensions. Methuen, London.
5. *Eisenhart, L. P.*: Contact transformations. Annals of Mathematics, ser. 2, vol. 30, pp. 211–249.
6. *Burstin, C.*: Beiträge zur mehrdimensionalen Differentialgeometrie. Monatshefte für Math. und Physik, vol. 36, pp. 97–130.
7. *Vranceanu, G.*: Sur l'indépendence des secondes formes fondamentales d'une V_m. Bul. Facul. de Stiinfe din Cernaufi, vol. 3, pp. 316–320.

8. *Vranceanu, G.*: Familles de variétes Riemannienes. Bul. Soc. de Stiinte din Cluz, vol. 4, pp. 434–443.
9. *Michal, A.*: Scalar extensions of an orthogonal ennuple of vectors. Amer. Math. Monthly, vol. 37, pp. 529–533.

1930. 1. *Weatherburn, C. E.*: Differential geometry of three dimensions. Vol. 2, University Press, Cambridge.
2. *Duschek, A.* and *Mayer, W.*: Lehrbuch der Differentialgeometrie. 2 vols., Teubner, Berlin.
3. *Eisenhart, L. P.*: Projective normal coordinates. Proceedings of Nat. Acad. of Sciences, vol. 16, pp. 731–740.
4. *Cartan, E.*: La théorie des groupes finis et continus et l'analysis situs. Memorial des Sciences Mathematiques, fasc. 42. Gauthier-Villars, Paris.
5. *Mattioli, G. D.*: Sulla determinazione della varieta riemannianne che ammettono gruppi semplicemente transitivi li movimenti. Rendiconti dei Lincei, ser. 6, vol. 11, pp. 369–371.
6. *Knebelman, M. S.*: On groups of motions in related spaces. Amer. Journ. of Math., vol. 56, pp. 280–282.
7. *Cartan, E.*: Notice historique sur la notion de parallelisme absolu. Math. Annalen, vol. 102, pp. 702–706.
8. *Vranceanu, G.*: Sulle condizione di rigidita di una V_m in un S_n. Rendiconti dei Lincei, vol. 11, pp. 375–389.

1931. 1. *Weyl, H.*: The theory of groups and quantum mechanics. Translated by H. P. Robertson, Methuen, London.
2. *McConnell, A. J.*: Applications of the absolute differential calculus. Blackie, London.
3. *Tucker, A. W.*: On generalized covariant differentiation. Annals of Math., vol. 32, pp. 451–460.
4. *Mayer, W.*: Beitrag zur Differentialgeometrie l-Dimensionsler Mannigfaltigkeiten die in Euklidischen Räumen eingebettet sind. Sitz. der Preuss. Akad. Wiss., vol. 27, pp. 3–12.
5. *Schouten, J. A.* and *van Kampen, E. R.*: Ueber die Krümmung einer V_m in V_n; eine Revision der Krümmungstheorie. Math. Annalen, vol. 105, pp. 144–159.
6. *Cutler, E. H.*: On the curvature of a curve in Riemann space. Transactions of Amer. Math. Soc., vol. 33, pp. 832–838.

1932. 1. *Veblen, O.* and *Whitehead, J. H. C.*: The foundations of differential geometry. Cambridge Tract 29. University Press, Cambridge.
2. *Eisenhart, L. P.*: Intransitive groups of motions. Proceedings of Nat. Acad. of Sciences, vol. 18, pp. 193–202.
3. *Robertson, H. P.*: Groups of motions in spaces admitting absolute parallelism. Annals of Mathematics, vol. 33, pp. 496–521.
4. *Cartan, E.*: Les espaces riemanniens symétriques. Verhand. des Intermat. Math. Kongresses, Zürich, pp. 152–161.
5. *Schoenberg, J. M.*: Some applications of the calculus of variations to Riemannian geometry. Annals of Mathematics, vol. 33, pp. 485–495.
6. *Kosambi, D. D.*: Géométrie differentielle et calcul des variations. Rendiconte dei Lincei, sec. 6, vol. 16, pp. 410–415.

1933. 1. *Eisenhart, L. P.*: Continuous groups of transformations. Princeton University Press.

2. *Eisenhart, L. P.*: Spaces admitting complete absolute parallelism. Bulletin of the Amer. Math. Soc., vol. 39, pp. 217–226.
3. *Robertson, H. P.*: Relativistic Cosmology. Reviews of Modern Physics, vol. 5, pp. 62–90.

1934. 1. *Thomas, T. Y.*: The differential invariants of generalized spaces. University Press, Cambridge.
2. *Graustein, W. C.*: The geometry of Riemann spaces. Transactions of the Amer. Math. Soc., vol. 36, pp. 542–585.
3. *Eisenhart, L. P.*: Separable systems of Stäckel. Annals of Mathematics, vol. 35, pp. 284–305.
4. *Thomas, T. Y.* and *Levine, J.*: Simple tensors and the problem of the invariant characterization of an N-tuply orthogonal system of hypersurfaces in a V_n. Annals of Mathematics, vol. 35, pp. 735–739.
5. *Cartan, E.*: Les espaces de Finsler. Hermann, Paris.
6. *Synge, J. L.*: On the derivation of geodesics and null geodesics, particularly in relation to the properties of spaces of constant curvature and indefinite line element. Annals of Mathematics, vol. 35, pp. 705–713.
7. *Thomas, T. Y.*: The reduction of degenerate quadratic forms. Proceedings of the Nat. Acad. of Sciences, vol. 20, pp. 215–219.
8. *Thomas, T. Y.*: On the variation of curvature in Riemann spaces of constant mean curvature. Annali di Matematica, vol. 13, pp. 227–238.
9. *Morse, M.*: The calculus of variations in the large. Colloquium Publications of the Amer. Math. Soc., vol. 18, New York.

1935. 1. *Mayer, W.*: Die Differentialgeometrie der Untermannigfaltigkeiten des R_n konstanter Krümmung. Transactions of the Amer. Math. Soc., vol. 38, pp. 167–309.
2. *Eisenhart, L. P.*: Groups of motions and Ricci directions. Annals of Mathematics, vol. 36, pp. 823–832.
3. *Cartan, E.*: La méthode du repère mobile. La théorie des groupes continue et les espaces généralises. Hermann, Paris.
4. *Eisenhart, L. P.*: Stäckel systems in conformal euclidean space. Annals of Mathematics, vol. 36, pp. 57–70.
5. *Tucker, A. W.*: Non-Riemannian subspaces. Annals of Mathematics, vol. 36, pp. 965–983.
6. *Myers, S. B.*: Riemannian manifolds in the large. Duke Math. Journ., vol. 1, pp. 39–49.
7. *Tompkins, C. B.*: Linear connections of normal space to a variety in euclidean space. Bulletin of the Amer. Math. Soc., vol. 41, pp. 931–936.
8. *Schouten, J. A.* and *Struik, D. J.*: Einführung in die neueren Methoden der Differentialgeometrie, vol. 1. Noordhoff, Groningen.

1936. 1. *Thomas, T. Y.*: Riemann spaces of class one and their characterization. Acta Math. vol. 67, pp. 169–211.
2. *Thomas, T. Y.*: On closed spaces of constant mean curvature. American Journal of Mathematics, vol. 58, pp. 702–704.
3. *Eisenhart, L. P.*: Simply transitive groups of motions. Monatsheften für Math. und Physik, vol. 43, pp. 448–462.
4. *Eisenhart, L. P.* and *Knebelman, M. S.*: Invariant theory of homogeneous contact transformations. Annals of Mathematics, vol. 37, pp. 747–765.

5. *Synge, J. L.*: On the connectivity of spaces of positive curvature. Quar. Journ. Math., vol. 7, pp. 316–320.

6. *Thomas, T. Y.*: Field of parallel vectors in the large. Compositio Mathematica, vol. 3, pp. 453–468.

7. *Levine, J.*: Groups of motions in conformally flat spaces. Bulletin of the Amer. Math. Soc., vol. 42, pp. 418–422.

1937. 1. *Allendoerfer, C. B.*: Einstein spaces of class one. Bulletin of the Amer. Math. Soc., vol. 43, pp. 265–270.

2. *Eisenhart, L. P.*: Riemannian spaces of class greater than unity. Annals of Mathematics, vol. 38, pp. 794–808.

3. *Thomas, T. Y.*: Extract from a letter by E. Cartan. American Journal of Mathematics, vol. 59, pp. 793–794.

4. *Cartan, E.*: L'extension du calcul tensoriel aux géométries non-affines. Annals of Mathematics, vol. 38, pp. 1–13.

5. *Allendoerfer, C. B.*: The imbedding of Riemann spaces in the large. Duke Math. Journ., vol. 3, pp. 317–353.

6. *Michal, A. D.*: General tensor analysis. Bulletin of the Amer. Math. Soc., vol. 43, pp. 394–401.

1938. 1. *Weatherburn, C. E.*: Riemannian geometry and the tensor calculus. University Press, Cambridge.

2. *Chern, S.*: On projective normal coordinates. Annals of Mathematics, vol. 39, pp. 165–171.

3. *Eisenhart, L. P.*: Fields of parallel vectors in Riemannian space. Annals of Mathematics, vol. 39, pp. 316–321.

4. *Myers, S. B.*: Arc length in metric and Finsler manifolds. Annals of Mathematics, vol. 39, 463–471.

5. *Mayer, W.* and *Thomas, T. Y.*: Fields of parallel vectors in non-analytic manifolds in the large. Compositio Mathematica, vol. 5, pp. 193–207.

6. *Levine, J.*: Metric spaces with geodesic Ricci curves. Bulletin of the Amer. Math. Soc., vol. 44, pp. 145–152.

7. *Fialkow, A.*: The Riemannian curvature of a hypersurface. Bulletin of the Amer. Math. Soc., vol. 44, pp. 253–257.

8. *Schouten, J. A.* and *Struik, D. J.*: Einführung in die neueren Methoden der Differentialgeometrie, vol. 2. Noordhoff, Groningen.

1939. 1. *Allendoerfer, C. B.*: Rigidity for spaces of class greater than one. American Journal of Mathematics, vol. 61, pp. 633–644.

2. *Berwald, L.*: Ueber Finslersche und Cartansche Geometrie II. Compositio Mathematica, vol. 7, pp. 141–176.

3. *Myers, S. B.* and *Steenrod, N. E.*: The group of isometries of a Riemannian manifold. Annals of Mathematics, vol. 40, pp. 400–416.

4. *Ficken, F. A.*: The Riemannian and affine differential geometry of product spaces. Annals of Mathematics, vol. 40, pp. 892–913.

5. *Fialkow, A.*: Conformal geodesics. Transactions of the Amer. Math. Soc., vol. 45, pp. 443–473.

6. *Weyl, H.*: On the volume of cubes. American Journal of Mathematics, vol. 61, pp. 461–472.

7. *Thomas, T. Y.*: Imbedding theorems in differential geometry. Bulletin of the Amer. Math. Soc., vol. 45, pp. 841–850.

8. *Thomas, T. Y.*: The decomposition of Riemann spaces in the large. Monatshefte für Math. und Physik, vol. 47, pp. 388–418.
9. *Levine, J.*: Metric spaces with geodesic Ricci curves. II. Bulletin of the Amer. Math. Soc., vol. 45, pp. 123–128.
10. *Grove, V. G.*: A tensor analysis for a V_k in a projective S_n. Bulletin of the Amer. Math. Soc., vol. 45, pp. 385–398.
11. *Fialkow, A.*: Totally geodesic Einstein spaces. Bulletin of the Amer. Math. Soc., vol. 45, pp. 423–428; vol. 48, pp. 167, 168.
12. *Thomas, T. Y.*: On the singular point locus in the theory of fields of parallel vectors. Bulletin of the Amer. Math. Soc., vol. 45, pp. 436–441.

1940. 1. *Eisenhart, L. P.*: An introduction to differential geometry. Princeton University Press.
2. *Berwald, L.*: Ueber Finslersche und Cartansche Geometrie. I, Mathematica, Chij, vol. 16.
3. *Bochner, S.*: Harmonic surfaces in Riemann metric. Transactions of the Amer. Math. Soc., vol. 47, pp. 146–154.
4. *Allendoerfer, C. B.*: The Euler number of a Riemann manifold. American Journal of Mathematics, vol. 61, pp. 633–644.
5. *Wong, Y.*: On the Frenet formulae of a V_m in a V_n. Quar. Journ. Math., vol. 11, pp. 146–160.

1941. 1. *Berwald, L.*: On Finsler and Cartan geometries. III. Annals of Mathematics, vol. 42, pp. 84–112.
2. *Hodge, W. V. D.*: The theory and application of harmonic integrals. University Press, Cambridge.
3. *Myers, S. B.*: Riemannian manifolds with positive mean curvature. Duke Math. Journ., vol. 8, pp. 401–404.
4. *Schwartz, A.*: The Gauss-Codazzi-Ricci equations in Riemannian manifolds. Journal of Mathematics and Physics, vol. 20, pp. 30–79.
5. *Coburn, N.*: Unitary spaces with corresponding geodesics. Bulletin of the Amer. Math. Soc., vol. 47, pp. 901–910.

1942. 1. *Fialkow, A.*: The conformal theory of curves. Transactions of the Amer. Math. Soc., vol. 51, pp. 435–501.
2. *Steenrod, N. E.*: Topological methods for the construction of tensors. Annals of Mathematics, vol. 43, pp. 116–131.

1943. 1. *Bompiani, E.*: Geometria delle equazioni e dei sistemi di equazioni differenziali ordinarie. Atti Reale Accad. d'Italia, pp. 1–35.
2. *Wong, Y.*: Family of totally umbilical hypersurfaces in an Einstein space. Annals of Mathematics, vol. 44, pp. 271–297.
3. *Wong, Y.*: Some Einstein spaces with conformally separable fundamental tensors. Transactions of the Amer. Math. Soc., vol. 53, pp. 157–194.
4. *Allendoerfer, C. B.* and *Weil, A.*: The Gauss-Bonnet theorem for Riemannian polyhedron. Transactions of the Amer. Math. Soc., vol. 53, pp. 101–129.
5. *Preissmann, A.*: Quelques propriétés globales des espaces de Riemann. Comm. Math. Helv., vol. 15, pp. 175–216.
6. *Wong, Y.*: A note on complementary subspaces in a Riemannian space. Bulletin of the Amer. Math. Soc., vol. 49, pp. 120–125.

1944. 1. *Einstein, A.* and *Bargmann, V.*: Bivector fields. Annals of Mathematics, vol. 45, pp. 1–14.
2. *Einstein, A.*: Bivector fields II. Annals of Mathematics, vol. 45, pp. 15–23.
3. *Chern, S.*: A simple intrinsic proof of the Gauss-Bonnet theorem for closed Riemannian manifolds. Annals of Mathematics, vol. 45, pp. 747–752.
4. *Fialkow, A.*: Conformal differential geometry of a subspace. Transactions of the Amer. Math. Soc., vol. 56, pp. 309–433.

1945. 1. *Hlavaty, V.*: Differential Liniengeometrie. Noordhoff, Groningen.
2. *Bompiani, E.*: Geometria degli spazi a connessione affine. Annali di Matematica, vol. 24, pp. 257–282.
3. *Chern, S.*: On the curvature integra in a Riemannian manifold. Annals of Mathematics, vol. 46, pp. 674–684.
4. *Walker, A. G.*: On completely harmonic spaces. Journal of the London Math. Soc., vol. 20, pp. 159–163.
5. *Thomas, T. Y.*: Absolute scalar invariants and the isometric correspondence of Riemann spaces. Proceedings of the Nat. Acad. of Sciences, vol. 31, pp. 306–310.
6. *Knebelman, M. S.*: On the equations of motion in a Riemann space. Bulletin of the Amer. Math. Soc., vol. 51, pp. 681–685.
7. *Chern, S.*: On Riemannian manifolds of four dimensions. Bulletin of the Amer. Math. Soc., vol. 51, pp. 964–971.
8. *Cartan, E.*: Les systemes differentials exterieurs et leurs applications géométriques. Hermann, Paris.

1946. 1. *Bochner, S.*: Vector fields and Ricci curvature. Bulletin of the Amer. Math. Soc., vol. 52, pp. 776–797.
2. *Chern, S.*: Some new viewpoints in differential geometry in the large. Bulletin of the Amer. Math. Soc., vol. 52, pp. 1–30.
3. *Walker, A. G.*: Symmetric harmonic spaces. Journal of the London Math. Soc., vol. 21, pp. 47–57.
4. *Cartan, E.*: Lecons sur la géométrie des espaces de Riemann, 2nd. ed., Gauthier-Villars, Paris.
5. *Schwartz, A.*: On higher normal spaces for V_m in V_n. American Journal of Mathematics, vol. 68, pp. 660–666.

1947. 1. *Vranceanu, C.*: Lecons de géométrie differentielle. Vol. 1. Bucarest.
2. *Bochner, S.*: Curvature in Hermitian metric. Bulletin of the Amer. Math. Soc., vol. 53, pp. 179–195.
3. *Berwald, L.*: Ueber Finslersche und Cartansche Geometrie IV. Annals of Mathematics, vol. 48, pp. 755–781.
4. *Bochner, S.*: On compact complex manifolds. Journal of the Indian Math. Soc., vol. 11, pp. 1–21.

1948. 1. *Allendoerfer, C. B.*: Global theorems in Riemannian geometry. Bulletin of the Amer. Math. Soc., vol. 54, pp. 249–259.
2. *Bochner, S.*: Curvature and Betti numbers. Annals of Mathematics, vol. 49, pp. 379–390.
3. *Eisenhart, L. P.*: Finsler spaces derived from Riemann spaces by contact transformations. Annals of Mathematics, vol. 49, pp. 227–254.

4. *Allendoerfer, C. B.*: Steiner's formulae on a general S^{n+1}. Bulletin of the Amer. Math. Soc., vol. 54, pp. 128–135.

5. *Levine, J.*: Invariant characterizations of two-dimensional affine and metric spaces. Duke Math. Journ., vol. 15, pp. 67–77.

1949. 1. *Taub, A. H.*: A characterization of conformally flat spaces. Bulletin of the Amer. Math. Soc., vol. 55, pp. 85–89.

2. *Eisenhart, L. P.*: Separation of the variables in the one-particle Schroedinger equation in 3-space. Proceedings of the Nat. Acad. of Sciences, vol. 35, pp. 412–418.

3. *Eisenhart, L. P.*: Separation of the variables of the two-particle wave equation. Proceedings of the Nat. Acad. of Sciences, vol. 35, pp. 490–494.

Index*

The numbers refer to pages

* This Index does not include the Appendices and the Additional Bibliography.

The results of paragraph 31 were based upon material presented by Professors Veblen and J. W. Alexander in their lectures before the appearance of the paper by Murnaghan. 1925, 7.